Biochemistry of Nucleic Acids II

Publisher's Note

The *International Review of Biochemistry* remains a major force in the education of established scientists and advanced students of biochemistry throughout the world. It continues to present accurate, timely, and thorough reviews of key topics of distinguished authors charged with the responsibility of selecting and critically analyzing new facts and concepts important to the progress of biochemistry from the mass of information in their respective fields.

Following the successful format established by the earlier volumes in this series, new volumes of the *International Review of Biochemistry* will concentrate on current developments in the major areas of biochemical research and study. New volumes on a given subject generally appear at two-year intervals, or according to the demand created by new developments in the field. The scope of the series is flexible, however, so that future volumes may cover areas not included earlier.

University Park Press is honored to continue publication of the *International Review of Biochemistry* under its sole sponsorship beginning with Volume 13. The following is a list of volumes published and currently in preparation for the series:

Volume 1: **CHEMISTRY OF MACROMOLECULES** (H. Gutfreund)
Volume 2: **BIOCHEMISTRY OF CELL WALLS AND MEMBRANES** (C. F. Fox)
Volume 3: **ENERGY TRANSDUCING MECHANISMS** (E. Racker)
Volume 4: **BIOCHEMISTRY OF LIPIDS** (T. W. Goodwin)
Volume 5: **BIOCHEMISTRY OF CARBOHYDRATES** (W. J. Whelan)
Volume 6: **BIOCHEMISTRY OF NUCLEIC ACIDS** (K. Burton)
Volume 7: **SYNTHESIS OF AMINO ACIDS AND PROTEINS** (H. R. V. Arnstein)
Volume 8: **BIOCHEMISTRY OF HORMONES** (H. V. Rickenberg)
Volume 9: **BIOCHEMISTRY OF CELL DIFFERENTIATION** (J. Paul)
Volume 10: **DEFENSE AND RECOGNITION** (R. R. Porter)
Volume 11: **PLANT BIOCHEMISTRY** (D. H. Northcote)
Volume 12: **PHYSIOLOGICAL AND PHARMACOLOGICAL BIOCHEMISTRY** (H. Blaschko)
Volume 13: **PLANT BIOCHEMISTRY II** (D. H. Northcote)
Volume 14: **BIOCHEMISTRY OF LIPIDS II** (T. W. Goodwin)
Volume 15: **BIOCHEMISTRY OF CELL DIFFERENTIATION II** (J. Paul)
Volume 16: **BIOCHEMISTRY OF CARBOHYDRATES II** (D. J. Manners)
Volume 17: **BIOCHEMISTRY OF NUCLEIC ACIDS II** (B. F. C. Clark)
Volume 18: **AMINO ACID AND PROTEIN BIOSYNTHESIS II** (H. R. V. Arnstein)
Volume 19: **BIOCHEMISTRY OF CELL WALLS AND MEMBRANES II** (J. C. Metcalfe)
Volume 20: **BIOCHEMISTRY AND MODE OF ACTION OF HORMONES II** (H. V. Rickenberg)

(Series numbers for the following volumes will be assigned in order of publication)
DEFENSE AND RECOGNITION II (E. S. Lennox)
BIOCHEMISTRY OF NUTRITION (A. Neuberger)
MICROBIAL BIOCHEMISTRY (J. R. Quayle)

Consultant Editors: H. L. Kornberg, Sc.D., F.R.S., Department of Biochemistry, University of Cambridge; and D. C. Phillips, Ph.D., F.R.S., Laboratory of Molecular Biophysics, Department of Zoology, University of Oxford

INTERNATIONAL REVIEW OF BIOCHEMISTRY

Volume 17

(Biochemistry of Nucleic Acids II)

Edited by

B. F. C. Clark, Sc.D.

Professor, Department of Chemistry
Aarhus University,
Aarhus, Denmark

UNIVERSITY PARK PRESS

Baltimore

UNIVERSITY PARK PRESS
International Publishers in Science and Medicine
233 East Redwood Street
Baltimore, Maryland 21202

Typeset by The Composing Room of Michigan, Inc.

Manufactured in the United States of America by
Universal Lithographers, Inc.,
and The Optic Bindery Incorporated.

Library of Congress Cataloging in Publication Data

Main entry under title:

Biochemistry of nucleic acids II.

   [International review of biochemistry; v. 17]
   Includes index.
   1. Nucleic acids.   I.   Clark, Brian Frederic
Carl.   II.   Series.
QP501.B527 vol. 17   [QP620]   574.8'732   78-22
ISBN   0-8391-1080-4

Consultant Editors' Note

The MTP *International Review of Biochemistry* was launched to provide a critical and continuing survey of progress in biochemical research. In order to embrace even barely adequately so vast a subject as "progress in biochemical research," twelve volumes were prepared. They range in subject matter from the classical preserves of biochemistry—the structure and function of macromolecules and energy transduction—through topics such as defense and recognition and cell differentiation, in which biochemical work is still a relatively new factor, to those territories that are shared by physiology and biochemistry. In dividing up so pervasive a discipline, we realized that biochemistry cannot be confined to twelve neat slices of biology, even if those slices are cut generously: every scientist who attempts to discern the molecular events that underlie the phenomena of life can legitimately parody the cry of *Le Bourgeois Gentilhomme,* "Par ma foi! Il y a plus de quarante ans que je dis de la Biochimie sans que j'en susse rien!" We therefore make no apologies for encroaching even further, in this second series, on areas in which the biochemical component has, until recently, not predominated.

However, we repeat our apology for being forced to omit again in the present collection of articles many important matters, and we also echo our hope that the authority and distinction of the contributions will compensate for our shortcomings of thematic selection. We certainly welcome criticism—we thank the many readers and reviewers who have so helpfully criticized our first series of volumes—and we solicit suggestions for future reviews.

It is a particular pleasure to thank the volume editors, the chapter authors, and the publishers for their ready cooperation in this venture. If it succeeds, the credit must go to them.

H. L. Kornberg
D. C. Phillips

Contents

Preface

Like its predecessor, this second volume on nucleic acids is concerned with the structure, biosynthesis, and genetic function of nucleic acids. However, the new volume is more selective in trying to review the more significant advances among the ever-expanding collection of research data published on the biochemistry of nucleic acids. The reader is also directed to other volumes of this series covering related topics in gene expression at a more functional level.

In the last three years, the first nucleic acid crystal structure has been determined. From it we have a detailed three-dimensional picture of yeast phenylalanine tRNA and an explanation for many of the constant parts of tRNAs, which are evident from the increasing number of established primary structures (now about 100). In addition, the intricate folding and specific interactions holding the tRNA together will probably be meaningful for understanding the tertiary structures of larger nucleic acids. A chapter on the structure of tRNA starts the volume and is followed by a chapter reviewing our current state of knowledge on how tRNA molecules are synthesized in the cell and on how the rare or minor bases in tRNA are put there. This fascinating subject has recently expanded rapidly but certainly is not yet completely understood—a good pointer to the fact that much more research effort is needed.

Another intriguing subject that has attracted the attention of many research workers in recent years is the structure and function of mRNA. The chapter included in this volume describes how mRNAs are "capped" at the 5'-terminal end by a modified base in a special linkage. How this cap is thought to be concerned in mRNA translation is reviewed. Furthermore, other mRNA complications, such as how the large nuclear species heterogeneous nuclear (Hn) RNA is related to mRNA and what is the effect of poly (A) attached to mRNA, are discussed in detail but the field is still not completely clarified.

Next are included two chapters on the recent extraordinary and productive technological advances in sequencing nucleic acids. In addition to describing the methodology, the chapters discuss the functional implication of the determined viral nucleic acid sequences. In recent years it has been very impressive that a total RNA sequence (about 3570 bases long) from RNA bacteriophage MS2 was determined, but this was followed in 1977 by the total DNA sequence (about 5375 bases long) of the DNA bacteriophage ϕX174. This latter achievement has been due to recently developed rapid sequencing techniques for DNA. This field is currently growing so rapidly that we have not yet come to understand much of the mass of sequence information now available. I should point out that the chapter describing the method of sequencing by the plus and minus method developed in Dr. F. Sanger's laboratory includes experimental details that are not yet published elsewhere, a unique feature of this volume.

This prime importance of gathering more structural information, even of a primary sequence nature, has been clearly indicated by novel and unexpected information about gene organization arising from bacteriophage DNA sequencing studies. Such studies have established that one piece of DNA can contain information for more than one gene read out of phase. Whether this is true in general or just for specially compact organizations of genes, such as in bacteriophage or viral genomes, remains to be substantiated.

Yet another topic in nucleic acids with an explosive growth rate is chromatic structure and function. This is also illustrated by the size of the chapter on this topic included herein, the length of which is necessary to do justice to the fact that the mass of data in the literature is beginning to allow sensible interpretations. The first level of structure appears now to be well understood, but higher orders of structure and functional states of chromatin remain to be investigated in the future. Finally, the last chapter is an up-to-date account of RNA synthesis and its control, largely concerning the more well-known aspects of prokaryotic RNA synthesis.

Clearly, a number of new developments are omitted, perhaps most obviously those concerning genetic manipulation, although some of the technology is included in the chapters on sequencing nucleic acids. In addition to considerations of time and space available in this volume, I feel that it is judicious to reserve this topic, together with new results on the major topics of ribosome structure and DNA replication, until the next volume, so that it can be covered both thoroughly and in the light of the most recent findings.

Finally, I would like to insert a word of caution to the reader who is anxious about that exact situation at his reading of this text—check the current literature. Our chapters are up to date at the time of our going to press, but none of the topics chosen can be considered to be in a stationary phase.

B. F. C. Clark

International Review of Biochemistry
Biochemistry of Nucleic Acids II, Volume 17
Edited by B. F. C. Clark
Copyright 1978   University Park Press   Baltimore

1
Structure of Transfer Ribonucleic Acid

J. E. LADNER
European Molecular Biology Laboratory, Heidelberg, Germany

The three-dimensional structure of tRNA has been amply reviewed recently (1–4). Consequently, this chapter does not attempt to duplicate this effort; instead it emphasizes the aspects of the structure which attract the attention of the author and seem of interest to biochemists.

Transfer ribonucleic acid (tRNA) is a polynucleotide chain of about 80 nucleotides. Its biological activities range from acting as a molecular interface between protein synthesis and genetic information to involvement in cell wall synthesis in bacteria (1–4).

The crystal structure of yeast tRNA^{Phe} has been refined to 2.5-Å resolution by Jack et al. (5) and Quigley and Rich (6). The detailed knowledge of the atomic structure presents a new basis for discussions of the function of tRNA. It must be remembered, however, that the crystal structure provides a view of the tRNA only in a "resting" conformation.

"Must this conformation change in order for the molecule to fulfill its many biological roles? And, if so, how does it change?" are questions for which we do not yet have complete answers. The methods being used to examine these problems include nuclear magnetic resonance studies, x-ray crystallography of interacting proteins and other tRNAs, chemical modification, and kinetic measurements.

STRUCTURE OF tRNA

The primary structure of RNA is the nucleotide sequence. The secondary structure is the formation of helical stretches by Watson-Crick base pairing, and the tertiary structure is the three-dimensional folding of the molecule.

The cloverleaf representation of tRNA indicates the secondary structure, one of three proposed by Holley et al. (7) when looking for stretches of complementary bases in the first sequenced tRNA. Since this first tRNA sequence, many more (1, 8) have followed, and they all can be arranged in the cloverleaf form (see Figure 1 for an example). The various parts of this representation are usually named for some common bases they contain or for a function they perform. The term ''stem'' refers to a stretch of helix with Watson-Crick pairing; ''arm'' refers to a helical stem together with a single-stranded loop at one end. The amino acid stem ends with -C-C-A on one side; the terminal adenosine has a 3'-OH instead of phosphate and this end is called the 3'- end of the sequence. Charged tRNA has the amino acid attached to the ribose of this terminal adenosine. At the beginning of the sequence, the first residue has a phosphate on its C-5' and this end of the sequence is referred to as the 5'-end. The loop of the D arm usually contains one or more dihydrouridine residues; the loop of the anticodon arm contains the three anticodon bases which indicate the amino acid for which the tRNA is specific. The extra loop is a stretch of residues between the anticodon and TψC arms (usually three to five nucleotides long, defining class 1 tRNAs) (1). There is also a group of tRNAs which has an extra arm, i.e., a helical stem plus a loop of 13 to 21 nucleotides, defining class 2 tRNAs (1). The loop of the TψC arm contains the highly conserved T-ψ-C sequence (thymidine-pseudouridine-cytidine).

The terms ''invariant'' and ''semi-invariant'' are also used extensively in this field. There are positions in the cloverleaf where specific bases, such as the terminal sequence -C-C-A, occur (Figure 1); these base positions, circled in Figure 1, are referred to as invariant. Other positions where the residue contains always a purine (adenine or guanine) or always a pyrimidine (uracil or cytosine) are termed semi-invariant and are shown in parentheses in Figure 1.

The major stabilizing interactions used to describe the tertiary structure are hydrogen bonding and base stacking. Hydrogen bonding means that an atom with a covalently attached proton is within a short distance, about 2.9 Å, of an atom which is sufficiently electronegative to share the hydrogen. It is usually described as an electrostatic interaction and is not very directional. Base stacking means that the planar, hydrophobic ring systems of spatially adjacent nucleic acid bases are overlapped.

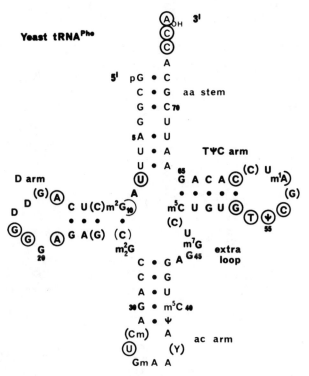

Figure 1. Cloverleaf structure of yeast tRNA^Phe (18). The invariant base positions are *circled*; semi-invariant bases are enclosed in *parentheses*. The standard nucleotide bases are shown as A, adenine; G, guanine; C, cytosine; and U, uracil. The modified bases are m²G, 2-methylguanine; D, dihydrouracil; m₂²G, 2,2-dimethylguanine; Cm, 2′-O-methylcytosine; Gm, 2′-O-methylguanine; Y, a highly modified guanine; m⁷G, 7-methylguanine; m⁵C, 5-methylcytosine; T, thymine; ψ, pseudouracil; m¹A, 1-methyladenine. In general, the letters refer to nucleosides, not just to the bases they contain. Herein they refer to bases where base interactions are considered. Heavy dots in stems indicate hydrogen bonding in Watson-Crick base pairs.

The tertiary structure of yeast tRNA^Phe has been described many times recently. The most detailed descriptions, including proposed hydrogen bonding, are in the recent papers describing 2.5-Å refinements (5, 6). A photograph of a 2.5-Å model built from the x-ray data with Kendrew skeletal parts is shown in Figure 2A. The accompanying diagram in Figure 2B illustrates the numerous tertiary base interactions holding the molecule in a folded T arrangement of the helices.

Simplified stereo views of the molecule and its various parts are shown in Figure 3, which should help the reader locate certain parts which may have functional significance in the total structure (see also Figure 2). A basic feature of the molecule is the TψC helix, which is stacked end-to-end on the amino acid helix to form a long helix. The D helix, residues 8 and 9, and the extra loop form a complex central core at approximately right angles to the long helix. The

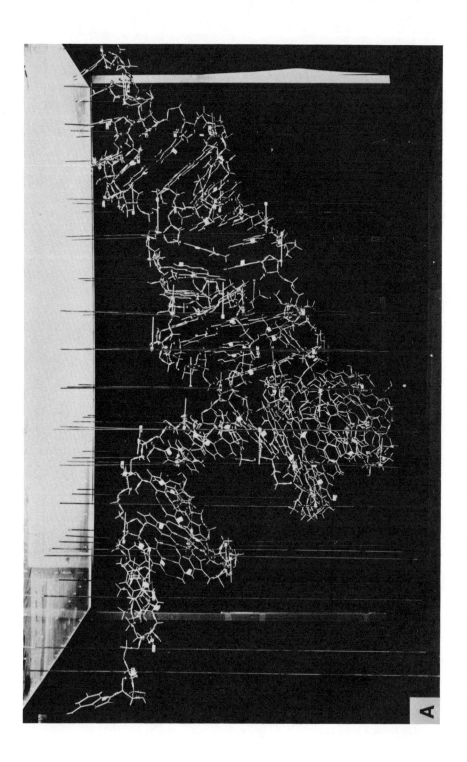

A

4

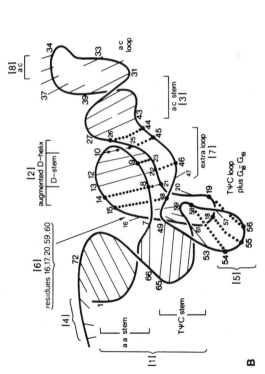

Figure 2. *A* (above), a photograph of a 2.5-Å model of the x-ray structure determined in Cambridge by Klug's group and built at the Department of Chemistry, Aarhus University, with the help of the author. *B*, an arrangement of the cloverleaf to show tertiary base interactions. Numbers in brackets refer to convenient substructures. Those not labeled in the picture are [1] long helix made of amino acid stem and T ψ C stem, [4] terminal -C-C-A, [5] T ψ C loop, [6] cluster of conserved bases. The letters ac stand for anticodon and aa stand for amino acid. Adapted from Ladner et al. (60).

5

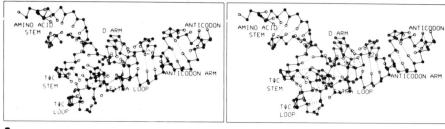

A

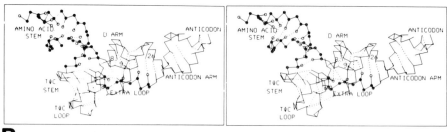

B

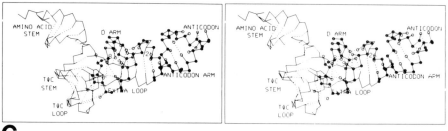

C

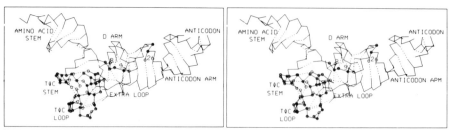

D

Figure 3. Stereo views of the yeast RNA^Phe structure. *Striped circles,* phosphate groups and centers of the ribose rings; *clear circles,* centers of the base ring systems. Covalent bonds are *solid*; hydrogen bonds are *dashed.* *A*, all groups; *B*, only groups of the amino acid stem and extra loop; *C*, only groups of the D arm and anticodon arm; *D*, only groups of the TψC arm and residues 8, 9, and 26; *E*, (*at right*), all the groups from a different viewpoint.

anticodon stem continues from this central core, but is tipped up about 20°, ending with the anticodon loop at the extreme tip of the molecule (see also Figure 2).

The central core is made of four chains, two from the D helix, residues 8 and 9, and the extra loop. Residues 8 and 9, sometimes affectionately referred to as the "ministreaker," form a fully extended stretch of nucleotide chain and span the length of the D helix. The invariant bases U8 and A14 hydrogen bond and stack on one end of the D helix, and A9 hydrogen bonds to A23 of the D helix pair U12·A23 to form a base triple. Two bases from the extra loop also pair with members of the D helix to form two more base triples: $m^2G10·C25·G45$ and $C13·G22·m^7G46$. In addition, two semi-invariant bases, C48 from the extra loop and G15 from the D loop, pair and stack on the pair U8·A14 to augment further the D helix. At the other end of the D helix, m_2^2G26 and A44 form a nonplanar hydrogen-bonded pair which forms the junction with the anticodon helix.

The D loop can be divided into three structural parts. The first part augments the D stem by stacking and hydrogen bonding and contains the invariant and semi-invariant residues A14, G15, and A21. The second contains variable residues, D16, D17, and G20 in yeast tRNA^{Phe}, which are on the outside of the molecule; it is easy to envisage that for other tRNAs different residues and different numbers of residues could be accommodated. These bases, along with U59 and C60 which project from the TψC loop, form a pocket of variable bases. In contrast, the third region of the D loop, the invariant G residues 18 and 19, with five invariant and semi-invariant residues of the TψC loop, forms a conserved stretch in which the loops reciprocally stabilize each other by stacking and

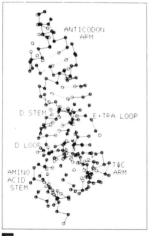
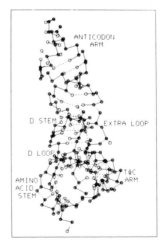

E

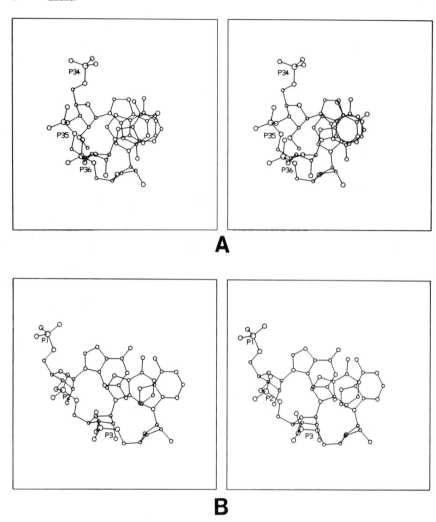

Figure 4. *A*, stereo view of the anticodon perpendicular to the plane of the central base. *B*, stereo view of an RNA-11 helix similarly projected on the plane of the central base.

hydrogen bonding. Three base pairs are formed by T54·m¹A58, G18·ψ55, and G19·C56.

The anticodon loop appears to be quite isolated from the rest of the molecule. There is no cross-strutting from the anticodon helix or loop to any other region. The loop is at the extreme tip of the molecule, and its arrangement is quite simple. The first two bases stack; then a sharp bend leads into a stack of five bases which begins with the "wobble" base of the anticodon. These bases are considerably more overlapped than the bases of a normal RNA helix (Figure 4).

The structure appears to be stabilized principally by stacking of the planar hydrophobic bases and by an extensive hydrogen-bonding system which makes considerable use of the 2'-OH groups of the riboses. These groups are used as both acceptors and donors. Detailed lists of the hydrogen-bonding network appear in the latest refinement papers by Jack et al. (5) and Quigley and Rich (6). Lacking exact criteria for hydrogen bonds, the slight differences in the lists are not surprising and are probably due to positional errors in the coordinates.

SOLVING CRYSTAL STRUCTURE OF YEAST PHENYLALANINE tRNA

The tertiary structure of yeast tRNA[Phe] has been determined by several groups. The monoclinic crystal form has been studied by Klug and co-workers in Cambridge, England, and by Sundaralingam's group at the University of Wisconsin; the orthorhombic crystal form has been studied by Rich and co-workers at the Massachusetts Institute of Technology. To an outsider, it may seem odd that so much independent effort has been put into the solving of the yeast tRNA[Phe] structure. However, so far, yeast tRNA[Phe] gives the most reproducible crystals capable of yielding high resolution x-ray data. Many different laboratories have attempted the crystallization of many different tRNAs from various sources and charging with several different amino acids. Finally, Kim et al. (9) obtained satisfactory crystals of yeast tRNA[Phe] in the orthorhombic form, and Ichikawa and Sundaralingam (10) and Ladner et al. (11) reported crystals of the monoclinic form.

The x-ray crystallographic determination of a structure generally progresses through several stages: 1) the production of reproducible crystals, 2) the introduction of heavy atoms into the crystals to provide isomorphous heavy atom derivatives, 3) the collection of x-ray data from the native and at least two derivatives, 4) the production of an electron density map and the fitting of an atomic model to the map, 5) the refinement of the model with the use of computer techniques, and 6) further collection of higher resolution native data and further refinement of the atomic model. The technique of x-ray diffraction of large molecules and the reason for the necessity of heavy atom derivatives have been presented in a readable manner for a novice to the field by Holmes and Blow (12) and Eisenberg (13), and briefly in a review of the tRNA structure by Kim (3). At present, the work on the structural elucidation of yeast tRNA[Phe] is at stage 6.

The atomic coordinates from the 2.5-Å refinements of the monoclinic form by Jack et al. (5) and of the orthorhombic form by Quigley and Rich (6) have been deposited in the Protein Data Bank. The published structures for the two crystal forms are the same, at least as far as the biochemist is concerned. It should, however, be appreciated that the refinement of the atomic coordinates and the collection of data to higher resolution are by no means wasted effort or pure scientific rivalry. This structure has provided the first detailed look at a polynucleotide in a crystal. It has shown that the hydrogen-bonding potential is more varied than just Watson-Crick type pairing, that the planar bases can form structural sheets which involve hydrogen bonding of three nucleotides, and that the ribose moieties can participate in forming structurally valuable hydrogen bonds between themselves, with the bases, and with phosphate oxygens. Furthermore, the presence of single-stranded loops, a region where four backbone chains run together, and regions of fairly standard double helix makes it of considerable interest to examine the torsion angles in as much detail as possible and to ask what restrictions can be placed on particular angles when predicting an unknown structure. The refined coordinates have revealed details about nucleic acid structure, including two angular correlations (5).

It will, of course, also be very interesting when a second tRNA structure is solved. The group working in Ebel's laboratory in Strasbourg (14) now has a complete set of 3-Å native data on yeast tRNA[Asp] and is attempting to use the molecular replacement method with yeast tRNA[Phe] coordinates to determine the structure. In general, it is assumed that all tRNAs have the same basic structure. The large number of known sequences can be made into the cloverleaf secondary structure. It is felt that with very conservative changes in the bonding, at least the sequences of class 1 (with no long extra arm) can be folded into the structure of yeast tRNA[Phe] (15, 16). Recently, a structure has been proposed for tRNAs of the second class (17), those with an extra arm (helix and loop). In this model, some of the triple base interactions are missing, and the helical stem of the extra arm emerges from the deep groove side of the D helix. The proposed structure must wait for a crystallographic analysis of a class 2 tRNA before it is deemed correct or not. It does seem somewhat surprising that the extra arm is not more integrated into the structure. One of the striking features of the yeast tRNA[Phe] structure is the predominance of base stacking. The proposed loss of the triple base interactions would decrease the very strong stacking of this region.

CONSERVATION OF CRYSTAL STRUCTURE

Two principal questions about the conservation of crystal structure are: 1) is it also the structure found in solution? and 2) do other tRNAs also have the same basic tertiary structure? The answer to both questions seems to be yes, although there may also be other conformations present.

One of the strongest arguments for the similar structure of other tRNAs comes from the prominent occurrence of invariant and semi-invariant residues in many of the tertiary interactions (Figure 5) which are concentrated in the complex central core. This point has been discussed in more detail by Klug et al. (15) and by Kim et al. (19). A further rationalization is that all the tRNAs must interact with some common proteins such as elongation factor Tu; consequently, it seems that a common basic structure would facilitate these interactions. On the other hand, the tRNAs must also express their individuality when reacting with the amino acyl-tRNA synthetases. It has been shown in companion studies of the chemical reactivity of yeast tRNA[Phe] by Robertus et al. (20), Rhodes (21), and Batey and Brown (22) that the availability of bases for chemical modification in

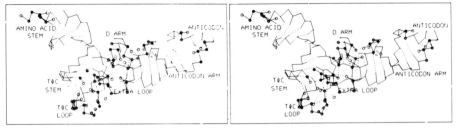

Figure 5. Stereo view with groups shown for the invariant and semi-invariant bases.

solution is in agreement with the crystal structure. The chemical modification work on other tRNAs has revealed a consistent pattern of exposed and buried residues as discussed in reviews by Brown (23) and Clark (1).

The chemical accessibility of adenine and cytosine residues in the bacterial initiator tRNA, *Escherichia coli* tRNA$_f^{Met}$, has been studied by Schulman and Pelka (24). This tRNA has an adenine residue in the position of G57 of yeast tRNA^{Phe}. It is reactive with 1 M chloroacetaldehyde at pH 6.0, which requires exposure of N1 and N6. The corresponding G57 in yeast tRNA^{Phe} is unreactive to kethoxal (25), which requires exposure of N1 and N2, to carbodiimide (21), which requires N1, and to tritium exchange (26) at C8. The authors correctly point out structural reasons for the nonreactivity of the guanine and use the reactivity of adenine in *E. coli* tRNA$_f^{Met}$ to infer a possible structural difference in tRNAs which have an adenine in this position. Although this may be true, one should be cautioned that the O6 position of G57 is exposed in the structure and the N1 appears available to attack by reagents which are not too bulky. There could, of course, be some weakening of the structure or some rearrangement. The two hydrogen bonds made by N2 of G57 to the riboses of G18 and G19 will be absent, but then the need to accommodate an internal NH$_2$ group also will be missing. There must be some distinction between initiator tRNA and the other tRNAs, but the distinction has been quite elusive to experimental probes so far. As suggested by Clark (1), the special function of *E. coli* tRNA$_f^{Met}$ is probably explained by the recognition of an initiation factor for some subtle feature of the structure. The initiation factor then assists in locating the tRNA in the correct ribosomal site. What is the feature being recognized by this initiation factor? It could be a difference in the structure of the TψC loop such as that suggested by Schulman and Pelka (24); it could be something else.

Nuclear magnetic resonance (NMR) has been used extensively lately to probe the structure of tRNA in solution. Kearns (27) reviewed the field in 1976. Most NMR work on biological molecules is done in D$_2$O in order to avoid a very large water peak. In D$_2$O, most of the spectrum of tRNA is quite broad because of the intrinsic line widths of the individual resonances and because of the occurrence of one or more resonances from every nucleotide in the same region. All tRNAs, however, contain some modified bases with methyl and methylene groups. Because there are relatively few per tRNA, these groups do provide some natural probes in the structure. In addition, in 1971 Kearns et al. (28) showed that in H_2O it was possible to see one proton from each Watson-Crick base pair in the low field region downfield from the very large water peak. This group of resonances, which comes from the proton on N1 of guanine or N3 of uracil when they are involved in hydrogen bonding, has received the most attention recently. These protons can only be observed in H_2O solution because they are exchanging with solvent protons; nevertheless, their involvement in hydrogen bonding gives them a sufficient lifetime to be detected by NMR. In D_2O, they are rapidly exchanged by deuterium atoms and consequently disappear from the proton spectrum. Two of the problems in NMR work are the assignment of the resonances and the integration of the spectrum to determine how many protons are represented by it.

At the moment, there is disagreement among the various NMR groups concerning the correct integration; for example, see the contribution of Bolton and Kearns and the reply by Reid in Nature's "Matters Arising" (29). Because of such internal disagreement and the large number of papers in the field, it is difficult to piece together a clear picture of the NMR work. However, the technique should be quite powerful and especially useful in looking for changes in conformation, whether they are attributable to interactions with proteins or the presence of ions. In reference to the comparison with crystal structure, it would be interesting to see spectra which were taken in the presence of the same concentrations of Mg^{2+} and spermine used in the crystallization. Polyamines are important in the protein biosynthesis system and probably play a role in maintaining the conformation of tRNA itself; for example, see the review by Sakai and Cohen (30). In a recent paper by Bolton et al. (31), they show that the low field spectrum of yeast tRNA[Phe] and other tRNAs does change with the concentration of Mg^{2+}; when more Mg^{2+} is present, more protons are seen in this region. The maximum number they find is 22.2 ± 1. The predicted number is 20 from the helices, plus 6 to 8 from tertiary interactions. In contrast, the spectra published by Robillard et al. (32) show no difference between no Mg^{2+} and 15 mM Mg^{2+}, and they claim 26 ± 1 resonances are included. Perhaps the NMR spectra do not show all the resonance intensity which the crystal study indicates should be present because the stability of some of the tertiary hydrogen bonds may not be great enough to give the proton a sufficient lifetime in the hydrogen-bonded environment to be detected. In the study of yeast tRNA[Asp] by Robillard et al. (33), they reject this explanation for not observing resonances from the D stem and suggest that at 35°C this helix is not formed. The sequence of yeast tRNA[Asp] is interesting in that the D stem written in the cloverleaf form contains two $A \cdot U$ pairs, one $G \cdot U$ pair, and one $G \cdot \psi$ pair. They do, however, tentatively assign resonances to the tertiary hydrogen bonds which would be equivalent to $G19 \cdot C56$ and $G46 \cdot G22$ in the yeast tRNA[Phe] structure. How these can be formed without the D helix presents an interesting model-building problem which may soon be answered by the solving of the three-dimensional structure by x-ray crystallography.

The low field spectra have demonstrated that in solution virtually all tRNAs have very similar secondary and tertiary hydrogen bonding involving the imino protons and consequently should have very similar three-dimensional structures. Daniel and Cohn (34) have examined the low field H_2O spectra and high field D_2O spectra of E. coli tRNA$_{f1}^{Met}$ and tRNA$_{f3}^{Met}$, which differ by only one residue (m^{7}G$\leftrightarrow$A). They have used this single residue change to assign positions for some of the tertiary structure interactions involving the imino protons and to assign positively the high field methyl resonance from m^{7}G. It is interesting, but not really surprising, that the single base change which would at least alter the C13· G22· m^{7}G46 triplet of the yeast tRNA^{Phe} structure also has sufficient effect on the tertiary structure to shift the resonances of other imino protons which are tentatively assigned to neighboring bases.

Robillard et al. (32) and Geerdes and Hilbers (35) have taken crystallographic atomic coordinates and tried to predict the low field region of the NMR spectrum. Although the predicted spectra do resemble the observed spectra, to say that they prove the solution structure to be virtually identical with the crystal structure seems somewhat too optimistic. The calculation of an NMR spectrum is rather difficult, so the inexact correspondence of the spectra does not prove that the solution and crystal structures are different. The calculations involve making assumptions about the perturbation of a large magnetic field on the wave functions which describe the ring systems. The resonances of the protons are quite sensitive to the exact environments; consequently, the inaccuracies in the calculation and in the coordinates both contribute to noticeable errors. Nevertheless, the calculations could be quite valuable as a guide for making assignments.

CONFORMATIONAL CHANGES

The determination of the three-dimensional structure has given no definitive answers about how the molecule performs its many functions. The two well-known regions of interaction, the -C-C-A end and the anticodon, appear free to perform their roles. Although it is still postulated by Yoon et al. (36) that the anticodon changes its conformation when pairing with the codon to allow the formation of a fourth base pair on the 5′ side of the anticodon, the evidence now favors retention of the anticodon stacked in the manner found in the crystals. Bald and Pongs (37) have measured high binding constants for U-U-C-purine nucleoside and U-U-C-m^{1}A, which can form only three base pairs. In addition, Grosjean et al. (38) have shown that the strong binding of two tRNAs with complementary anticodons does not involve any slow reorganization of the molecule.

However, evidence is building for a conformational change induced in the charged Phe-tRNA^{Phe} by the binding of the anticodon to its complementary codon when the Phe-tRNA^{Phe} forms a complex with the elongation factor Tu and GTP. Schwarz et al. (39) use equilibrium dialysis and binding of the complement to T-ψ-C-G to demonstrate that the binding of the complement is significantly increased by the codon-anticodon binding when the charged tRNA is bound to 30 S subunits in the presence of elongation factor Tu and GTP. The results are interesting but should be considered with some caution because of the high concentrations used and more basically because the interpretation of oligonucleotide binding is ambiguous. One of the striking features of the tRNA strucuture is the presence of base pairings other than Watson-Crick type. The usual interpretation of oligonucleotide binding experiments, however, assumes only Watson-Crick complementary binding. Robertson et al. (40) use yeast tRNA^{Phe} in which ethidium bromide has been substituted for the base next to the anticodon; in other experiments, the substitution has been at the positions of D16/17. They then follow the fluorescence of the ethidium bromide during the complex formation and find that their results can be explained in terms of a conformational change involving the D loop (and by implication the TψC loop also because these loops interact). This change is induced when the ribosome complex is formed in the presence of poly (U), which can act as the messenger RNA in this case.

The disruption of the structure of the tRNA observed in the crystal form in order to expose the T-ψ-C-G sequence for complementary binding would seem rather great. In addition, the formation of standard Watson-Crick pairs provides no explanation for the modified bases thymine and pseudouracil. It has been shown by Erdmann et al. (41) that the unmodified oligonucleotide U-U-C-G does not activate the stringent factor-dependent synthesis of guanosine tetra- and pentaphosphates whereas T-ψ-C-G does. This result suggests a specific biological role for the modified bases and may be an indication that the T-ψ-C-G is involved in an interaction more complex than Watson-Crick pairing, although it is quite conceivable that some conformational change does occur.

Thompson and Stone (42) studied the error rejection of noncognate charged tRNA (aa-tRNA) in the reaction of poly (U)-programmed ribosomes in the presence of elongation factor Tu and GTP. They interpret their results as favoring a two-step discrimination involving an initial, reversible binding of the ternary complex (aa-tRNA:elongation factor Tu:GTP) to the ribosome and a second proofreading step, after the GTP hydrolysis but preceding the irreversible binding of aa-tRNA to the position where it accepts the growing peptide chain (A-site).

If it is reasonable to consider these recent results together, it seems likely that, with respect to the tRNA, the groups are describing different aspects of a common conformation change. The exposure or interaction of the common T-ψ-C-G sequence could be a signal that the molecule involved is indeed aa-tRNA. The proofreading step is simply the formation at the anticodon of an incomplete triplet interaction which is insufficient to cause the conformational change which would allow further interaction via the T-ψ-C-G sequence, which could also be involved in the correct positioning of the molecule into the A-site. The initial step in the binding could be a binding between the ternary complex and the ribosome which does not directly involve the messenger; the second proofreading step then produces a conformational change of the T-ψ-C-G region if successful. Whether the first step involves any conformational changes in the tRNA or elongation factor is an open question. What is the extent of the conformational change of the T-ψ-C-G region? How does the binding of the anticodon become propagated down to the T-ψ-C-G region? Is the interaction with T-ψ-C-G the formation of Watson-Crick pairs or is it a different form of interaction? What, if any, is the role of Mg^{2+}? All these questions await future experiments.

Prior to these latest reports, the ternary complex was examined by NMR by Shulman et al. (43). Unfortunately, the complex is so large that the spectrum is considerably broader than that of tRNA alone. These workers studied the imino proton region and found that the helical stems remained intact, but the spectrum is far too broad to say anything about any tertiary interactions. Maybe in the light of these latest results the NMR will be repeated in the presence of $(Up)_7U$.

Of course, it would be desirable to trap the tRNA in the altered conformation and solve the three-dimensional structure. This unfortunately seems a long way off. However, the solving of the structure of fragments of elongation factor Tu is under way by Sneden et al. (44), Gast et al. (45), and by x-ray groups at Aarhus University (Denmark) and the University of Virginia (46). The native crystalline Tu-GDP is also being studied by Leberman et al. (47). When the structure of both elongation factor Tu and tRNA are known in one conformation perhaps it will be possible to get some clue about the nature of possible conformational changes.

It is still not clear whether the structure of charged tRNA is the same as that of the uncharged form. A recent NMR study by Kan et al. (48) reports the loss of one or two protons from the imino region of the spectrum in H_2O, but no difference in the methyl and methylene resonances in D_2O between the charged and uncharged tRNA. In order to ensure the chemical stability of the charged form, the tRNA was altered to produce an amide linkage to the amino acid. A previous NMR study by Wong et al. (49) reported no differences in the two forms.

The interaction with the synthetase presents another possible situation in which a conformational change might occur. Willick and Kay (50) have used circular dichroism to study the interaction of $tRNA_2^{Glu}$ with its synthetase. They found that both the synthetase and the tRNA probably undergo conformational changes, but caution that the change they observed could be accounted for by only subtle changes in the helical regions and would not require disruption of hydrogen bonding. In addition, they interpret their results to indicate the formation of a hydrogen bond between the first base in the anticodon and the enzyme. Again, a detailed atomic structure of a synthetase may answer some of the questions. Irwin et al. (51) have reported a chain tracing of tyrosyl tRNA synthetase. From studies of the mechanism of the synthetase-tRNA interaction, Krauss et al. (52) have found that the specificity of the reaction is improved by the presence of Mg^{2+}, which is not directly involved in the complex formation but instead guarantees the establishment of the tertiary structure of the tRNA. They also conclude that there is a conformational transition involved in the binding which is essential for the recognition process.

The positions of three possible strong Mg^{2+} binding sites in the crystal structure have been published by Jack et al. (53). These three sites are: 1) between the phosphate oxygens of U8 and A9, 2) between phosphate oxygens of G20 and A21, and 3) involved with a phosphate oxygen of G19, N7, and O6 of G20 and O4 of U59. It is quite possible that, as suggested by Stein and Crothers (54), the sensitivity of the tRNA structure to Mg^{2+} may be an indication that selective ribosome-mediated removal of Mg^{2+} could trigger unfolding of the tRNA, which would then return to the folded form upon return of Mg^{2+}. It is pointed out by Jack et al. (53) that their third site does involve the region where the TψC and D loops interact, an observation which is in agreement with the kinetic unfolding experiments of Stein and Crothers (54) and also in agreement with the chemical modification results of Rhodes (55). In these experiments, fingerprints of T_1 ribonuclease digests of radioactively labeled tRNA after chemical modification with varying amounts of Mg^{2+} showed that two sites are protected by Mg^{2+}, one in the region of residues 8 and 9 and the other near residues 59 and 60 of the TψC loop. Leroy et al. (56) have just published a study of the effect of ions on the folding of tRNA as followed by emission and fluorescence spectroscopy. Their results suggest that the molecular organization brought about by divalent ions starts in the central region with the anticodon being affected only as the ion concentration is increased. Further interesting results are being reported by Rigler and Ehrenberg (57, 58). By using tRNA[Phe] with and without fluorescent probes in positions 16, 17, and 37, they find evidence for a dynamic equilibrium between three conformations, with the populations of these conformations depending on the ion concentrations. Bina-Stein and Stein (59) also discuss the significance of Mg^{2+} binding in terms of an allosteric model for the function of tRNA.

CONCLUSION

Evidence seems to be mounting to support the idea that tRNA is indeed a dynamic molecule which, with some nudging from its various protein partners and by changes in ion concentrations, can change its conformation to reveal itself first as a tRNA and then as a specific tRNA. Present experimental evidence indicates that any major conformational transitions last only while the molecule is in contact with a protein or other RNA binding system, returning to a folded, stacked, and hydrogen-bonded structure when released.

ACKNOWLEDGMENTS

The author acknowledges many useful discussions and exchanges during her collaboration with A. Klug and his group at the Medical Research Council Laboratory of Molecular Biology in Cambridge. Also, special thanks are given to Dr. R. C. Ladner for criticism and for assistance in producing the stereo views, to Dr. A. Jack for plotting the pictures, and to Mr. C. Christensen for printing them. Acknowledgment is also given to Prof. B. F. C. Clark for discussions and for making his manuscripts available before publication.

REFERENCES

1. Clark, B. F. C. (1978). *In* S. Altman (ed.), Transfer RNA. MIT Press, Cambridge.
2. Rich, A., and Raj Bhandary, U. L. (1976). Annu. Rev. Biochem. 45: 805.
3. Kim, S. H. (1976). Prog. Nucleic Acid Res. Mol. Biol. 17:181.
4. Sigler, P. B. (1975). Annu. Rev. Biophys. Bioeng. 4:477.
5. Jack, A., Ladner, J. E., and Klug, A. (1976). J. Mol. Biol. 108:619.
6. Quigley, G. J., and Rich, A. (1976). Science 194:796.
7. Holley, R. W., Apgar, J., Everett, G. A., Madison, J. T., Marguisee, M., Merill, S. H., Penswick, J. R., and Zamir, A. (1965). Science 147:1462.
8. Barrell, B. G., and Clark, B. F. C. (1974). Handbook of Nucleic Acid Sequences. Joynson Bruvvers Ltd., Oxford.
9. Kim, S. H., Quigley, G., Suddath, F. L., and Rich, A. (1971). Proc. Natl. Acad. Sci. USA 68:841.
10. Ichikawa, T., and Sundaralingam, M. (1972). Nature New Biol. 236:174.
11. Ladner, J. E., Finch, J. T., Klug, A., and Clark, B. F. C. (1972). J. Mol. Biol. 72:99.
12. Holmes, K., and Blow, D. (1965). In D. Glick (ed.), Methods of Biochemical Analysis, Vol. 13, p. 113. John Wiley and Sons, New York.
13. Eisenberg, D. (1970). In P. Boyer (ed.), The Enzymes, 3rd Ed., Vol. 1, p. 1. Academic Press, New York.
14. Thierry, J. C., private communication.
15. Klug, A., Ladner, J. E., and Robertus, J. D. (1974). J. Mol. Biol. 89:511.
16. Kim, S. H., Sussman, J. L., Suddath, F. L., Quigley, G. J., McPherson, A., and Rich, A. (1974). Proc. Nati. Acad. Sci. USA 71:4970.
17. Brennan, T., and Sundaralingam, M. (1976). Nucl. Acids Res. 3:3235.
18. Raj Bhandary, U. L., and Chang, S. A. (1968). J. Biol. Chem. 243:598.
19. Kim, S. H., Sussman, J. L., Suddath, F. L., Quigley, G. J., McPherson, A., Wang, A. H. J., Seeman, N. C., and Rich, A. (1974). Proc. Natl. Acad. Sci. USA 71:4970.
20. Robertus, J. D., Ladner, J. E., Finch, J. T., Rhodes, D., Brown, R. S., Clark, B. F. C., and Klug, A. (1974). Nucl. Acids Res. 1:927.
21. Rhodes, D. (1975). J. Mol. Biol. 94:449.
22. Batey, I. L., and Brown, D. M. (1977). Biochim. Biophys. Acta 474:378.
23. Brown, D. (1974). *In* P. O. P. T'so (ed.), Basic Principles in Nucleic Acid Chemistry, Vol. 2, ch. 1. Academic Press, London.
24. Schulman, L. H., and Pelka, H. (1976). Biochemistry 15:5769.
25. Litt, M. (1971). Biochemistry 10:2223.
26. Gamble, G. C., and Schimmel, P. R. (1974). Proc. Natl. Acad. Sci. USA 71:1356.
27. Kearns, D. R. (1976). Prog. Nucleic Acid Res. Mol. Biol., Vol. 18, p. 91. Academic Press, New York.
28. Kearns, D. R., Patel, D. J., and Shulman, R. G. (1971). Nature (Lond.) 229:338.
29. Bolton, P. H., and Kearns, D. R. (1976). Nature (Lond.) 262:423.
30. Sakai, T. T., and Cohen, S. S. (1976). Prog. Nucleic Acid Res. Mol. Biol., Vol. 17, p. 15. Academic Press, New York.
31. Bolton, P. H., Jones, C. R., Bastedo-Lerner, D., Wong, K. L., and Kearns, D. R. (1976). Biochemistry 15:4370.
32. Robillard, G. T., Tarr, C. E., Vosman, F., and Berendsen, H. J. C. (1976). Nature (Lond.) 262:363.

33. Robillard, G. T., Hilbers, C. W., Reid, B. R., Gangloff, J., Dirheimer, G., and Shulman, R. (1976). Biochemistry 15:1883.
34. Daniel, W. E., and Cohn, M. (1976). Biochemistry 15:3917.
35. Geerdes, H. A. M., and Hilbers, C. W. (1977). Nucl. Acids Res. 4:207.
36. Yoon, K., Turner, D. H., Tinoco, I., von der Haar, F., and Cramer, F. (1976). Nucl. Acids Res. 3:2233.
37. Bald, R., and Pongs, O., personal communication.
38. Grosjean, H., Söll, D. G., and Crothers, D. M. (1976). J. Mol. Biol. 103:499.
39. Schwarz, U., Menzel, H. M., and Gassen, H. G. (1976). Biochemistry 15:2484.
40. Robertson, J. M., Kahan, M., Wintermeyer, W., and Zachau, H. G. (1977). Eur. J. Biochem. 72:117.
41. Erdmann, V. A., Lorenz, S., Sprinzl, M., and Wagner, R. T. (1976). In N. C. Kjeldgaard (ed.), Control of Ribosome Synthesis, p. 427. Munksgaard, Copenhagen.
42. Thompson, R. C., and Stone, P. J. (1977). Proc. Natl. Acad. Sci. USA 74:198.
43. Shulman, R. G., Hilbers, C. W., and Miller, D. L. (1974). J. Mol. Biol. 90:601.
44. Sneden, D., Miller, D. L., Kim, S. H., and Rich, A. (1973). Nature (Lond.) 241:530.
45. Gast, W. H., Leberman, R., Schulz, G. E., and Wittinghofer, A. (1976). J. Mol. Biol. 106:943.
46. Clark, B. F. C., personal communication.
47. Leberman, R., Wittinghofer, A., and Schulz, G. E. (1976). J. Mol. Biol. 106:951.
48. Kan, L. S., Tso's, P. O. P., Sprinzl, M., von der Haar, F., and Cramer, F. (1976). Biophys. J. 16:11a.
49. Wong, Y. P., Reid, B. R., and Kearns, D. R. (1973). Proc. Natl. Acad. Sci. USA 70:2193.
50. Willick, G. E., and Kay, C. M. (1976). Biochemistry 15:4347.
51. Irwin, M. J., Nyborg, J., Reid, B. R., and Blow, D. M. (1976). J. Mol. Biol. 105:577.
52. Krauss, G., Riesner, D., and Maass, G. (1976). Eur. J. Biochem. 68:81.
53. Jack, A., Ladner, J. E., Rhodes, D., Brown, R. S., and Klug, A. (1977). J. Mol. Biol. 111:315.
54. Stein, A., and Crothers, D. M. (1976). Biochemistry 15:160.
55. Rhodes, D., personal communication.
56. Leroy, J. L., Guéron, M., Thomas, G., and Favre, A. (1977). Eur. J. Biochem. 74:567.
57. Rigler, R., and Ehrenberg, M. (1976). Q. Rev. Biophys. 9:1.
58. Rigler, R., personal communication.
59. Bina-Stein, M., and Stein, A. (1976). Biochemistry 15:3912.
60. Ladner, J. E., Jack, A., Robertus, J. D., Brown, R. S., Rhodes, D., Clark, B. F. C., and Klug, A. (1975). Proc. Natl. Acad. Sci. USA 72:4414.

International Review of Biochemistry
Biochemistry of Nucleic Acids II, Volume 17
Edited by B. F. C. Clark
Copyright 1978   University Park Press   Baltimore

2
Transfer RNA Biosynthesis

S. ALTMAN
Yale University, New Haven, Connecticut

This research was supported by a grant from the United States Public Health Service, Bethesda, Maryland.

The biosynthesis of transfer RNA occurs in several distinct stages: 1) transcription of the tRNA gene; 2) enzymological cleavage and trimming of the tRNA precursor RNA, yielding a molecule containing the mature tRNA sequence; and 3) minor nucleotide biosynthesis within the polymeric precursor or unmodified tRNA substrate. Stage 1 clearly occurs before the others, but events which might be included in stages 2 or 3 cannot always be described in a well defined temporal sequence. This chapter deals primarily with events which can be included in step 2 as defined above, but also gives a summary of the other events in tRNA biosynthesis. In addition, all those tRNA precursors thus far identified are described. Because several reviews of tRNA biosynthesis have appeared during the last few years (1–4), this chapter focuses on recent advances and refers the reader to earlier publications for a more detailed discussion of some other results. Most of the detailed new information regarding tRNA biosynthesis has been derived from work with *Escherichia coli,* but recent results in eukaryotic systems, including yeast, silkworms, and mammalian tissue culture cells, are also described.

tRNA GENE ORGANIZATION

Transfer RNA genes in *E. coli* have been mapped primarily through the recognition of nonsense and missense suppressor mutations (5–7). In some cases, the actual products of the suppressing genes have been shown to be tRNAs with altered anticodons which allow recognition of nonsense or missense codons. Genetic mapping experiments have located tRNA genes at various positions around the *E. coli* map (8), showing that not all tRNAs are clustered and thus cannot be transcribed together as one long polycistronic transcript (unlike bacteriophage T4, which is discussed below). Biochemical experiments in which tRNAs have been hybridized to fragments of *E. coli* DNA have also shown that tRNA genes are not all clustered (9). These results give some indication of the maximum size of some tRNA gene transcripts, but give no information concerning the way in which the final level of a particular tRNA species within the cell is controlled.

A most significant new result concerning tRNA gene organization was obtained when it was shown that tRNA species were located in rRNA cistrons which had been transposed into phages (10). These results, demonstrated originally by hybridizing tRNA to transducing phage DNA, have been extended to show that about 4 tRNA species hybridize to the approximately 6–10 rRNA cistrons in *E. coli* (11). To date, 3 of the tRNAs found in the spacer region between 16 S and 23 S sequences (Figure 1) have been identified: $tRNA_2^{Glu}$, $tRNA_1^{Ile}$, and $tRNA_{1b}^{Ala}$. It has been possible to treat the 30 S rRNA precursor molecule from *E. coli* (12, 13) with crude cell extracts which process the precursor to yield several cleavage products, among them the tRNA species (11). Thus far, it is not certain how many copies of each species occur in any one spacer region, or whether more than one species can exist in any one spacer region.

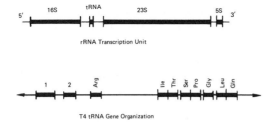

Figure 1. Organization of rRNA transcription unit of *E. coli* and tRNA genes in bacteriophage T4. *Top*, the organization of an *E. coli* rRNA transcription unit. tRNA hybridizes to a DNA restriction fragment in the spacer region between the 16 S and 23 S sequences. The total length of the unit shown is about 5,000 nucleotide pairs. Adapted from Lund et al. (10) with permission of MIT Press. *Bottom*, the organization of T4 tRNA genes as determined by hybridization mapping of the various RNA species to DNA restriction fragments. The abbreviations refer to the tRNA species found in this cluster; the relative size of the region can be judged from the size of the tRNA genes. In addition, there are other RNA species of unknown functions (bands 1 and 2) coded in this region. Adapted from Velten et al. (21) with permission of Elsevier/North Holland Biomedical Publishers.

Because the spacer tRNAs are major species within the cell, their abundance is possibly attributable to the influence of the same strong promoter control that governs rRNA. This theory does not take into account any possible regulatory features of the post-transcriptional processing pathways.

Mapping of bacteriophage-encoded tRNAs by deletion mutant techniques has shown that T4 tRNAs are clustered in one segment of the T4 genome (14–18) (Figure 1), whereas T5 tRNAs are located at several loci (19, 20). Recently, with the availability of DNA restriction fragments and tRNA hybridization techniques, a detailed map of the T4 genes and a partial map of T5 tRNA genes have been constructed (20, 21). From the organization of the single cluster of tRNA genes in T4, it seems that several hundred nucleotides can separate some of the mature tRNA sequences. However, control of the transcription of these genes and at least two other non-tRNA RNA species in this region seems to be governed by one promoter region (22). Because attempts to isolate one long polycistronic transcript have failed, processing probably occurs before transcription of the entire region is completed, as suggested for the processing of the large rRNA transcription unit in wild type *E. coli* (12, 13).

Other new data regarding tRNA gene organization have come from studies of eukaryotes for which the biochemical characterization of tRNA genes has progressed rapidly in the past few years. Most of these data tend to confirm the initial observations that tRNA genes in eukaryotes are always transcribed as monocistronic transcripts, some 20 or so nucleotides longer than the mature tRNA sequence (1).

Careful studies of the reannealing kinetics of yeast tRNA species to yeast DNA fragments indicate that the tRNA sequences are separated from each other, in segments of the genome where they are clustered, by spacer regions which may vary in size up to 10 times that of the mature tRNA sequence (23). Similar data are available for *Drosophila melanogaster* (24) and *Xenopus laevis* (25).

Genetic studies have shown that yeast suppressor genes (presumably tRNA structural genes) can be scattered throughout the yeast genome (26). Cell organelles, such as chloroplasts and mitochondria, cannot be classified as pro- or eukaryotes. Hybridization data indicate that chloroplasts may have 26 cistrons, enough to code for the full complement needed for protein synthesis (27). Mitochondria from yeast (28) and other eukaryotes probably have close to 20 tRNA cistrons (reviewed in ref. 4).

Recently, putative precursors to tRNA species have been isolated from yeast and from silkworms (29–31). These species are about 90–100 nucleotides in size. Thus, as yet there is no evidence in any eukaryotic organism that more than one tRNA sequence is included in any transcript of tRNA genes. The evidence strongly points to tRNA transcripts containing only one tRNA sequence as a reflection of the nature of the gene organization and control. The difference between tRNA gene control in pro- and eukaryotic organisms presents one of the most interesting open questions in studies of tRNA biosynthesis.

Nature of tRNA Gene Transcripts

Control of the functioning of RNA polymerase in the transcription of tRNA and rRNA has been reviewed in recent papers (32, 33) and is not discussed in great detail in this section (see also chapter 7 of this volume). Instead, the role of DNA promotor sequences in governing tRNA gene transcription and the nature of the transcripts are considered. In *E. coli,* certain major tRNA species have their transcription controlled by promoters for the transcription units of rRNA species (10, 11). The promoter governing the transcription of T4 tRNAs also controls transcription of other RNA species and possibly that of the lysozyme gene as well (22). Few data are available regarding the details of RNA polymerase interaction with these sequences.

The best information on promoter sequences of tRNA genes per se comes from studies of the *E. coli* tRNA^{Tyr} genes. By enzymatically elongating synthetic gene sequences hybridized to transducing phage DNA in the 3′ to 5′ direction beyond the initiation of transcription, Khorana and his colleagues have identified a DNA sequence which they have called a promoter region as it is close to the site of initiation of transcription and has a symmetric structure, a feature shared by other promoter sequences (34–36). The identical sequence precedes both tRNA$_1^{Tyr}$ and tRNA$_2^{Tyr}$ gene precursor sequences, although the genes for the two species are widely separated on the *E. coli* map and about 3 times more tRNA$_2^{Tyr}$ is made than tRNA$_1^{Tyr}$. It seems inevitable that additional segments of DNA, further controlling the interaction of RNA polymerase with the tRNA^{Tyr} sequences, will be found, otherwise an explanation for the large difference in mature tRNA levels between the two species must rely solely on post-transcriptional control mechanisms. If the promoter sequence for every tRNA is outside the transcribed precursor sequence and if these promoter sequences are all similar to that for the tRNA^{Tyr} genes, which would allow uniform control of RNA polymerase initiation of transcription, could it not be possible that the extra

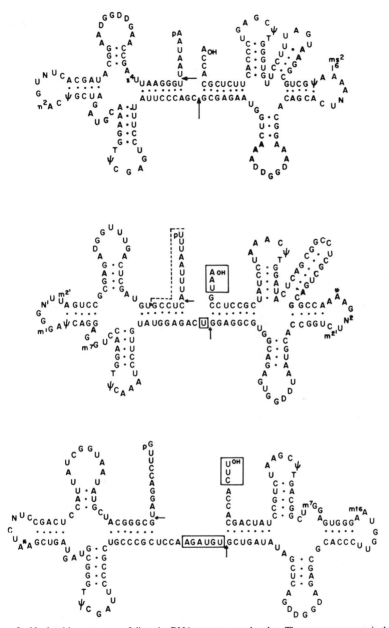

Figure 2. Nucleotide sequences of dimeric tRNA precursor molecules. The *upper sequence* is that of the Gln–Leu precursor encoded by T4 (59). The *arrows* pointing toward the sequence represent RNase P cleavage sites. The *middle sequence* is that of the T4 Pro–Ser precursor (57) and the *lower sequence* that of the Gly–Thr (46) precursor of *E. coli*. Nucleotides at the 3′-terminal not found in the mature tRNAs are *boxed*. The *arrows* indicate the RNase P cleavage sites. In the T4 Pro–Ser precursor, the RNase T₁-generated fragment which is resistant to RNase P cleavage is outlined by the *dotted line*.

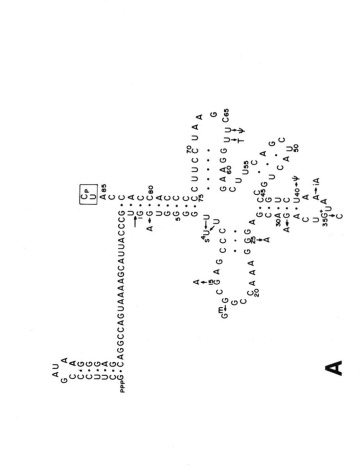

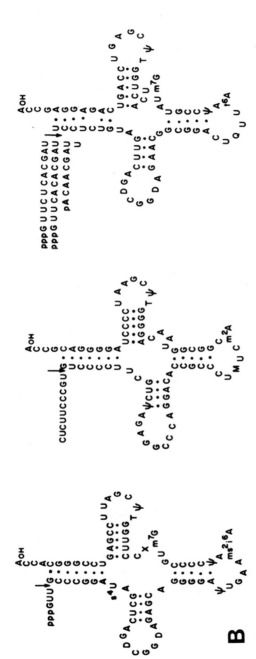

Figure 3. Nucleotide sequences of monomeric tRNA precursor molecules of *E. coli. A*, nucleotide sequence of the tRNA₁^Tyr precursor (43) molecule showing modifications found in the mature molecule, the position and nature of various mutations decreasing tRNA yield (see Table 1), and the anticodon su₃⁺ mutation (C35) which does not alter tRNA yield. The *arrow* pointing inward is the site of RNase P cleavage, and extra 3′-terminal nucleotides are *boxed. B*, nucleotide sequences of three monomeric tRNA precursors which were isolated by affinity chromatographic methods from an RNase P thermosensitive host. The *arrows* point to the sites of RNase P cleavage. *Left*, precursor to tRNA^Phe (47); *center*, precursor to tRNA₂^Glu (47); *right*, precursors to tRNA^Asp (i.e., four species were isolated (48)). No extra nucleotides were found at the 3′-termini of these molecules. The mature tRNA sequences are drawn with modifications intact.

5'-segment of precursors is the region of the gene where other controlling elements exist? The heterogeneous nature of these sequences (Figures 2 and 3) could reflect the ability of the cell to vary the level of different tRNA species through controls of the rate of transcription. Alternatively, every tRNA species could have a different promoter sequence which has evolved to provide the right level of initiation of transcription for the individual species. The heterogeneity in sequence and length of the extra 5'-proximal nucleotides of the precursor molecules could then be regarded as vestiges of ancient events, no longer having any meaningful function. Additional data regarding the control of transcription of tRNA genes have been derived from in vitro studies of tRNA gene transcription.

The appearance of mature tRNAs transcribed in vitro in systems directed by transducing phage DNA has been used as an assay for factors involved in production of these tRNAs (32, 37–41). Apparently, optimum tRNA production requires the RNA polymerase holoenzyme and is enhanced by inclusion of the transcription termination factor, ρ, in the mixture, although the role of ρ does not seem to aid transcription termination at a particular site (see also Chapter 7). Khorana and his colleagues have identified symmetrical elements in the nucleotide sequence adjacent to the 3'- terminus of $tRNA_1^{Tyr}$. The relationship of these elements to natural sites of transcription termination and/or the function of ρ is unknown (34, 35, 42).

The tRNA gene transcripts made in vitro from transducing phage DNA carrying these genes can be used as substrates to study the enzymology of processing. At least three nucleases seem to be involved in the trimming of tRNA from the very long initial transcript from the DNA (38–41). Two of these seem to be RNase P (43, 44) and a 3'–5'-exonuclease. Bikoff and Gefter (40) and Bikoff et al. (41) identified two more endoribonucleolytic activities which can cleave a long transcript from phage $\phi 80su_3^+$ containing the tRNA sequence, but these enzymic activities do not yet seem to have any direct correlation with functions identified in vivo. In most respects, these in vitro studies have tended to confirm independent studies of nuclease trimming action of the initial gene transcripts in vivo.

NUCLEOTIDE SEQUENCES OF tRNA PRECURSOR MOLECULES

Additional data regarding the organization of tRNA genes and transcriptional control of tRNA gene expression can be derived from an examination of the nucleotide sequences of tRNA precursor molecules.

Escherichia coli tRNAs

Two methods have been used to isolate those precursors to E. coli tRNAs which have had their sequences completely determined: a) isolation of precursors to tRNAs amplified in vivo by transcription from transducing phages (45, 46) and b) affinity chromatography techniques using characteristics of particular anticodons (47, 48). (With the use of these latter techniques, it is possible to adsorb to a tRNA-cellulose column specifically those tRNA precursors with codons complementary (in an antiparallel fashion) to the purified tRNA species used to make the tRNA-cellulose material. In addition, precursor tRNAs containing the modified base Q (which contains an accessible diol) in their anticodons selectively bind to borate-substituted cellulose derivatives.) The sequences derived from these studies are shown in Figures 2 and 3. It is immediately apparent that there is no primary sequence homology in the "extra" nucleotides and, furthermore, that the length of the extra regions is highly variable, even in the few cases in which primary transcripts have been identified. In addition, indirect studies of extra sequences in several other precursors showed that the 5'-proximal extra sequence may be highly variable in size. Those precursors not having a terminal triphosphate cannot be identified as primary gene transcripts and must then be products of cleavage events.

From an examination of these sequences it is clear that the enzyme involved in processing precursors at the 5'-termini of the mature tRNA cannot be sequence specific. Consequently, the possible regulatory roles of the extra sequences can be only speculated. The composition and size of the various extra 5'-sequences may reflect controlling elements for the fine tuning of the production of each isoaccepting class of tRNAs.

The complete nucleotide sequences of only a few precursors derived from *E. coli* are known. As noted above, some of these precursors were isolated by using cells infected with transducing phage carrying and amplifying the tRNA genes. Furthermore, in all cases, mutations in either the precursors themselves or in host cell enzymes resulted in the slowing down of the processing events, giving additional aid in the isolation procedures. If a cell with thermosensitive RNase P is incubated briefly at the restrictive temperature with ^{32}P and the RNA products are separated by electrophoresis in acrylamide gels, many precursor species can be identified (49–53). Some of these molecules are up to 600 or more nucleotides in length and contain as many as 5–7 tRNA sequences (54–56). Some of the very large molecules contain identical species repeated in tandem, indicating that several tRNA genes are present in *E. coli* in several copies and are located in clusters which are transcribed as a unit. The source of these deductions has been the partial sequence analysis of these large molecules, in particular, stoichiometric analysis of derived oligonucleotides containing the suspected anticodon sequences. Thus, although Figures 2 and 3 present completely known sequences, the data shown do not reflect the nature of all known transcription units or the variety of partial information regarding the nature of the large precursor species.

Bacteriophage T4 tRNAs

The complete sequence is known for four T4 tRNA precursors which have been isolated from two dimeric precursors (Figure 2) (57–61). At least four more T4 tRNAs can be isolated, and it is thought that all eight of these tRNAs are transcribed as a unit from the T4 genome (22). No precursor species containing

Table 1. Location in tRNA moieties of precursor molecule of mutations which affect processing events

tRNA precursor	Mutation	Affected function
E. coli Tyr$_1$	G15 → A15	RNase P cleavage
	G25 → A25	RNase P cleavage
	G31 → A31	RNase P cleavage
	G31 → U31	Probably RNase P cleavage
	G46 → A46	Probably RNase P cleavage
	G1 → A1	Not known; no precursor identified
	G2 → A2	Not known; no precursor identified
	C(−4) → U(−4)[a]	Not known; enhanced tRNA production
T4 Gln[b]–Leu	C11 → U11	Nucleotide modification; partial loss of RNase P and C-C-A repair at glutamine moiety
	G40 → A40	Nucleotide modification; partial loss of RNase P and C-C-A repair at glutamine moiety
	C62 → U62	Nucleotide modification; partial loss of RNase P and C-C-A repair at glutamine moiety
T4 Gln–<u>Leu</u>[b]	C72 → U72	Nucleotide modification; RNase P cleavage of both moieties
T4 Pro–<u>Ser</u>[b]	G67 → A67	Nucleotide modification; 3′ → 5′ exonuclease at serine terminus
	C70 → U70	Nucleotide modification; 3′ → 5′ exonuclease at serine terminus
	C75 → U75	Nucleotide modification; 3′ → 5′ exonuclease at serine terminus

Data, taken from ref. 43, 57, 59, 62–68.

[a] This mutation is four nucleotides from the 5′ terminus of the mature tRNA sequence and is in the 5′ extra sequence.

[b] Underlined moiety contains the mutation noted.

more than two tRNA sequences has been isolated, so partial processing of the long transcript must occur before transcription is completed. Examination of the sequences reiterates the conclusions drawn from the *E. coli* tRNAs. There is no primary sequence homology in the extra sequences and their size, too, is variable. One significant difference exists between the T4 Pro–Ser precursor and all other precursors thus far examined. The T4 Pro–Ser species lack -C-C-A in their initial gene transcripts. This sequence must be inserted into the tRNAs as part of their maturation process as is discussed further in the following section.

Knowledge of the primary sequences of both *E. coli* and T4 tRNA precursors is extremely useful because many mutants of these tRNAs disrupt the biosynthetic process. It is possible to correlate the phenotypical defects with the

Table 2. Second site mutation of tRNA$_1$^{Tyr} that restores normal
biosynthesis

Original mutation	Second site mutation
G31 → U31	C16 → U16
	C45 → U45
	C41 → A41
G31 → A31	C16 → U16
	C45 → U45
	C41 → U41
G25 → A25	C11 → U11
G46 → A46	C54 → U54
G1 → A1	C81 → U81
G2 → A2	C80 → U80

Data from refs. 4, 62–66.

position of the mutations in the tRNAs (and precursors) and to make inferences
regarding structure-function relationships (Table 1). It is worthwhile pointing out
that, in the absence of various genetic data, it might be concluded that T4 tRNAs
are transcribed as dimeric sequences. Thus, inferences regarding gene organization drawn from a description of precursor primary sequences may be misleading, a fact which should be remembered when considering the results presented
in the next section. For example, of the known mutations which occur in the
tRNA^{Tyr} gene, only those apparently not affecting secondary or tertiary structure (anticodon changes) do not affect precursor tRNA processing (Table 1)
(62–65). The nature of the features of the tertiary structure important for processing must be complex because the second mutations U16 and U45 enhance, in a
less than obvious fashion, the processing of precursors containing the U31 or
A31 mutations (Table 2). It has been proposed that the enhancement might occur
because the tRNA moiety is prevented from folding into an "incorrect" structure
by the second site mutations (66). Most second site mutations which restore
tRNA production do so, apparently, by restoring hydrogen-bonding arrangements.
　　Several mutations in the TψC loop region of the serine moiety of the Pro-
Ser T4 dimeric precursor slow down the action of the 3'–5'-exonuclease activity
(Table 1) (67). Therefore, this area of the molecule might be important for
substrate recognition by this enzyme. In other enzymes needed for processing,
studies of T4 precursors and their mutations are less illuminating. McClain and
his colleagues have shown that sites scattered throughout the tRNA moieties can
affect the functioning of the nucleotide modification enzymes and can reduce the
effectiveness of RNase P and the enzyme that incorporates C-C-A$_{OH}$ at the 3'
terminus. Mutations in the glutamine moiety of the Gln–Leu precursor affect
only RNase P processing of the glutamine moiety, whereas mutations in the
leucine moiety alter RNase P processing of both tRNA moieties (Table 1) (68).

Eukaryotic tRNAs

To date, preliminary data concerning the pure precursor species isolated from yeast (T. McCutchan and D. Söll, personal communication) and silkworms (R. Garber, personal communication) show that extra sequences are present at both ends of these molecules, that the tRNAs are at least partly modified, and that the extra sequences in precursor tRNAs homologous to known precursors from *E. coli* are significantly different from them. It is of great interest to determine the exact nature of these extra sequences because they may shed light on the following questions: a) are the extra sequences conserved throughout evolution for homologous tRNAs (reflecting tRNA species-specific control mechanisms)? b) are the extra sequences in any one organism conserved (reflecting some aspect of the gene organization in each organism)?

ENZYMOLOGICAL PROCESSING OF tRNA PRECURSORS

Endonucleases

Only one enzyme, RNase P (43, 44), has been shown to perform an obligatory function in the endonucleolytic processing of tRNA precursor molecules. Not only can it be demonstrated that this enzymic activity cleaves precursor molecules in vitro to generate their correct 5'-termini, but thermosensitive lethal mutants of *E. coli,* affecting RNase P function, also exist. Cell extracts made from these mutants and lacking RNase P function contain endonucleolytic activities which can cleave multimeric precursors (50, 54), but the role of these activities, named RNase P_2 (50) and RNase O (54), in vivo is unclear because no mutants have been isolated which affect these functions. Whether or not RNases P_2 and O are indeed the same or different activities has not yet been determined. Nor has it been rigorously demonstrated that these latter activities are distinct from RNase III (69) (another *E. coli* endoribonuclease which has a variety of substrates) or that the cleavages they perform are distinct from the secondary cleavages of RNase III (70). (To determine unequivocally that RNases P_2 and O are distinct activities from RNase III, it is above all necessary to demonstrate that their action is not affected by the addition of double-stranded RNA to the reaction mixtures in vitro. Considerations based on salt effects on the reaction or elution molarities from certain ion exchange media may be misleading because RNase III responds differently to different ionic conditions. One can imagine that RNase III may be needed for some steps in the processing of tRNA precursors from the large rRNA precursor molecule (12, 13) and that it could, under certain conditions, also cleave other tRNA precursor molecules, but not in the same fashion as RNase P.) The characteristics of RNase P are described below in some detail, although some discussion is also devoted to the putative RNases P_2 and O and their possible role in tRNA processing.

RNase P   Several of the tRNA precursor molecules of known sequence have been used as substrates in vitro for *E. coli* extracts containing RNase P activity (43, 46–48, 57, 59). In every case, through fingerprint analysis of the cleaved product, it has been demonstrated that RNase P cleaves the precursor molecule to generate the correct 5′ terminus of the tRNA sequence contained therein. Precursors to at least 35, if not all, tRNAs accumulate at restrictive temperatures in certain thermosensitive *E. coli* mutants (Figure 4). RNase P, purified from extracts of these mutants, is thermosensitive in vitro, and RNase P from one particular mutant also has an altered isoelectric point (71). Thus, there is strong evidence that the RNase P enzyme is essential in the maturation of all species of tRNA.

As mentioned previously, an examination of precursor nucleotide sequences shows that RNase P cannot be recognizing primary structure in its search for the appropriate cleavage site in various substrate molecules. Indeed, Guthrie et al. (72) have shown that RNase P does not attack a fragment of the T4 Pro–Ser precursor (Figure 2), thereby implicating larger features of the substrate molecule in the recognition process. In addition, as mentioned above, mutations exist in several locations of the various tRNA precursors (Figures 2 and 3*A*) which affect processing by RNase P (Tables 1 and 2). Loci far removed in space from the cleavage site can alter the enzyme-substrate recognition process. The simplest explanation is that RNase P recognizes a conformational feature shared by all precursor molecules, namely their tRNA moiety (assuming that the solution structures of the tRNA moieties are quite similar) (73). It is not yet clear what the necessary details of the region immediately surrounding the point of cleavage must be. On the 5′-side of the cleavage site, as few as three nucleotides are adequate for recognition of the tRNA^{Tyr} precursor. Most of the 5′-proximal extra sequence of tRNA^{Tyr} is unnecessary for RNase P action (65). On the 3′-side, i.e., the aminoacyl-accepting end of the tRNA molecule, which in the three-dimensional structure is very close to the 5′-end of the tRNA sequence, matters seem more complicated. McClain has shown that T4 precursors, which must have -C-C-A inserted into their sequences for maturation, cannot be cleaved efficiently by RNase P until this sequence is present (74–76). On the other hand, in vitro studies of the requirement for an intact -C-C-A sequence in tRNA^{Tyr} for efficient cleavage do not indicate a rigorous requirement for the presence of this sequence (71). Furthermore, tRNA^{Tyr} precursor isolated from thermosensitive cells under restrictive conditions, a good RNase P substrate tRNA, still has an untrimmed 3′-terminus. Thus, the effect of an intact -C-C-A sequence or untrimmed 3′-terminus upon RNase P cleavage may vary with the individual precursor molecule under consideration.

Finally, certain mutants of tRNA^{Tyr}, which affect the hydrogen bonding of the amino acid stem very close to termini of the mature tRNA sequence, do not yield precursor under any conditions yet tried. These mutants also produce very little mature tRNA. It has been proposed that the possible disruption of the hydrogen bonding of the amino acid stem of the tRNA moiety of the precursor makes the molecule much more susceptible to degradative exonucleases (77). These exonucleases compete with the normal processing enzymes and reduce the

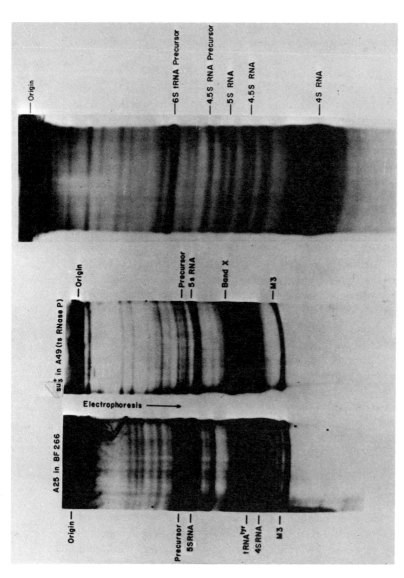

Figure 4. Autoradiographs of acrylamide gel separations of RNA species extracted from *E. coli* under different conditions for tRNA precursor separation. *Left,* RNA extracted from a host normal in RNase P function infected with φ80su⁺₃-A25. The A25 mutation slows the action of RNase P on the tRNA precursor. *Center,* RNA extracted from host cell temperature sensitive for RNase P function, infected at the restrictive temperature with φ80su⁺₃. Note the underproduction of mature RNA and the many bands in the *upper portion* of the gel. These bands represent multimeric tRNA precursors. *Right,* RNA extracted from cells as shown in the center panel except that the cells are uninfected. Thus, there is no enrichment for any one particular tRNA species. Note again the many unlabeled bands throughout the gel which are mostly tRNA precursor species. Adapted from Altman et al. (65) with permission of Brookhaven National Laboratory; and Bothwell et al. (80) with permission of *Journal of Biological Chemistry.*

amount of precursor traveling along the normal pathway to maturation (77). (It is well established, that 3'-5'-exonucleases such as snake venom phosphodiesterase stop at regions of secondary structure but proceed readily under conditions in which this structure is denatured (78)). When such mutations (e.g., A1 or A2 in Table 1 (79)) occur in the first gene of the tRNA^{Tyr} doublet on the transducing phage ϕ80, not only is the yield of the first gene product less than 10% of normal, but the second gene yield of the tRNA is also reduced by about 50% (79). Mutations in the same region of the first gene which do not alter the hydrogen bonding of the amino acid stem do not decrease mature tRNA yield as drastically.

In vitro RNase P not only cleaves tRNA precursor molecules but also cleaves the precursor to *E. coli* 4.5 S RNA (80) and bacteriophage ϕ80-induced M3 RNA (81, 82). In these reactions, the enzyme seems to make use of both primary and secondary structure in recognizing its cleavage site, as does RNase III in the recognition of some of its substrates (83). The possible biological roles of these other RNase P substrates are not known, but the precursor to 4.5 S RNA does accumulate under restrictive conditions in RNase P thermosensitive mutants.

RNases P$_2$ and O When some of the multimeric molecules are exposed to extracts of thermosensitive *E. coli* which have been heat-treated, a residual endonucleolytic activity can be detected which reduces the molecular weight of these molecules (50, 51, 54, 56). Fingerprint analysis shows that the cleavages are always in intercistronic regions. If these substrates are subsequently exposed to active RNase P, tRNA-sized molecules are generated.

In some cases, in vitro treatment of the large precursors with RNase P generates intermediates of various sizes rather than mature tRNA molecules. Subsequent treatment with RNase P$_2$ or O is needed as a prerequisite for final processing by RNase P. These experiments in vitro have been interpreted as evidence that at least one additional endonuclease, besides RNase P, is needed for the maturation of large precursor molecules in vivo, and that the process of size trimming may be a well defined combination of cleavage events with different precursors (see Figure 5). That is, perhaps because of conformational restrictions, either RNase P$_2$ (or O) or P must act before the other enzyme in order to expose further cleavage sites in the multimeric precursors.

It should be strongly emphasized, however, that test tube conditions may not always be relevant to those in vivo. Ordering a sequence of enzymic reactions is best done by analyzing such a pathway with mutants which inactivate various steps (as has been done for one T4 precursor by McClain; see below). Characterization of RNases P$_2$ and O is still incomplete, and although it seems likely that under certain conditions such endonucleases may participate in the processing of tRNAs, their involvement can only be rigorously demonstrated with the isolation of lethal mutants affecting these functions.

Evidence exists for at least one other mode of attack on precursor molecules in vivo. Two precursor molecules, those of tRNA^{Asp} and tRNA^{Tyr} , when iso-

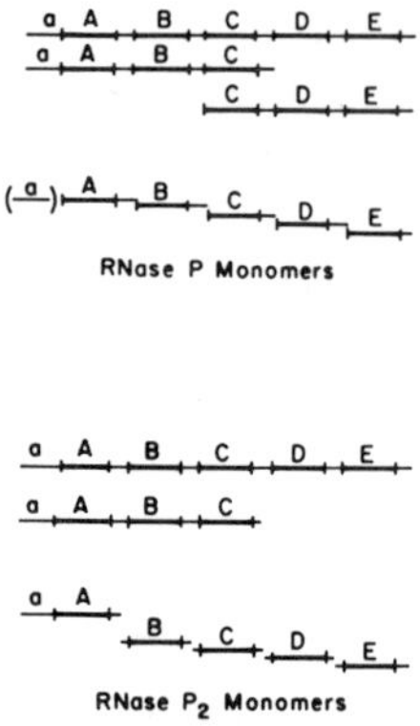

a A B C D E
a A B C
      C D E

(a) A B C D E

RNase P Monomers

a A B C D E
a A B C

a A
      B C D E

RNase P₂ Monomers

Figure 5. Schematic plan for steps in the processing of large *E. coli* tRNA precursors. *Top,* the hypothetical action of RNase P on a pentameric precursor molecule. In this case, it is suggested that the first cleavage splits the large molecule into two parts and subsequent cleavages generate all of the correct 5' termini. The resulting monomeric precursors must still have their 3' termini trimmed. *Bottom,* the hypothetical action of RNase P₂ on the large precursor. This latter enzymic activity cleaves in intercistronic regions, generating fragments which must be acted on further by RNase P and a 3' terminus trimming enzyme. If RNase P₂ does indeed play a role in normal tRNA precursor maturation, processing of large precursor molecules occurring through a combination of the steps shown in the upper and lower parts of the diagram can be envisioned. That is, cleavages by RNase P might be preceded or followed by cleavages by RNase P₂, each cleavage facilitating the next in the sequence. Different large precursor molecules might be processed by a different combination and ordered sequences of RNase P and RNase P₂. Adapted from Schedl et al. (56) with permission of MIT Press.

lated from a host with thermosensitive RNase P (48, 49, 79), can be found incorrectly processed in low molar yields. That is, the tRNA sequence is found with one additional nucleotide, $_{OH}$Up, at the 5'-terminus. None of the known tRNA processing enzymes generate 5'-hydroxyl groups. Perhaps the thermosensitive RNase P has some residual aberrant activity, and the products then generated are susceptible to phosphatase activity.

Exoribonucleases

Crude extracts of *E. coli* have the ability in vitro to remove nucleotides from the 3'-terminus of radioactive precursors prepared either in vivo or in vitro (38–41, 43, 52). At least four of the 3'-terminal nucleotides of tRNA^{Tyr} precursor can be removed by an activity present in crude extracts of *E. coli*. Several exoribonucleases found in *E. coli* might be responsible for this action. Schedl found one such activity, P₃, which copurified with RNase II (84), a 3'–5'-exoribonuclease. At least one other apparently distinct activity, identified by Bikoff et al. (40, 41), performs this same exonucleolytic function. Shimura and his collaborators (54) have also described another exonuclease, RNase Q, which carries out the same exonucleolytic function. It is possible that several exonucleases exist in *E. coli* which can perform the precursor trimming in vitro but that in vivo only one of these performs this particular function under normal conditions. One candidate is the so-called BN exonuclease, which is absent in *E. coli* strain BN. In this strain,

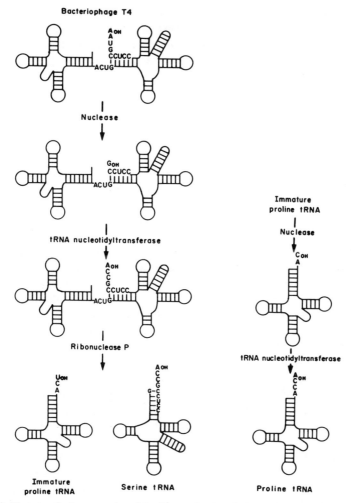

Figure 6. Five steps in the processing of the T4-encoded Pro–Ser tRNA precursor. The first step is trimming of the 3'-terminus to remove nucleotides not found in the mature tRNA^{Ser} sequence. This terminus is then repaired by the C-C-A enzyme (nucleotidyltransferase); subsequently, RNase P cleaves the precursor at the sites indicated. Trimming and repair of the tRNA^{Pro} 3'-terminus then occur. Reproduced from Seidman et al. (57) with permission of Academic Press Inc. (London) Ltd.

T4 precursors cannot be processed to mature tRNAs, because extra nucleotides at the 3'-termini are not removed from them (74). These data have facilitated the ordering of five distinct steps in the maturation of some T4 tRNAs (Figure 6). The BN mutation may not be in the structural gene for an exonuclease but clearly prevents the action of an exonuclease responsible for the trimming of at least some T4 precursors. It is not known whether this exonuclease is identical with those labeled variously as RNase II, P_3, or similarly, Q, of Sakano and Shimura (54).

An additional, less interesting role for 3'-exonucleases in tRNA biosynthesis is the degradative role. The 5'-proximal fragments released by RNase P cleavage of precursors and incorrect precursors, such as that to the mutant tRNA^{Tyr} A1, must be degraded to mononucleotides, and 3'-5'-exonucleases can perform this function.

tRNA Nucleotidyltransferase (Action at -C-C-A$_{OH}$ Terminus)

All tRNAs contain the 3'-terminal sequence -C-C-A$_{OH}$, yet some tRNA gene transcripts, notably those encoded by genes for T4 tRNA^{Pro} and tRNA^{Ser}, do not have this sequence intact. Some mechanism must exist for providing the T4 transcripts with this sequence. An enzyme which can add the terminal residues containing cytidines and adenosine to transcripts lacking these nucleotides has been purified from both E. coli and rat liver (85). It is clear that before addition of -C-C-A to T4 tRNA gene transcripts, nucleotides found in the transcript beyond the 3'-end of the base-paired acceptor stem must be removed by exonuclease action. In addition to the obvious need for the CCA enzyme (nucleotidyltransferase) to complete the necessary sequence on pretrimmed tRNA gene transcripts, it has been postulated that it also plays a role in the repair of mature tRNAs (86). By suitably labeling nucleotides or tRNAs or both the constant in vivo turnover of the 3'-terminal adenosine of mature tRNAs can be demonstrated. In a sense, then, the CCA enzyme is a repair enzyme and, indeed, is often called the CCA repair enzyme.

Cells lacking the CCA enzyme (or having reduced levels) do not support the normal production of mature T4 tRNAs because the -C-C-A terminus is added very slowly, if at all, to these molecules (87). The accumulation of some tRNAs lacking the -C-C-A terminus in such a strain, and also the accumulation of precursor molecules to which the -C-C-A had not been added, have been observed. This is evidence that cleavage of certain tRNA precursors by RNase P might require the presence of the -C-C-A sequence for efficient action by the endoribonuclease. This expectation has been confirmed in vitro by Schmidt et al. (76) in experiments comparing the rates of RNase P cleavage of two bacteriophage tRNA precursor molecules, identical in sequence except at their 3' termini. In vitro, however, the discrimination between the two species is not as rigorous as the in vivo results suggest. However, in vitro experiments are usually done with great excess of enzyme present and at concentrations of attendant carrier tRNA which are probably less than concentrations in vivo. This latter fact is important because the ability of RNase P to discriminate between precursors, either with or without the -C-C-A terminus, depends on the concentration of mature tRNA, which is a competitive inhibitor of RNase P in the reaction mixture (71). The presence of the -C-C-A sequence in a tRNA precursor may affect the cleavage rate of RNase P, but not its specificity.

Deutscher has recently found that the CCA enzyme may play a role in T4 phage maturation (88). Phage which have grown in cells with low levels of CCA enzyme are incapable of producing more than 10% normal bursts when plated on normal hosts. It seems that DNA replication or packaging or both is disturbed in the second cycle infection which implies that some entity, possibly tRNA, is packaged in phage heads and plays some role in directing normal phage DNA replication. In phage produced in the mutant cells, the packaged tRNA is defective and thus leads to aberrant DNA replication in the next round of replication.

Nucleotide Modification

tRNA gene transcripts are made by RNA polymerase with the use of the four nucleotides containing adenosine, cytidine, uridine, and guanosine. Yet all mature tRNAs participating in protein synthesis have a high content of rare or minor nucleotides (89, 90). The nucleotide modifications are carried out enzymically with an intact tRNA polymer as substrate. The action of at least one modifying enzyme, notably (m^5U)methyltransferase, is under stringent control as one might anticipate (T. Ny and G. Björk, personal communication), because tRNA gene transcription itself is under the same control.

Some of the nucleotide modification reactions have been performed in vitro by using partially purified enzyme preparations and either chemically or naturally undermodified tRNA or partially processed precursor molecules as substrates (89–91).

Nucleotide modifications in tRNA precursors do not appear to be necessary for the size maturation of the tRNA gene transcripts. Several modification-deficient mutants of *E. coli* have been characterized but in none is any particular tRNA gene product missing (although the function of certain mature tRNAs may be affected) (92, 93). Because the full range of modification-deficient mutants is not available, in vitro experiments must indicate whether or not modification is necessary for processing. In fact, tRNA synthesized enzymically in vitro can be processed to tRNA-sized gene products in the absence of any modifying enzymes (32, 37–39, 41–42, 91, 94).

With the use of precursor tRNAs made in vivo, RNase P cleavage can be carried out normally without fully modified T, Gm, iA, and s^4U in the substrate (91). An analysis of several different precursors has led to the notion that certain modifications are carried out at a faster rate than others on precursor tRNAs (91). It also seems that some modifications, such as Gm or pseudouridine when found in anticodon loops, may be carried out efficiently only on the mature tRNA of certain species in vivo. The fact that T4 precursors are usually isolated fully modified may be explained by their very long lifetimes in vivo (perhaps attributable to slow cleavage by endonucleases awaiting the action of 3'- trimming activities). Thus, even those modification reactions which may proceed at a relatively slow rate on *E. coli* precursors have time to go to completion on T4 precursors. In addition, a multimeric *E. coli* precursor which lacks Gm can be cleaved in vitro by using crude cell extracts, and one of the monomeric products subsequently has the modification. The monomeric product is then processed further in vitro to generate a mature species (54). The Gm modification enzyme probably can act only on the monomeric substrate. Because *E. coli* precursor tRNAs are normally isolated from strains in which their half-lives have been artificially lengthened, it is not possible to say definitely that certain modifications always take place at the precursor level. Rather, it can be said that most modifications of *E. coli* precursors and certainly those of T4 precursors can occur at the precursor level.

To study the functions of the modifications in mature tRNAs, it is usually necessary to study tRNAs from which they are absent. Such experiments have shown that pseudouridine is implicated in the regulation of certain operons when this modification appears in the anticodon loop (93). Various methylations or other aliphatic modifications exert some effect on ribosome binding or aminoacylation efficiencies (95, 96). Thiolation and other modifications of uracil undoubtedly alter codon-anticodon interactions when these are found in the first (wobble) anticodon position (97–100). Most of the effects are partial losses of activity or alterations of the kinetics of certain reactions. It is as if the cell uses the nucleotide modifications to fine tune some aspects of protein synthesis.

Processing of tRNA Gene Transcript in Eukaryotes

It has long been established that cytoplasmic extracts of eukaryotic cells, specifically, mammalian cells, yeast, and silkworms, contain an activity which is capable of reducing the molecular weight of the homologous precursor species to that of tRNA size (1). No firm identification has been made of the cleavage sites of eukaryotic substrates because of the unavailability of radiochemically pure precursor to a single tRNA species.

tRNA processing enzymes have been partially purified from eukaryotic systems by experiments based on the following argument. Because enzyme-substrate specificity of tRNA processing enzymes seems to lie in structural conformation, if it is assumed that these structures have been preserved throughout evolution, a prokaryotic tRNA precursor exposed in vitro to a eukaryotic cell extract should be accurately and efficiently cleaved. This approach, initially with the use of the *E. coli* tRNA[Tyr] precursor as a substrate probe, proved fruitful in the identification of RNase P-like activities in mammalian cells (101), chick embryo thigh muscle (E. J. Bowman, personal communication), and silkworms (R. L. Garber, personal communication). An endoribonuclease activity capable of generating the correct 5'-terminus of the mature tRNA sequence has been purified from each organism. 3'-Exonucleolytic activity has also been identified in each case. Although the experiments with the *E. coli* precursor have been instructive in identification of these activities and in demonstration of the conservation of such enzymic mechanisms, it must be shown that these enzymes, partially purified through the use of a prokaryotic substrate, can also process their homologous eukaryotic precursor tRNAs. This has been done with extracts of KB cells (101) and silkworms (R. L. Garber, personal communication). It has not yet been possible to show that the RNase P-like activity in KB cells actually generates the correct 5'-termini of eukaryotic tRNAs because no individual KB tRNA precursor species has been isolated in radiochemically pure form. Recently, however, some success has been achieved in studies on silkworms.

When silkworms (or isolated silk glands) are labeled radioactively with a short (6 hr) pulse of ^{32} P$_i$ during the fifth instar just prior to silk fibroin synthesis, precursor tRNAs can be easily isolated by phenol extraction and electrophoresis in acrylamide gels. Two-dimensional gel electrophoresis yields precursor species, according to the criterion that a high degree of homology exists between their fingerprints and those of certain mature tRNA species (R. Garber, personal communication). These individual precursor species can serve as substrates for a partially purified, RNase P-like activity which has been isolated from silk glands by using the *E. coli* tRNA^{Tyr} precursor as substrate in purification assays. Proof that cleavage occurs at the 5′-termini of the mature tRNA sequences awaits the total sequence analysis of individual precursor substrates and their cleavage products.

Certain gaps in our knowledge of eukaryotic precursor molecules make it impossible to state that only two enzymes, RNase P-like activity, and a 3′-5′-exonuclease are needed for the size processing of these monocistronic molecules. It is necessary to show that individual eukaryotic precursor species have 5′-triphosphate termini; otherwise, it may be that they are already partially processed when isolated. In addition, our knowledge of 3′-transcription termination steps is virtually nonexistent; therefore, statements regarding the extent of processing in vivo prior to precursor isolation cannot be definitive. On the other hand, in vitro studies of transcription from isolated chromatin by RNA polymerase III have shown that putative tRNA precursor species, made under conditions which do not seem to allow any significant processing events to take place, are of approximately the same size as those isolated in vivo and that the heterogeneous collection of molecules is enriched in 5′-triphosphate termini (102–106). Thus, it may indeed be proven that only two enzymes are needed to accomplish the size processing of eukaryotic tRNA precursor molecules.

CONTROL OF GENE EXPRESSION

Regulation of Transcription
The control of transcription itself necessarily depends on the mechanism of action of RNA polymerase and the accessibility of template DNA to that enzyme. In *E. coli,* the rate of transcription of tRNA and rRNA genes seems to be tightly coupled to the cell's growth rate, but the coupling mechanism is not completely understood (107). The coupling mechanism may act through agents which decrease the tight binding of RNA polymerase to promoter regions (regions of DNA where transcription can start) (reviewed in ref. 33), but the way in which tRNA gene transcription is directly regulated under normal cell growth conditions is not known definitely (see also chapter 7).

Factors Affecting RNA Polymerase-Promoter Interaction   The binding of *E. coli* RNA polymerase holoenzyme to DNA promoter sequences must be characterized, under a given set of conditions, by a particular binding constant for each promoter sequence. The strength of this interaction can be altered by several factors, including salt concentration, temperature, and the presence or absence of certain nucleotide and protein cofactors (33). In *E. coli,* at least, regulation of transcription of tRNA genes is accomplished in part by the stringent response to amino acid starvation, i.e., a drop in the intracellular amino acid concentration. During this response, ppGpp and pppGpp (108) accumulate in the cell and, through interactions with RNA polymerase holoenzyme, weaken the affinity of the enzyme for tRNA and rRNA promoter sites. Although it is well established that various inducible protein factors, particularly a certain one in- duced after T4 infection (33), can associate with holoenzyme and alter its speci- ficity, there is no evidence that a similar mechanism is used to alter the rate of transcription of tRNA genes, in particular, in uninfected cells. However, under certain conditions with transducing phage DNA as template, elongation factors important in protein synthesis, EF-Tu.Ts, can influence the level of transcription of a tRNA gene (109). Thus, in addition to use of the nucleotide derivatives ppGpp and pppGpp, transcription control can be coupled to translational pro- cesses through protein factors involved in protein synthesis.

In eukaryotes, the DNA template may not be uniformly accessible to RNA polymerase because of the nonuniform architecture of the chromosome (i.e., different chromosomal proteins bound at different places). The unfolding of local chromosomal structure may be affected during various developmental events (hormone action, acetylation of histones, or the phosphorylation of certain pro- teins) which may lead to the exposure of promoter sites. In eukaryotes, an additional control over transcription has been introduced through the diversity of the RNA polymerase enzymes themselves. There are three different RNA polymerases, of which one (RNA polymerase III) seems to be specifically used for the transcription of tRNA genes, among others (105). The controls exerted over RNA polymerase action in *E. coli* could also apply, in principle, to the eukaryotic RNA polymerase III. Thus, eukaryotes have evolved an additional mechanism for controlling classes of gene transcription: different enzymes are responsible for the transcription of different classes of genes. Accurate control of the amounts of functional tRNAs needed for specific protein synthesis, however, may be more easily governed by post-transcriptional nucleotide modification of specific tRNAs or by degradation of specific tRNAs or their precursors.

Regulation of Amount of tRNA Gene Product   Regulation of tRNA gene product occurs at two obvious levels, through control of the amount of transcript made and through control of the amount of transcript actually processed into a usable molecule by the cell. The latter type of control is exemplified in the previous discussion of post-transcriptional processing enzymes and their sub- strate specificity. tRNA molecules, in which the secondary or tertiary structure or both have been altered by mutation, are not produced in normal amounts in growing cells because the processing steps are delayed. This control can be exerted at the level of 3′-exonuclease action, -C-C-A addition, or RNase P cleavage. The decrease in processing rates leads not only to the accumulation of precursor but also to the degradation of larger amounts of gene transcript during the waiting time. Apparently, processing enzymes and enzymes which recognize these RNAs as unstable molecules prone to complete degradation constantly compete for precursor molecules (77). Normally, the processing enzymes work so efficiently that the bulk of the transcribed sequences are made into mature products. However, any mutation which decreases the rate of specific substrate cleavage renders a precursor susceptible to degradation by exonucleases for a longer period of time. Any small adjustment in degradative enzyme function leads to an alteration in the amount of final product produced.

CONCLUSION

Certain enzymological processes, once established in the simplest organisms, seem to be preserved throughout evolution. For example, tRNA structures are essentially the same in *E. coli* and in human cells, and the specificity of an enzyme such as RNase P remains unchanged in its recognition requirement for a tRNA-like conformation in the two organisms. What has changed, or rather evolved, in complexity is the nature of control of tRNA gene expression. The accessibility of RNA polymerase to eukaryotic tRNA genes, their transcription as monocistronic units, and their more complex modifications and infinite variations in isoacceptor levels during developmental processes seem to be responses to the need for intricate control mechanisms for modulating the expression of all genes in higher organisms.

ACKNOWLEDGMENTS

I wish to acknowledge the many contributions to work in my laboratory made by A. Bothwell, E. Bowman, R. Garber, R. Koski, and B. Stark. A. M. Körner provided invaluable assistance in the preparation of this manuscript, which was completed in spring, 1977.

REFERENCES

1. Burdon, R. H. (1971). Prog. Nucleic Acid Res. Mol. Biol. 11:33.
2. Schaefer, K., and Söll, D. (1974). Biochemie 56:795.
3. Altman, S. (1975). Cell 4:21.
4. Smith, J. D. (1976). Prog. Nucleic Acid Res. Mol. Biol. 16:25.
5. Smith, J. D. (1972). Annu. Rev. Genet. 6:235.
6. Hartman, P. E., and Roth, J. R. (1973). Adv. Genet. 17:1.
7. Berg, P. (1973). Harvey Lect. 247.
8. Bachmann, B. J., Low, K. B., and Taylor, A. L. (1976). Bacteriol. Rev. 40:116.
9. Brenner, D. J., Fournier, M. Y., and Doctor, B. P. (1970). Nature (Lond.) 227:448.
10. Lund, E., Dahlberg, J. E., Lindahl, L., Jaskunas, R., Dennis, P. P., and Nomura, M. (1976). Cell 7:165.
11. Lund, E., and Dahlberg, J. E. (1977). Cell. 11:247.
12. Dunn, J. J., and Studier, F. W. (1973). Proc. Natl. Acad. Sci. USA 70:3296.

13. Nikolaev, N., Silengo, L., and Schlessinger, D. (1973). Proc. Natl. Acad. Sci. USA 70:3361.
14. Daniel, V., Sarid, S., and Littauer, U. Z. (1970). Science 167:1682.
15. Scherberg, N. H., and Weiss, S. B. (1970). Proc. Natl. Acad. Sci. USA 67:1164.
16. Wilson, J. H., Kim, J. S., and Abelson, J. (1972). J. Mol. Biol. 71:547.
17. McClain, W. H., Guthrie, C., and Barrell, B. G. (1972). Proc. Natl. Acad. Sci. USA 69:3703.
18. Wilson, J. H., and Abelson, J. N. (1972). J. Mol. Biol. 69:57.
19. Chen, M. J., Locker, J., and Weiss, S. B. (1976). J. Biol. Chem. 251:536.
20. Hunt, C., Hwang, L. T., and Weiss, S. B. (1976). J. Virol. 20:63.
21. Velten, J., Fukuda, K., and Abelson, J. (1976). Gene 1:93.
22. Kaplan, D. A., and Nierlich, D. P. (1975). J. Biol. Chem. 250:934.
23. Feldmann, H. (1976). Nucleic Acids. Res. 3:2379.
24. Grigliatti, T. A., White, B. N., Tener, G. M., Kaufman, T. C., Holden, J. J., and Suzuki, D. T. (1973). Cold Spring Harbor Symp. Quant. Biol. 38:461.
25. Clarkson, S. G., and Birnstiel, M. L. (1973). Cold Spring Harbor Symp. Quant. Biol. 38:451.
26. Gilmore, R. A., and Mortimer, R. K. (1966). J. Mol. Biol. 20:307.
27. Schwartzbach, S. D., Hecker, L. I., and Bennett, W. E. (1976). Proc. Natl. Acad. Sci. USA 73:1984.
28. Reijnders, L., and Borst, P. (1972). Biochem. Biophys. Res. Commun. 47:126.
29. Blatt, B., and Feldmann, H. (1973). FEBS Lett. 37:129.
30. Chen, G. S., and Siddiqui, M. A. Q. (1973). Proc. Natl. Acad. Sci. USA 70:2610.
31. Tsutsumi, K., Majima, R., and Shimura, K. (1974). Jap. J. Biochem. 76:1143.
32. Daniel, V., Beckmann, J. S., Grimberg, J. I., and Zeevi, M. (1976). In N. Kjeldgaard and O. Maaloe (eds.), Control of Ribosome Synthesis, pp. 268–278. Munksgaard, Copenhagen.
33. Travers, A. (1976). Nature (Lond.) 263:641.
34. Sekiya, T., and Khorana, H. G. (1974). Proc. Natl. Acad. Sci. USA 71:2978.
35. Loewen, P. C., Sekiya, T., and Khorana, H. G. (1974). J. Biol. Chem. 249:217.
36. Sekiya, T., Contreras, R., Kupper, H., Landy, A., and Khorana, H. G. (1976). J. Biol. Chem. 251:5124.
37. Ikeda, H. (1971). Nature (Lond.) 234:198.
38. Grimberg, J. I., and Daniel, V. (1974). Nature (Lond.) 250:320.
39. Daniel, V., Grimberg, J. I., and Zeevi, M. (1975). Nature (Lond.) 257:193.
40. Bikoff, E. K., and Gefter, M. L. (1975). J. Biol. Chem. 250:6240.
41. Bikoff, E. K., LaRue, B. F., and Gefter, M. L. (1975). J. Biol. Chem. 250:6248.
42. Kupper, H., Contreras, R., Landy, A., and Khorana, H. G. (1975). Proc. Natl. Acad. Sci. USA 72:4754.
43. Altman, S., and Smith, J. D. (1971). Nature (New Biol.) 233:35.
44. Robertson, H. D., Altman, S., and Smith, J. D. (1972). J. Biol. Chem. 247:5243.
45. Altman, S. (1971). Nature (New Biol.) 229:19.
46. Chang, S., and Carbon, J. (1975). J. Biol. Chem. 250:5542.
47. Vogeli, G., Grosjean, H., and Söll, D. (1975). Proc. Natl. Acad. Sci. USA 72:4790.
48. Vogeli, G., Stewart, T. S., McCutchan, T., and Söll, D. (1977). J. Biol. Chem. 252:2311.
49. Schedl, P., and Primakoff, P. (1973). Proc. Natl. Acad. Sci. USA 70:2091.
50. Schedl, P., Primakoff, P., and Roberts, J. (1974). Brookhaven Symp. Biol. 26:53.
51. Schedl, P. (1975). Ph.D. dissertation, Stanford University, Stanford, California.
52. Sakano, H., Yamada, S., Ikemura, T., Shimura, Y., and Ozeki, H. (1974). Nucleic Acids Res. 1:355.

53. Ikemura, T., Shimura, Y., Sakano, H., and Ozeki, H. (1975). J. Mol. Biol. 96:69.
54. Sakano, H., and Shimura, Y. (1975). Proc. Natl. Acad. Sci. USA 72:3369.
55. Ilgen, C., Kirk, L. L., and Carbon, J. (1976). J. Biol. Chem. 251:922.
56. Schedl, P., Roberts, J., and Primakoff, P. (1976). Cell 8:581.
57. Seidman, J. G., Barrell, B. G., and McClain, W. H. (1975). J. Mol. Biol. 99:733.
58. Barrell, B. G., Seidman, J. G., Guthrie, C., and McClain, W. H. (1974). Proc. Natl. Acad. Sci. USA 71:413.
59. Guthrie, C. (1975). J. Mol. Biol. 95:529.
60. Guthrie, C., Seidman, J. G., Altman, S., Barrell, B. G., Smith, J. D., and McClain, W. H. (1973). Nature (New Biol.) 246:6.
61. Nierlich, D. P., Lamfrom, H., Sarabhai, A., and Abelson, J. N. (1973). Proc. Natl. Acad. Sci. USA 70:179.
62. Abelson, J. N., Barnett, M. L., Landy, L., Russell, R. L., and Smith, J. D. (1970). J. Mol. Biol. 47:15.
63. Smith, J. D., Barnett, L., Brenner, S., and Russell, R. L. (1970). J. Mol. Biol. 54:1.
64. Anderson, K. W., and Smith, J. D. (1972). J. Mol. Biol. 69:349.
65. Altman, S., Bothwell, A. L. M., and Stark, B. C. (1974). Brookhaven Symp. Biol. 26:12.
66. Smith, J. D. (1974). Brookhaven Symp. Biol. 26:1.
67. McClain, W. H., Barrell, B. G., and Seidman, J. G. (1975). J. Mol. Biol. 99:717.
68. McClain, W. H., and Seidman, J. G. Nature (Lond.) 257:106.
69. Robertson, H. D., Webster, R. E., and Zinder, N. D. (1968). J. Biol. Chem. 243:82.
70. Dunn, J. J. (1976). J. Biol. Chem. 251:3807.
71. Stark, B. (1977). Ph.D. thesis, Yale University, New Haven, Connecticut.
72. Guthrie, C., Seidman, J. G., Comer, M. M., Bock, R. M., Schmidt, F. J., Barrell, B. G., and McClain, W. H. (1974). Brookhaven Symp. Biol. 26:106.
73. Chang, S. E., and Smith, J. D. (1973). Nature (New Biol.) 246:165.
74. Seidman, J. G., Schmidt, F. J., Foss, K., and McClain, W. H. (1975). Cell 5:389.
75. Seidman, J. G., and McClain, W. H. (1975). Proc. Natl. Acad. Sci. USA 72:1491.
76. Schmidt, F. J., Seidman, J. G., and Bock, R. M. (1976). J. Biol. Chem. 251:2440.
77. Altman, S., and Robertson, H. D. (1973). Mol. Cell. Biochem. 1:83.
78. Miller, J. P., Hirst-Burns, M. E., and Philipps, G. R. (1970). Biochim. Biophys. Acta 217:176.
79. Ghysen, A., and Celis, J. E. (1974). Nature (Lond.) 29:418.
80. Bothwell, A. L. M., Stark, B. C., and Altman, S. (1976). Proc. Natl. Acad. Sci. USA 73:1912.
81. Pieczenik, G., Barrell, B. G., and Gefter, M. L. (1972). Arch. Biochem. Biophys. 152:152.
82. Bothwell, A. L. M., Garber, R. L., and Altman, S. (1976). J. Biol. Chem. 251:7709.
83. Robertson, H. D., and Dunn, J. J. (1975). J. Biol. Chem. 250:3050.
84. Singer, M. F., and Tolbert, G. (1965). Biochemistry 4:1319.
85. Deutscher, M. P. (1973). Prog. Nucleic Acid Res. 13:51.
86. Deutscher, M. P., and Evans, J. A. (1977). J. Mol. Biol. 109:593.
87. Deutscher, M. P., Foulds, J., and McClain, W. H. (1974). J. Biol. Chem. 249:6696.
88. Morse, J. W., and Deutscher, M. P. (1976). Biochem. Biophys. Res. Commun. 73:953.
89. Söll, D. (1971). Science 173:293.
90. Nishimura, S. (1972). Prog. Nucleic Acid Res. Mol. Biol. 12:49.

91. Schaefer, K., Altman, S., and Söll, D. (1973). Proc. Natl. Acad. Sci. USA 70:3626.
92. Bjork, G. R., and Isaakson, L. A. (1970). J. Mol. Biol. 51:83.
93. Singer, C. E., Smith, G. R., Cortese, R., and Ames, B. N. (1972). Nature (New Biol.) 238:72.
94. Yang, S., Reinitz, E. R., and Gefter, M. L. (1973). Arch. Biochem. Biophys. 157:55.
95. Gefter, M. L., and Russell, R. L. (1971). J. Mol. Biol. 39:145.
96. Bjork, G. R., and Neidhardt, F. C. (1975). J. Bacteriol. 124:99.
97. Colby, D. S., Schedl, P., and Guthrie, C. (1976). Cell 9:449.
98. Agris, P. F., Söll, D., and Seno, T. (1973). Biochemistry 12:4331.
99. Sekiya, T., Takeishi, K., and Ukita, T. (1969). Biochim. Biophys. Acta 182:411.
100. Feinstein, S. I., and Altman, S. (1977). J. Mol. Biol. 112:453.
101. Koski, R. A., Bothwell, A. L. M., and Altman, S. (1976). Cell 9:101.
102. Marzluff, W. F., Jr., Murphy, E. C., and Huang, R. C. C. (1974). Biochemistry 13:3689.
103. McReynolds, L., and Penman, S. (1974). Cell 1:139.
104. Sklar, V. E. F., and Roeder, R. G. (1977). Cell. 10:405.
105. Roeder, R. G. (1976). In M. Chamberlin and R. Losick (eds.), RNA Polymerases, pp. 285-329. Cold Spring Harbor Press, Cold Spring Harbor.
106. Marzluff, W. F., Jr., and Huang, R. C. C. (1975). Proc. Natl. Acad. Sci. USA 72:1082.
107. Kjeldgaard, N., and Maaløe, O. (eds.). (1976). Control of Ribosome Synthesis. Munksgaard, Copenhagen.
108. Cashel, M., and Gallant, J. (1974). In M. Nomura, A. Tissieres, and P. Lengyel (eds.), Ribosomes, pp. 733-745. Cold Spring Harbor Press, Cold Spring Harbor.
109. Horvitz, H. R. (1974). J. Mol. Biol. 90:739.

International Review of Biochemistry
Biochemistry of Nucleic Acids II, Volume 17
Edited by B. F. C. Clark
Copyright 1978   University Park Press   Baltimore

3
Methylation and Polyadenylation of Heterogeneous Nuclear and Messenger RNA

F. M. ROTTMAN
Michigan State University, East Lansing, Michigan

The literature covered in this review was current through November, 1976.

During the past several years, it has become increasingly apparent that control of gene expression in higher organisms occurs at the post-transcriptional level (1). At the same time, mRNA molecules have become the object of intense chemical and physical characterization in an attempt to correlate structural and biological properties and to decipher the key role that these molecules play in genetic expression. Until recently, mRNA species were poorly defined, reflecting both their heterogeneous size and cellular diversity. With the discovery of 3'-terminal polyadenylic acid sequences (poly(A)) in most mRNAs, it became feasible to purify rapidly mRNA molecules collectively. Specific mRNAs, purified from specialized tissues, have also been isolated and shown to function as templates in the cell-free synthesis of single proteins. Thus, it is now possible to define the structure of mRNA molecules with greater precision and to search for modifications in the primary structure of mRNA that might function in the post-transcriptional regulation of genetic information.

Heterogeneous nuclear RNA (HnRNA) is another class of RNA molecules that must be considered in this context. This class of complex molecules is generally believed to contain sequences that are the nuclear precursors to cytoplasmic mRNA. The relatively small amount of HnRNA, in addition to its extensive diversity in size and nucleotide sequence, makes it equally difficult to characterize. Nonetheless, analysis of HnRNA structure indicates that these molecules are also modified post-transcriptionally, much like mRNA. Comparison of post-transcriptional modifications in HnRNA and mRNA can provide a clearer understanding of the proposed precursor-product relationship between these two RNA species.

This chapter concentrates on post-transcriptional modifications that occur in the primary structure of both HnRNA and mRNA: methylation and capping at the 5'-terminus and polyadenylation at the 3'-terminus. Although the biological role of these modifications is not completely understood, their importance as objects of study is 2-fold: first, they provide a unique approach in the exploration of precursor-product relationships between HnRNA and mRNA, and, second, they offer the cell another possible control site for regulating gene expression.

Several reviews on polyadenylation of mRNA and processing of HnRNA and mRNA (1–5) have summarized the earlier work in these areas and should be consulted for further details. This chapter emphasizes current developments in these areas as well as recent work on HnRNA and mRNA methylation.

RELATIONSHIP BETWEEN HnRNA AND mRNA

The generation of cytoplasmic mRNA molecules from large nuclear precursors was first proposed by Scherrer and Marcaud in 1968 (6). In their model, a specific nucleotide sequence present in HnRNA is processed by a series of nucleases which catalyze cleavages and give rise to a smaller mature mRNA sequence, much the same as ribosomal RNA arises from a larger precursor (4). It is difficult to demonstrate such a model directly for a specific mRNA molecule. Much of the difficulty centers around the true size of nuclear precursors and the assertion that many large nuclear RNAs are in fact artifacts of RNA aggregation. Under stringent conditions of denaturation, nuclear ovalbumin (7) and globin (8) mRNA sequences are nearly the same size as their mature cytoplasmic mRNA counterparts. However, there are reports claiming that large nuclear globin sequences do exist (9), and this situation is by no means settled. In adenovirus sequences present in transformed cells, large nuclear precursors have been documented (10) and seem to be obligatory for generation of cytoplasmic adenovirus mRNA.

It should be mentioned that the failure to find high molecular weight nuclear precursors does not prove that these sequences do not exist in the cell; it may merely reflect extremely rapid processing events in the nucleus. Silk fibroin mRNA was first thought to be the same size in both nucleus and cytoplasm because no high molecular weight nuclear species could be demonstrated (11). Subsequent analysis, however, indicated that lowering the temperature under which the larvae were labeled from the normal 24°C to 11.5°C apparently reduced the rate of nuclear processing and permitted the observation of a larger nuclear precursor (12), raising the interesting possibility that the biological half-life of certain high molecular weight precursors may be extremely short, especially in those cells dedicated primarily to the synthesis of a single protein. Processing of some eukaryotic mRNA precursors may be a very rapid event, possibly occurring even before transcription is completed. This model has a precedent in bacterial systems in which T7 mRNA processing begins prior to completion of transcription (13).

Several laboratories have looked for the presence of pppNp 5'-termini on HnRNA in an attempt to characterize HnRNA by other than physical methods and thereby circumvent the argument of artifactual aggregation. Such structures, which should be present only on nascent transcripts, have been described in HnRNA. Bajszar et al. (14) first reported finding pppAp and pppGp on the 5'-termini of HnRNA. Another study involving alkaline digestion of Ehrlich ascites cell HnRNA (15) indicated that as much as 70% of poly(A)-containing HnRNA transcripts contained pppNp termini, whereas in L cells the distribution of HnRNA 5'-termini was found under certain labeling conditions to be 20% pppN, 18% ppN, 42% pN, and 20% cap structures (16). These chemical studies should permit a more rigorous analysis of the size of nascent HnRNA transcripts. The correspondence between the molecular weight of HnRNA as determined by pppN content and size on denaturing gels was reasonably close, indicating that under conditions employed in these studies the majority of poly(A)-containing HnRNA are small (<28 S) and does not arise from processing events that would sacrifice the 5'-end (15).

Analysis of HnRNA turnover within the nucleus of L cells (17) suggested that only approximately 2% of the HnRNA enters the polysomes as mRNA. Similar results were obtained earlier with sea urchin HnRNA, in which it was estimated that approximately 90% of the HnRNA was degraded within the nucleus (18). This apparently low survival rate of nuclear RNA sequences has raised many questions concerning the efficiency of processing and the role, if any, of those sequences that do not become cytoplasmic mRNA. Comparison of nuclear and cytoplasmic RNA sequences by hybridization studies indicates that a significant number of nuclear sequences are confined to the nucleus and are not represented in cytoplasmic mRNA (19), whereas all cytoplasmic sequences are also found in the nucleus. As pointed out by Lewin (5), consideration of the rates of HnRNA synthesis and cytoplasmic mRNA turnover presents a paradox. The extremely high rate of nuclear RNA synthesis is not paralleled by a corresponding accumulation of cytoplasmic mRNA. This raises the possibility, first discussed by Price et al. (20), that the cytoplasmic mRNA may be the product of a unique or special subclass of HnRNA that survives the processing mechanism, the remaining HnRNA serving another purpose or simply being degraded. Consequently, at least two possibilities for the generation of cytoplasmic mRNA sequences from HnRNA precursors must be considered. First, perhaps only a portion of a large HnRNA molecule is retained and processed, the remainder being used for another purpose or merely discarded. Second, perhaps a small subclass of HnRNA molecules, possibly similar in size to mature mRNA, is selectively processed to cytoplasmic mRNA. The extent to which either or both of these mechanisms function in mRNA processing is presently unknown.

In an early model of mRNA processing proposed by Darnell et al. (1), the 3'-terminus of a large HnRNA molecule gives rise to a smaller mature mRNA sequence with retention of the poly(A) sequence. This model predicts that mRNA sequences would be found principally in large HnRNA and at the 3'-terminal end of these molecules. Subsequent studies designed to test this model have produced ambivalent results, and the size of HnRNA sequences that give rise to mRNA is presently unclear. Comparison of HnRNA size with size of the DNA genome in several species suggested that, within a given taxonomic order, the size of HnRNA increases with increasing genome size (21), thus adding further complexity to this problem. Steady state-labeled HnRNA from HeLa cells was shown by cDNA probes to contain message sequences in some large HnRNA species that were greater than 45 S (22). Other studies, however, suggested that mRNA arises principally from small molecular weight HnRNA. By using pppN termini to follow the size of HnRNA, it was concluded that most cytoplasmic mRNA could result from a mere 5'-terminal modification of a small molecular weight fraction of HnRNA (15). In addition, the unique subclass of HnRNA that was postulated to give rise to mRNA was found to be of smaller size than the bulk of HnRNA (20). Perhaps most of the cytoplasmic messages are derived from short HnRNA molecules with a smaller number coming from large nuclear precursors.

At issue are central questions on the nature of the rapidly turning over nuclear species: whether or not these sequences are joined to surviving structural mRNA sequences, and whether the sequences contained within this rapidly degraded species represent some structural genes. In this connection, Egyhazi (23) has described an RNA molecule that is made in the Balbiani rings of the salivary gland and can be isolated as a 75 S molecule from both the nucleus and the cytoplasm. Some of these molecules, presumably containing mRNA sequences, are released into the nucleoplasm and subsequently into the cytoplasm as intact 75 S molecules. The remaining approximately 80% of these 75 S molecules are degraded within the nucleus. It seems likely that the degraded and retained sequences are identical, although this has not been proved directly. Thus, at least in this organism, it would seem that sequences containing information for the synthesis of proteins are being degraded within the nucleus. Such a mechanism would suggest an additional point of cellular control, functioning at the processing level between transcription and transport into the cytoplasm.

Studies on the biological half-life of HnRNA have resulted in estimates ranging from as low as 3 min to as high as 23 min, the latter representing a uniform decay rate of L cell HnRNA (17). There are several mRNA classes in the cytoplasm, each with separate decay rates ranging from 1 hr to 24 hr, depending on the specific cell under examination (24–27). In HeLa cells, there seem to be three separate classes of mRNA, each with a different stability (26), whereas in L cells the turnover of mRNA has been described by a single decay rate (25). These complicated kinetics, along with problems associated with RNA precursor pool equilibration and depletion, make comparative studies on the precursor-product relationship between HnRNA and mRNA extremely difficult. Consequently, the specific nature of the nuclear precursors to mRNA, both as nascent transcripts and immediate precursors to cytoplasmic mRNA, remains uncertain. Following an mRNA sequence coding for a single protein from transcription in the nucleus to translation on ribosomes should help to clarify this point. Detailed examination of such a sequence in both nucleus and cytoplasm for 5'-terminal pppN and cap structures (see below) should provide a clearer understanding of precursor size and the overall nature of mRNA processing.

POLYADENYLATION

Poly(A) in mRNA

The first post-transcriptional modification of mRNA described was the addition of polyadenylic acid (for reviews see refs. 2 and 3). The importance of this modification lies not only in its biological role within the cell but also as a tool for the efficient purification of mRNA from other cellular RNAs. Several methods for mRNA isolation have been devised which involve retention of the mRNA via its 3'-terminal poly(A) fragment. Perhaps the two procedures most commonly used are retention on columns of either oligo(dT)-cellulose or poly(U)-Sepharose. Oligo(dT)-cellulose is particularly useful for mRNA isolation because it is stable to dilute alkali, a treatment which can be used to recycle the column and remove residual nucleic acid as well as any contaminating nuclease. However, oligo(dT) columns often exhibit a small but significant nonspecific binding of RNA that is activated by alkaline treatment. This nonspecific binding is usually not a problem because of the large excess of non-mRNA species present in preparations applied to the column. It has also been observed that oligo(dT) residues shorter than six nucleotides do not retain poly(A) under ionic conditions normally employed for mRNA retention (28). By contrast, poly(U)-Sepharose binds poly(A)-containing mRNA efficiently, but sporadic elution of some poly(U) from the column can potentially interfere with subsequent translation of the mRNA in cell-free systems.

Originally it was thought that all mRNAs contained poly(A), with the exception of histone mRNAs. Subsequently, it was shown that as much as 30% of total cytoplasmic mRNA lacked poly(A) and that this was not the result of cleavage from mRNA sequences originally containing poly(A) (29). A specific mRNA coding for the protein casein is present both with and without poly(A) (30). Poly(A) segments have also been found in mitochondrial RNA (31), being shorter than the poly(A) sequence on cytoplasmic mRNA.

Although the length of poly(A) on cytoplasmic mRNA can vary, it generally contains 150–200 nucleotides. Three size classes of poly(A) have been reported on globin mRNA (32). Observations that poly(A) seems to become shorter with time (32, 33) suggest that the length of the poly(A) terminus may be related to the age of the mRNA in the cytoplasm.

Biological Role of Poly(A) in mRNA   For a possible biological role of poly(A), Darnell et al. (1) proposed that poly(A) may function in the processing and transport of mRNA molecules from nucleus to cytoplasm. This model was based largely on the kinetics of nuclear and cytoplasmic poly(A) labeling and also on inhibitor studies with the nucleoside analogue, 3'-deoxyadenosine. However, the transport and translation of mRNA molecules lacking poly(A) indicate that poly(A) is not an absolute requirement for mRNA function (29, 34). For instance, examination of casein mRNAs (30) revealed that a large part of the mRNAs coding for casein does not contain poly(A). Furthermore, this lack of poly(A) was not a result of its removal from older poly(A)-containing mRNAs, but instead represented newly synthesized mRNAs which were not polyadenylated. Similar conclusions were reached in studies on the mRNA of sea urchin embryos (35). Both studies imply that polyadenylation is not required for the proper processing of mRNA.

The possible involvement of poly(A) in translation has also received much attention. Obviously, poly(A) is not an absolute requirement for translation because, as noted above, many functional mRNAs lack this modification. Messenger RNA naturally lacking poly(A) is found on polysomes (30, 35), and the presence of poly(A) does not insure in vivo translation. Some 40% of the cytoplasmic poly(A)-containing mRNA found in myeloma cells is not bound to polysomes, and a significant fraction of these sequences cannot be chased into polysomes (36). Early studies using in vitro cell-free protein synthesizing systems, which compared poly(A)-containing mRNA with mRNA from which poly(A) was removed, indicated that the poly(A) was not required for cell-free translation (37, 38).

However, more subtle effects were noted in subsequent in vitro experiments. After removal of the poly(A) terminus from globin mRNA by polynucleotide phosphorylase, the initial rate of amino acid incorporation directed by this mRNA was nearly identical with that of poly(A)-containing mRNA (39). After longer times, however, the incorporation decreased more rapidly with mRNA lacking poly(A), suggesting that this mRNA was less stable in the in vitro system. Similar studies were performed in vivo in which Xenopus oocytes were injected with deadenylated rabbit globin mRNA (40). At early times, mRNA lacking poly(A) behaved in the same way as the natural poly(A)-containing mRNA, but at later times globin synthesis fell off markedly with the mRNA lacking poly(A). These experiments have led to a proposed role for poly(A) in which poly(A) enhances the functional stability of the mRNA in the cytoplasm (40), implying that poly(A) must somehow attenuate the degradative system responsible for the turnover of cytoplasmic mRNA. Extensions of these experiments in Xenopus oocytes have documented that the poly(A) tail also enhances the stability of cytoplasmic mRNA, as measured by hybridization with cDNA probes (41). A clearer role for poly(A) in these studies could be documented by reconstruction of the initial properties of the mRNA molecule through restoration of the poly(A) sequence. Huez et al. (42) re-established the stability of globin mRNA in Xenopus oocytes from which the poly(A) was previously removed by readenylation with a poly(A) addition enzyme from Escherichia coli. The minimum effective length of added poly(A) for stability in the Xenopus oocytes system was 30 nucleotides. Although it seems quite certain that poly(A) enhances mRNA stability both in vitro and in vivo, the mechanism of this enhancement is not clear. Presumably, poly(A) affects the conformation of the remainder of the mRNA molecule and thereby influences either its translation or its susceptibility to nucleases.

Poly(A) has been reported to interact with other sequences in the mRNA (43) as well as with a small RNA species, translational control RNA (44). In addition, a protein has been isolated from an mRNA-protein complex which appears to interact specifically with the poly(A) tail (45). No functional significance has been clearly demonstrated for any of these poly(A) interactions.

Poly(A) in HnRNA

The existence of poly(A) in HnRNA sequences permitted the isolation of a subclass of nuclear RNA molecules containing this modification and provided a potential method for following mRNA processing. Only about 20% of HnRNA molecules is polyadenylated under certain experimental conditions, and their possible role as precursors to cytoplasmic mRNA remains unsettled (5). At issue is the conservation of poly(A)-containing HnRNA sequences and the extent to which the poly(A) sequence is an appropriate tag for following HnRNA conversion to mRNA.

If poly(A) sequences are used by the cell to designate which HnRNA molecules exit into the cytoplasm, minimum intranuclear turnover of poly(A) and little cytoplasmic formation of poly(A) would be expected. Jelinek et al. (46) have shown that the kinetics of poly(A) labeling indicates that an early synthesis of nuclear poly(A) is followed by its subsequent appearance in the cytoplasm. Not all of the poly(A) made in the nucleus, however, appears in the cytoplasm, a conclusion based on both kinetic analysis (47) and hybridization studies with cDNA probes (48). Furthermore, this kinetic analysis is complicated by the description of a class of cytoplasmic mRNA molecules with short half-lives (26, 27) which could contain some of the sequences that rapidly disappear from the nucleus. Complementary DNA probes have also been made against nuclear and cytoplasmic poly(A)-containing RNA sequences from *Xenopus* liver in order to explore the relationship between these two classes of RNA (49). It was estimated from these studies that more than 10^4 different poly(A)-containing nuclear sequences were degraded intranuclearly and did not enter the cytoplasm. Similar hybridization data on nuclear and cytoplasmic RNA from *Drosophila* (19) also indicate that some poly(A)-containing HnRNA does not reach the cytoplasm. Thus, it seems that polyadenylation is not required of all mRNA sequences for transport to the cytoplasm, nor does the presence of poly(A) ensure that a sequence containing this modification is transported into the cytoplasm.

Enzymic Addition of Poly(A)

Edmonds and Abrams (50) first isolated a poly(A) polymerase activity from mammalian nuclei that was capable of catalyzing the addition of poly(A) to the 3'-end of RNA. Whether or not this enzyme is responsible for in vivo addition of poly(A) to the RNA still remains an open question. More recently, this activity has been studied in isolated nuclei (51) and chromatin (52) in an attempt to analyze both the synthesis of nuclear precursor RNA and poly(A) addition in vitro. The poly(A) tails of RNA formed in isolated nuclei at high ATP concentrations were 180–200 nucleotides long, similar to the length found in most eukaryotic mRNAs.

In addition to nuclear polyadenylation, it seems that this post-transcriptional modification also occurs, to a limited degree, in the cytoplasm. Elongation of the mRNA poly(A) terminus takes place in the cytoplasm of mammalian cells (53). This cytoplasmic activity, in contrast to that in the nucleus, is essentially insensitive to inhibition by cordycepin and catalyzes the addition of only 7–8 adenylate residues. If both cytoplasmic elongation and removal of poly(A) occur on the same mRNA molecule, it implies that poly(A) sequences in the cytoplasm are in a dynamic state and that some balance between these two processes results in the maintenance of poly(A) within a certain size range. Recent kinetic data are consistent with a model in which not all cytoplasmic poly(A) is derived from a conserved nuclear pool (54). These kinetic data do not eliminate, however, a model in which a subclass of poly(A)-containing HnRNA enters the cytoplasm very rapidly.

The kinetics of radioisotope incorporation has also been used to show that nuclear poly(A) is probably added post-transcriptionally as the last step in mRNA synthesis (47, 55). Furthermore, this poly(A) cannot be a transcriptional product because pyrimidine tracts corresponding to the length of poly(A) are not found in DNA from animal cells (56, 57).

Mammalian cell HnRNA does contain a short stretch of oligo(A), approximately 25 AMP residues long (58). This oligo(A), in contrast to poly(A), is located internally within the HnRNA molecule, and in a mixture of HnRNAs seems to be bounded on its 3'-side by an oligonucleotide of common composition. It also differs from terminal poly(A) in that its synthesis is inhibited by low levels of actinomycin D but is resistant to cordycepin, suggesting a transcriptional origin. An interesting function for this oligo(A) sequence has been proposed (59) in which the nuclear oligo(A) serves as the primer for post-transcriptional polyadenylation. In this model, the HnRNA molecule is cleaved at or near the oligo(A) sequence, thus generating an initiation point for poly(A) polymerase-catalyzed extension.

Sequences Adjacent to Poly(A) in HnRNA and mRNA

Molloy and Darnell (60) examined the nucleotides on the 5'-side of HeLa cell poly(A) in both mRNA and HnRNA and found a pyrimidine oligonucleotide composed of 2 uridylic acid residues and 1 cytidylic acid residue. These results deal with nucleotide composition rather than sequence and, therefore, were consistent with, but did not prove, a common 5'-sequence adjacent to poly(A). The sequence of this pyrimidine oligonucleotide has recently been examined further in L cells and was found to be heterogeneous in total L cell mRNA (61), a result which is inconsistent with a model for poly(A) generation in which initiation of polyadenylation occurs at a common sequence for all mRNAs. Further sequence analysis of this region in several specific eukaryotic mRNAs is presently in progress and should help to establish the extent of similarities or differences in various mRNAs (62, 63). Recent work on the untranslated region of mRNA located near poly(A) indicates the presence of a common AU-rich hexanucleotide in several mRNAs (64).

Several laboratories have studied much longer nucleotide sequences adjacent to poly(A) in HnRNA and mRNA by hybridization techniques employing cDNA probes. Of particular interest is the estimate of the relative number of such poly(A)-containing HnRNA sequences that are confined to the nucleus. In reports dealing with RNA from three rather divergent species—HeLa cells (22), *Xenopus* (49), and *Drosophila* (19)—the estimates of sequences confined to the nucleus were 30%, 33%, and 30%, respectively. This remarkable similarity may be coincidental, or it may reflect an overall similarity in the general nature of processing events functioning in these divergent organisms.

METHYLATION OF mRNA AND HnRNA

Description of Methylation in mRNA

The absence of methylation in eukaryotic mRNA was an assumption of long standing and was based primarily on extrapolation from bacterial studies which indicated less than 1 methyl group in 3,500 nucleotides (65). In fact, lack of methylation was used for some time as one criterion in the characterization of mRNA (66). Previously, any small amount of methylation observed in mRNA or HnRNA was in the range of expected rRNA or tRNA contamination and was consequently discounted. With the discovery of 3'-terminal poly(A) sequences in mRNA, it became possible to purify readily eukaryotic mRNA from rRNA and tRNA, both of which are heavily methylated.

Perry and Kelley (67) reported the presence of methylation in poly(A)-containing L cell mRNA at a level of 2.2 methyl groups per 1,000 nucleotides, or approximately one-sixth the level found in rRNA. Alkaline hydrolysis of this mRNA produced methylnucleotides in both mono- and oligonucleotide fractions. Working with Novikoff poly(A)-containing polysomal RNA, Desrosiers et al. (68) also described the existence of methylation in eukaryotic mRNA. Methylnucleosides obtained from Novikoff mRNA were found to contain the four common $2'-O$-methylnucleosides plus N^6-methyladenosine and an unknown nucleoside, later identified as 7-methylguanosine. This mRNA methylnucleoside composition was unique in that it differed significantly from rRNA and tRNA.

At the same time, several laboratories (69–71) described the presence of methylnucleotides in mRNA molecules derived from viruses that infect animal cells. Furthermore, the enzymic activity capable of methylating the viral mRNA was located within the purified virus particle. Given the limited viral genome, this result suggests an important role for methylation in viral expression. The location of these enzymes within the virion also facilitated the subsequent characterization of methylating enzymes.

5'-Terminal Cap Structures

Description of Cap Structures   A significant amount of the methylated nucleotides remained as an oligonucleotide when both cellular and viral mRNA were digested with alkali. This alkali-stable oligonucleotide had a high negative charge and possessed several properties which were not consistent with a simple structure. The oligonucleotide could not be phosphorylated by polynucleotide kinase, and treatment with alkaline phosphatase removed only a single phosphate, a result that is inconsistent with the presence of pppNp 5'-terminal structures. Furthermore, its high negative charge and alkali stability could not be explained solely on the basis of repeating adjacent 2'-O-methylnucleotides because methylnucleoside composition data allowed for a maximum of only two such nucleosides per cellular mRNA molecule.

Significant to the structural elucidation of this methylated oligonucleotide was the report by Ro-Choi et al. (72) describing a sequence in a low molecular weight nuclear RNA in which 2,2,7-trimethylguanosine was joined by a 5'-5'-diphosphate bond to an adjacent 2'-O-methylnucleoside. Many of the data that had been accumulating to this point from both cellular mRNAs (67, 68) and viral mRNAs (69–71) were consistent with a similar, but not identical, structure. Such a 5'-terminal oligonucleotide, in which a 7-methylguanylic acid residue linked by 5'-5'-pyrophosphate linkage to either one or two adjacent 2'-O-methylnucleotides, was proposed as a general structure for the 5'-terminus of eukaryotic mRNA molecules (73). These 5'-termini have been called "cap" structures (Figure 1) and have since been found in a wide range of organisms. They seem to be ubiquitous, being present in the mRNA of several lower forms, including yeast (74, 75), the slime mold *Dictostelium* (76), silk worm (77), and

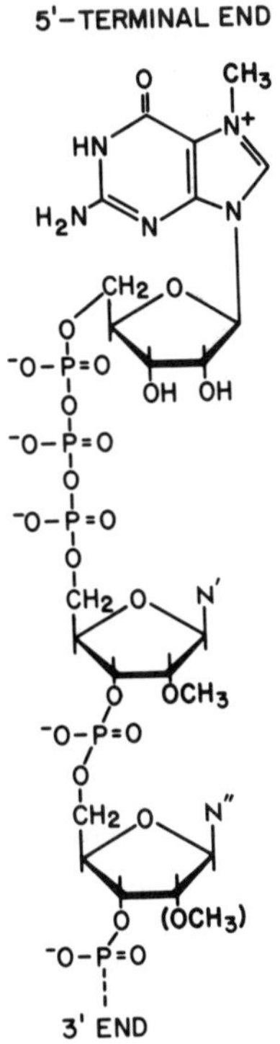

Figure 1. The 5'-terminal end of eukaryotic mRNA. The nucleoside 7-methylguanosine is attached by an unusual 5'-5'-pyrophosphate linkage containing three phosphates to a 2'-O-methylnucleoside with the base indicated as N'. This terminus is called cap 1. An additional adjacent 2'-O-methylnucleoside, with the base indicated as N'', is present in cap 2 structures.

sea urchin (78). 5'-Terminal caps have been found in all animal cell mRNAs examined to date, as well as in most mRNAs from viruses infecting both animals and plants. A partial summary of the cap structures found in these mRNAs is given in Tables 1 and 2. Cap structures are unique to mRNA, HnRNA, and several species of low molecular weight nuclear RNA. There are several mRNA molecules which are exceptional and lack cap structures, including polio virus mRNA (110, 111), encephalomyocarditis virus (113), and the plant virus, satellite tobacco necrosis virus (114).

Isolation and Characterization of Cap Structures   Both the 5'-5'-pyrophosphate bond and 2'-O-methyl groups result in nucleotide linkages that are stable to either alkaline hydrolysis or cyclizing nucleases, such as T_2. Treatment of methyl-labeled mRNA with either dilute alkali or T_2 produces two labeled fractions, the 5'-terminal oligonucleotide cap structure and a methylated mononucleotide derived from within the mRNA molecule (cf. Figure 2). The oligonucleotide can be separated from methyl-labeled mononucleotide by several methods, including chromatography on DEAE-cellulose or DEAE-Sephadex in the presence of 7 M urea (71, 115), paper electrophoresis (116, 117), high pressure liquid chromatography on weak anion exchange resins (87), or by binding of the oligonucleotide via the free *cis*-hydroxyls of m^7G to dihydroxyborate cellulose (118).

Subsequent analysis of isolated cap structures indicated that caps can be present in three basic forms: cap 0, a cap structure which is methylated only at 7-methylguanosine (m^7G); cap 1, which contains an additional single 2'-O-methylnucleotide at N' (cf. Figure 1); and cap 2, which contains 2'-O-methylnucleotides at both N' and N" (87, 90, 117, 119). Intact cap 1 and cap 2 structures can be resolved on DEAE-urea columns (84, 90, 119). Partial digestion products of caps, N'mpN and N'mpN"mpN, derived from cap 1 and cap 2, respectively, can be used to determine the ratio of cap structures in an mRNA preparation (120).

Isolated cap structures can be systematically degraded to determine the level of methyl labeling in the m^7G, N', and N" positions of the cap. Treatment with periodate, followed by β elimination, releases 7-methylguanine from the end of an isolated cap structure or an intact mRNA molecule (87, 119). Alternatively, nucleotide pyrophosphatase attacks the 5'-5' linkage releasing 7-methylguanylic acid (72). This enzyme also releases 7-methylguanylic acid from intact mRNA, indicating that the cap structure is in a relatively exposed position within the mRNA and accessible to the enzyme (120). Although enzymic treatment results in a quantitative and reproducible release of m^7G for methylnucleoside analysis, it is not useful if the mRNA is used in subsequent studies requiring an intact molecule because contaminating nuclease in the commercial enzyme preparation produces extensive degradation. Terminal 5'-polyphosphates remaining after m^7G removal are susceptible to hydrolysis by bacterial alkaline phosphatase.

Nuclease P_1 cleaves phosphodiester bonds, including those adjacent to 2'-O-methylnucleosides in cap structures, but does not attack the 5'-5' linkage

Table 1. Eukaryotic mRNA methylation

Organism	Cap 1 (m⁷GpppN'm)	Cap 2 (m⁷GpppN'mpN"m)	Internal methylation	Reference
Slime mold	$m^7GpppG^A_{Am}$	None	None	76
Yeast	m^7GpppG^A	None	None	74
	N'm present	None	None	75
Insect				
Silkworm silk fibroin	GpppN[a]	m⁷GpppAmUm	Not determined	77
Tobacco hornworm oocyte mRNA	m⁷Gppp[b]		Not determined	79
Brine shrimp	N'm present	Not determined	Not determined	80
Duck globin mRNA	m⁷Gppp[b]	N"m present	None	81
Rabbit reticulocyte mRNA	N'm present	Not determined	Not determined	82
Mouse kidney	N'm = Am, m⁶Am, Gm, Um, Cm	N"m present (?)	Present	83
Mouse L cells	N'm = m⁶Am, Am, Gm, Um, Cm	N"m = Um, Cm, Am, Gm	m⁶A	84
Mouse myeloma (MPC-11-662)	N'm = m⁶Am, Am, Gm, Um, Cm	N"m = Um, Cm, Am, Gm	m⁶A	85
Hamster kidney cells (BHK-21)	N'm = m⁶Am, Am, Gm, Um, Cm	N"m present (?)	m⁶A, m⁵C	86
Novikoff hepatoma cells (N1S1)	N'm = m⁶Am, Am, Gm, Um, Cm	N"m = Um, Cm, Am, Gm	m⁶A	87
Monkey kidney cells (BSC-1)	N'm = m⁶Am, Am, Gm, Um, Cm	Not determined	m⁶A	88
HeLa cells	N'm = m⁶Am, Am, Gm, Um, Cm	N"m = Um, Cm, Am, Gm	m⁶A	89
	N'm = Am, Gm, Um, Cm	N"m = Um, Gm, Cm	m⁶A	90

[a] mRNA is capped but not methylated.
[b] m⁷G is present in inverted 5'-5' linkage; the rest of mRNA was not examined.

Table 2. Viral RNA methylation

	Type of Genome	5′-Terminus of virion	Cytoplasmic mRNA cap	Internal methylation	Reference
Plant viruses					
Brome mosaic virus	ss-RNA (+)	m^7GpppG	Not determined	Not determined	91
Tobacco mosaic virus	ss-RNA (+)	m^7GpppG	Not determined	Not determined	92, 93
Alfalfa mosaic virus	RNA	m^7GpppG	Not determined	Not determined	94
Avian viruses					
Rous sarcoma virus (Prague B)	ss-RNA (+)	$m^7GpppGm$	Not determined	Not determined	95
Avian sarcoma virus-B77	ss-RNA (+)	$m^7GpppGm$	Not determined	14–16 m^6A[a]	96
Newcastle disease virus	ss-RNA (−)	Not determined	m^7GpppG[a]	Not determined	97
Insect viruses					
Cytoplasmic polyhedrosis virus	ds-RNA	$m^7GpppAm$	Not determined	None	98
Mammalian viruses					
Vaccinia virus[b]	DNA		$m^7Gppp^{Am}_{Gm}$[a]	None	99

Simian virus 40[c]	DNA		$m^7Gppp_{Gm}^{Am}$	m⁶A present	88
Adenovirus[c]	DNA	$m^7GpppGm$	$m^7Gppp_{Am}^{m^6Am}$, $m^7Gppp_{Am}^{m^6Am}N''m$	m⁶A, m⁵C present	100, 101
Reovirus[b]	ds-RNA	$m^7GpppGm$	$m^7GpppGm$, $m^7GpppGmCm$	None	70, 102
Vesicular stomatitis virus[b]	ss-RNA (−)	ppA	m^7Gpppm^6Am, $m^7Gpppm^6Am_{Am}^{m^6Am}$[d]	None	103, 104
Influenza virus[c]	ss-RNA (−)	ppA	$m^7Gppp_{Am, Gm}^{m^6Am}$	3 m⁶A	105
Moloney murine leukemia virus[c]	ss-RNA (+)	$m^7GpppGm$	Not determined	15–23 m⁶A	106
Sindbis virus[b]	ss-RNA (+)	m^7GpppA	m^7GpppN	m⁵C	107, 108
Feline leukemia virus[c]	ss-RNA (+)	No cap	Not determined	10 m⁶A	109
Poliovirus[c]	ss-RNA (+)	pUp, pAp	pUp	Not determined	110, 111
Encephalomyocarditis virus	ss-RNA (+)	No cap	Not determined	None	112

[a] mRNA synthesized in vitro by virion core.
[b] Replicates in cytoplasm.
[c] Replicates in nucleus.
[d] RNA synthesized in vitro by virion core is $m^7GpppAm$.

5'-end   3'-end

$$m^7G^{5'}ppp^{5'}N'mpN''mpNp--------m^6A------m^6A--------ApApApApApApApA$$

Figure 2. Post-transcriptional modifications of eukaryotic mRNA. The base-methylated nucleoside N^6-methyladenosine (m^6A) is located within the mRNA molecule between the 5'-terminal cap structure and the 3'-terminal poly(A) segment. m^7G, 7-methylguanosine; N'm and N''m, the 2'-O-methylnucleosides of cap 1 and cap 2, respectively.

(cf. Figure 2). Thus, nuclease P_1 treatment of a cap 2 structure, $m^7G^{5'}$ $ppp^{5'}N'mpN''mpN'''$, produces $m^7G^{5'}ppp^{5'}N'm$, commonly called the cap core, plus $pN''m$ and pN'''. Resolution of these P_1-generated fragments permits a comparison between the labeling at N''m and the core. The remaining core structure can be hydrolyzed with nucleotide pyrophosphatase and alkaline phosphatase to yield free nucleosides, including M^7G and N'm. A combination of these degradative methods, coupled with subsequent resolution of the various fragments, has been used to analyze the methylation patterns in mRNA and HnRNA.

Methylation at N' Position of Cap Structures   Most 5'-terminal cap structures in mRNA from higher organisms contain a 2'-O-methyl group on the N' nucleoside. Exceptions to this generalization include yeast mRNA (74), some *Dictostelium* mRNAs (76), plant viral RNAs such as tobacco mosaic virus (92, 93) and brome mosaic virus (91), and one animal virus, Newcastle disease virus (97). *Dictostelium* mRNA is different in that mRNA species derived from this organism contain a mixture of cap structures, of which only about 20% is methylated at the N' position (76). Although it seems that N'm is most often a purine-containing nucleoside, especially in lower organisms and viruses, pyrimidine nucleosides are also found at this position. This may indicate that such pyrimidine-containing termini are the result of cleavage at sites within larger precursors, rather than the result of simple capping of purine-initiated nascent transcripts.

Another methylated nucleoside in addition to the four common 2'-O-methylnucleosides has been found at the N' position of cap structures. This dimethylated nucleoside, N^6, 2'-O-dimethyladenosine, has been described at the N' position in cap structures in mRNAs from both animal cells and viruses (84, 100, 117, 121, 122).

Methylation at N'' Position of Cap Structures   Total cytoplasmic poly(A)-containing mRNA isolated from mouse myeloma cells (117), HeLa cells (90, 119), L cells (84), and Novikoff cells (87) contains a mixture of cap 1 and cap 2 structures. Methylation at the N'' position to produce cap 2 occurs primarily, but not exclusively, on pyrimidines. Under certain labeling conditions with L-[^{3}H-methyl]methionine, nearly 50% of the methylation in this position in Novikoff mRNA is in purines (120). Animal viral mRNA sequences contained within the intact virion are not methylated at N''. Silk fibroin mRNA, on the other hand, contains only cap 2 structures, indicating that modification at the N'' position is not restricted to higher organisms (77). The relationship between cap 1 and cap 2 structures and their biogenesis is discussed in a later section.

Multiply methylated nucleosides, such as N^6, $2'$-O-methyladenosine, have not been found in the N'' position of eukaryotic mRNA (89). All of the Am present at N'' appears to be devoid of further methylation (120).

Internal Methylation of mRNA

In addition to the methylnucleotides present in $5'$-terminal cap structures, mRNA contains other methylnucleotides located internally between the $5'$-cap and the $3'$-terminal poly(A) sequence. After digestion with alkali, a significant portion of the methyl label present in mRNA was found in the mononucleotide fraction (67). The properties of this mononucleotide were consistent with methylation of the heterocyclic base rather than the sugar moiety of the nucleotide, and it was subsequently identified as N^6-methyladenylic acid (68). Under certain conditions of labeling with L-[^{3}H-methyl]methionine, three of the approximately six methyl groups found in each Novikoff mRNA molecule were present in N^6-methyladenylic acid (87). Although this methylated nucleotide has been found in a number of mRNAs (68, 88, 89, 100, 117), it is conspicuously absent in the mRNA of several lower organisms, including yeast (74), *Dictostelium* (76), and silk fibroin mRNA (77). N^6-methyladenylic acid appears to be generally found in animal cell mRNA, with the exception of globin mRNA (81), which raises the possibility that this modification may be unique to animal cells and viruses infecting animal cells. Several animal viral mRNA sequences have also been found to contain N^6-methyladenosine (m^6A), including adenovirus (90, 100), simian virus 40 (88) avian sarcoma (96, 123), and feline leukemia virus (109). Feline leukemia viral RNA, isolated from intact virus particles, seems to be unique in that it contains approximately 10–14 m^6A residues per viral RNA, but it does not contain the usual $5'$-cap, possibly reflecting the defective nature of this viral isolate (109).

The number of m^6A residues in an mRNA molecule seems to be roughly proportional to the size of the RNA. When adenovirus-specific mRNA (100) or L cell cytoplasmic mRNAs (124) were fractionated on gradients, the larger mRNA species had a significantly higher content of m^6A. Similarly, with Novikoff cells increased times of labeling with L-[^{3}H-methyl]methionine shortened the average size of cytoplasmic mRNA and reduced the content of m^6A per mRNA molecule (125).

The general distribution of m^6A residues within mRNA sequences has been studied by subjecting poly(A)-containing mRNA to partial alkaline hydrolysis and comparing the methylnucleotide content of fragments containing the poly(A) sequence with those lacking poly(A). With this approach, it has been estimated that greater than 90% of the m^6A in adenovirus mRNA sequences is not located in the $3'$-terminal one-third of the molecule (100). Earlier studies indicated the essential absence of m^6A in the poly(A) segment itself (87, 115).

More precise information on the distribution of m⁶A residues within HeLa cell mRNA has been obtained by Wei et al. (89). A combination of nucleases and phosphatase was used to examine the nucleotides adjacent to internal m⁶A. Only two sequences were detected, Gpm⁶ApC and Apm⁶ApC. This surprisingly simple pattern, in which cytidylic acid was always located on the 3'-side and either adenylic or guanylic acid on the 5'-side of the m⁶A residue, suggests a considerable degree of sequence specificity in the post-transcriptional formation of m⁶A. A recent study noted that the distribution of internal m⁶A in avian sarcoma virus (126) was nearly identical with that reported earlier for HeLa mRNA. It would indeed be interesting if this mRNA modification occurs at similar nucleotide sequences in all mRNA molecules.

Another nucleoside methylated in the heterocyclic ring moiety, 5-methylcytidine, is present internally in three different mRNAs. The percentage of internal methylnucleoside present in the mRNA from cultured hamster kidney cells (86), adenovirus (100), and Sindbis virus (107) as 5-methylcytidine (m⁵C) was 20%, 10–20%, and nearly 100%, respectively, the remainder being m⁶A. Less than 3% of the internal methylnucleoside in Novikoff mRNA is present as m⁵C (127).

Several early indications of internal NmpN and NmpNmpN residues in mRNA were most likely attributable to contamination with small amounts of rRNA. This is not a trivial problem and probably results from a specific association between mRNA and rRNA because mRNA purified by repeated passage over oligo(dT)-cellulose columns remains significantly contaminated with 18 S rRNA (87). Perhaps this association of mRNA with 18 S rRNA is related to the specific interaction observed earlier between similar bacterial RNA species (128). Association of this rRNA and mRNA can be disrupted by a brief heat treatment prior to application to the oligo(dT)-cellulose column (87).

Although the existence of internal m⁶A and m⁵C in mRNA has been documented and the nucleotides immediately adjacent to m⁶A residues have been defined, these studies do not indicate the location of these post-transcriptional modifications relative to the coding sequence in mRNA. Two central issues concerning internal mRNA methylation still remain unclear. The first is whether m⁶A and m⁵C residues are contained within the region of the mRNA that is translated or, alternatively, are located in a nontranslated and possibly repeated sequence of the message. The second relates to the possible biological role of internal methylation. As discussed below, a potential biological role for the 5'-terminal cap seems to be developing, but a reasonable hypothesis for the function of internal methylnucleosides has not yet appeared. Internal methylation is not a universal prerequisite for mRNA translation because numerous mRNAs lack this modification. It may be involved in processing of nuclear precursors, and, in this context, it is noteworthy that viral mRNAs synthesized in the nucleus generally contain m⁶A, whereas those synthesized in the cytoplasm do not.

Methylation in HnRNA

The discovery of methylation and cap structures in cytoplasmic mRNA raised the question of the existence of comparable modifications in HnRNA. In their early report on mRNA methylation, Perry and Kelley (67) noted a small but significant level of methylation in L cell HnRNA. Subsequently, it was shown that L cell HnRNA did contain cap structures as well as internal m^6A (84). Similar results were obtained with HeLa cell HnRNA (129) and in nuclear precursors containing simian virus 40 (88) and adenovirus sequences (100). Of particular interest was the finding of only cap 1 structures in the HnRNA of L cells; no methylation was observed in the N″ position. In addition, compositional analysis of the methylnucleosides in the HnRNA cap indicated a remarkable similarity to cytoplasmic mRNA cap 1 structures, but not to those of cap 2 (84). The presence of caps on HnRNA molecules and the similarity in cap 1 structures in HnRNA to those found in a subclass of mRNA suggest the possibility that some cytoplasmic mRNA molecules may be derived from the 5′- end of nascent nuclear transcripts. Some nascent transcripts, probably containing pppN 5′-termini (14–16), may be rapidly processed to produce capped termini. This does not imply, however, that all capped HnRNA molecules represent nascent nuclear transcripts. In any event, the presence of caps on the 5′-termini of large HnRNA species opens the possibility that larger nuclear precursors are processed with retention of the 5′-terminus.

Internal m^6A (84) and m^5C (100) are also present in HnRNA molecules, indicating that these modifications are nuclear events. As mentioned earlier, apparently only those viral RNAs synthesized in the nucleus have internal methylnucleotides. As in cytoplasmic mRNA, longer HnRNA molecules have a higher number of internal m^6A residues per RNA molecule than do shorter molecules (100). Thus, it seems that all post-transcriptional modifications found in cytoplasmic mRNA, with the exception of methylation at N″, can occur at the precursor level when the molecule still resides in the nucleus.

Kinetics of mRNA and HnRNA Methylation

The finding of only cap 1 in HnRNA and its striking similarity in methylnucleoside composition to one class of cap structures in mRNA suggested the possible preservation of 5′-capped sequences during processing from nucleus to cytoplasm. Perry and Kelley (124) pulse-labeled L cell HnRNA with L-[³H-methyl]methionine and followed the progress of cap transfer from the nucleus to cytoplasm. After 3 hr, essentially all of the caps in >50 S HnRNA which had been labeled with a pulse were chased from the nucleus into the cytoplasm. At early time periods, there was good recovery of nuclear-labeled caps in cytoplasmic mRNA, consistent with minimum cap turnover within the nucleus (124). The kinetics of labeling of cap 2 structures in these studies suggested that they were being made in the cytoplasm. In similar studies, Friderici et al. (120)

examined the kinetics of labeling of Novikoff cytoplasmic mRNA with L-[³H-methyl]methionine. After a short labeling period of 20 min, most of the label in cytoplasmic mRNA was in cap 2 structures; more specifically, >80% was found in the N″ position of cap 2. With longer labeling times, m⁷G and N′m also became radioactive. A model for mRNA methylation was suggested in which m⁷G, N′m, and internal m⁶A are all products of nuclear methylation events, whereas formation of cap 2 structures by methylation at N″ is a subsequent event occurring in the cytoplasm. Apparently, the double methylation at N′ forming N⁶-2′-O-methyladenosine can also occur in the nucleus because it has been found in HnRNA (84). The kinetic studies on Novikoff mRNA revealed that the methylnucleoside composition of cap 1 structures changed with time of labeling. At longer labeling times, the Cm content of N′m increases relative to Am, Um, and Gm, possibly reflecting the variation in cytoplasmic stability of several classes of mRNAs. With no causal relationship implied, those molecules with greater stability may happen to be enriched with Cm. Similarly, as steady state labeling is approached, cap 2 structures are enriched until approximately half the mRNAs contain this terminus (120).

Comparison of the methylnucleoside composition of cap 1 structures and the core of cap 2 structures, formed by the removal of N″m with P_1 nuclease, was used to analyze the precursor-product relationship between cap 1- and cap 2-terminated mRNA molecules. If conversion of cap 1 into cap 2 involves selection of a specific subclass of cap 1-containing mRNA molecules, this might be expected to result in a different composition of N′m in these two cap structures. No such differences were observed in Novikoff mRNA (120), whereas some variation in N′m composition was observed in L cells (124). However, different labeling conditions were employed in these two studies, pulse chase being used for L cells and continuous steady state labeling for Novikoff cells. Consideration of selective mRNA turnover and continuous entry of new molecules into the cytoplasmic mRNA pool make studies on total poly(A)-containing mRNA difficult to interpret.

In considering the cytoplasmic generation of cap 2 structures, it should be noted that viral RNAs normally containing only cap 1 within the virion can be isolated with cap 2 termini when prepared from the cytoplasm of infected cells. Reovirus RNA isolated from virions or synthesized in vitro by viral enzymes contains a cap 1 structure, m⁷GpppGmpC (116). Upon infection of the host cell, cytoplasmic viral RNA with a cap 2 structure, m⁷GpppGmpCmp, is formed (102). The biological necessity for this subsequent modification of a viral transcript by host machinery is not known. Recent studies on cap structures in noninfected L cell mRNA indicate an enrichment of cap 2-containing mRNAs associated with polysomes (124).

In Vitro Capping and Methylation of mRNA

In many respects, viruses represent ideal systems for studying methylation and biological function of mRNA. Early studies with several viruses indicated that the enzymic machinery required for RNA capping and methylation was contained within the virion (69–71, 130). Not only is this a unique advantage in the isolation and purification of the enzymes catalyzing these modifications, but also viruses can be used to provide a source of unmodified mRNA for other studies. The studies on virion-associated methylation have demonstrated an S-adenosylmethionine-dependent in vitro capping and methylation of viral RNA sequences to produce RNA terminated with the general structure, m^{7}GpppN'mp. S-adenosylhomocysteine is a powerful inhibitor of the methylation reaction and prevents methylation, but not capping. There appear to be two basic mechanisms for capping or addition of the terminal guanylate residue. The first involves addition of GMP, derived from GTP, to an RNA molecule terminated with 5'-diphosphate (e.g., ppNpNp - - -). In this reaction, only the α-phosphate of GTP is incorporated into the cap structure. This reaction has been studied in considerable detail in systems derived from vaccinia (131) and reovirus (132). The second mechanism involves the addition of GDP, or both α- and β-phosphates of the GTP donor, to an acceptor molecule terminated with a single 5'-phosphate (e.g., pNpNp - - -). This reaction is used in the formation of capped vesicular stomatitis viral sequences (133). The second reaction may be of interest as a model for cellular mRNA capping because VSV mRNA is made as a large molecular weight nuclear precursor and is subsequently processed to smaller molecules (133). Cleavage of precursors would generate monophosphorylated termini which could then be capped. However, recent studies by Schibler and Perry (16) on the 5'-termini of HnRNA have shown that approximately 20% of the HnRNA molecules is terminated by ppN - - - , the remainder containing pN - - - -, pppN - - - - , and cap structures. Furthermore, the base composition of the N' position in HnRNA caps is most nearly reflected by the composition of ppN - - - - termini. These results are most consistent with a capping mechanism similar to that described in vaccinia (131) and reovirus (132), in which addition occurs to a ppN - - - - terminus. Presumably, this ppN - - - terminus could be generated from a nascent 5'-end or from an internal cleavage. As already pointed out (16), this would require an as yet undescribed enzyme activity for generating ppN - - - - ends from internal cleavage sites.

Capping and methylation enzymes have been extensively purified from vaccinia virus cores (134, 135). Although purified to near homogeneity, separation of the guanylyltransferase and (guanine-7)-methyltransferase activities was not possible. The purified enzyme was also capable of capping the synthetic RNA homopolymers poly(A) and poly(G), indicating a rather low level of acceptor specificity. As with viral RNA acceptors, the synthetic RNAs must also contain a 5'-terminal diphosphate. Methylation of the sugar moiety of N' to form N'm is believed to result from a separate methylase which as yet has not been purified. Formation of N"m in cap 2 structures, as discussed earlier, is presumed to occur in the cytoplasm. Neither this enzyme nor the nuclear enzyme responsible for formation of internal m⁶A formation has been isolated.

Capping of reovirus RNA is believed to occur early in transcription, shortly after initiation (132). In this reaction, the first phosphodiester bond formed yields pppGpC. A nucleotide pyrophosphatase removes the terminal phosphate to produce ppGpC. Guanylyltransferase then catalyzes the transfer of GMP to this dinucleotide acceptor, forming the capped structure GpppGpC. Methylation at the 7-position of the terminal guanosine also occurs at this level (132). Although this mechanism seems likely for the in vitro generation of reovirus RNA termini, it should be mentioned that both enzyme activities can nevertheless function on long, intact RNA molecules in vitro.

In vitro studies with isolated nuclei have been used to study the capping and methylation reaction in HeLa (136) and L cell nuclei (137). Large molecular weight RNA molecules, as well as cap 1 and cap 2 structures, were synthesized in nuclear preparations from L cells (137). The finding of cap 2 structures might not have been expected if, as suggested by in vivo studies, the formation of N″m in cap 2 is a cytoplasmic event. However, the authors point out that their nuclear preparation had residual cytoplasmic tags and one class of cap 2 structures made in this system was labeled only in the N″m position (137).

The possible turnover of cap structures in mRNA has been examined. The stability of caps apparently parallels that of the entire RNA molecule, indicating no separate turnover of the capped portion of the message (124). An enzymic activity has been detected in HeLa cells that cleaves the pyrophosphate linkage of the cap, producing pm⁷G and ppN (138). This enzyme is specific for cap structures, but does not hydrolyze a cap that is located at the end of a long RNA chain, functioning only on the free cap structure itself or when it is present in an oligonucleotide less than 10 nucleotides long. Another enzyme activity has been isolated from tobacco cells that can selectively cleave the m⁷G from an intact, capped viral RNA (139). This enzyme should be very useful in studying the biological function of the cap.

Possible Biological Role of mRNA and HnRNA Methylation

Role in Translation   The first report of a possible biological function for cap structures involved the cell-free translation of viral RNA (140). Reovirus and VSV RNA were synthesized in vitro in the absence of methyl donors to generate viral RNA messages lacking the 5′-terminal m⁷G. Methylation of the RNA could occur, however, in the cell-free translation system from wheat germ. In order to demonstrate a complete dependence on methylated cap structure, it was necessary to include in this system the methylation inhibitor S-adenosylhomocysteine. Translation of nonmethylated RNAs in the wheat germ cell-free system was very inefficient in comparison to RNAs containing the intact cap structure (140). Subsequently, these studies were extended to include globin mRNA. Because it was not possible to isolate this mRNA devoid of the cap, the terminal m⁷G was removed by periodate oxidation and β elimination (82). Globin mRNA which was treated in this fashion also showed decreased template activity in the wheat germ extracts. It has been suggested that periodate treatment of mRNA may cause some nonspecific damage of mRNA (141) and thus affect translation by a mechanism other than removal of m^7G from the 5'-terminal cap. With proper precautions, however, periodate oxidation apparently can be used to modify only the m^7G residue. For example, periodate treatment of satellite tobacco necrosis viral RNA, an RNA lacking a cap structure, did not inhibit its translation in wheat germ (142).

Stimulation of translation by the 5'-cap is thought to occur at the level of initiation. When several reovirus RNAs, each containing cap structures at various stages of completion, were used, the formation of a 40 S ribosome-mRNA complex depended on an intact 5'-terminal cap (143). The nature of this complex was studied in more detail by using brome mosaic viral RNA and wheat germ ribosomes (144). Ribonuclease treatment of this ribosome-mRNA complex produced a ribosome-protected fragment corresponding to the 5'-terminal end of coat protein mRNA. Included within this sequence was the cap, m^7GpppG, separated by only 10 nucleotides from the AUG initiator codon. Wheat germ ribosomes have also been used to protect fragments of reovirus RNA (145). The protected 5'-terminal fragments from mixed reovirus mRNAs ranged in length from 31–65 nucleotides and contained cap structures. These oligonucleotides rebound to ribosomes after dissociation and purification. Further reduction in size of the fragments to a length of 7–10 nucleotides by treatment with T_1 nuclease eliminated its ribosomal binding capacity (143).

Additional evidence concerning the role of the cap in translation was provided by another approach in which the mononucleotide 7-methylguanylic acid was used to inhibit effectively mRNA translation (146) of capped mRNAs such as globin, TMV, and HeLa cell poly(A)-containing RNA, but it did not inhibit tobacco necrosis viral RNA, an RNA molecule known to lack the cap. Presumably, 7-methylguanylic acid acts as a structural analogue of the cap and interferes with initiation, because formation of the 80 S initiation complex was inhibited, whereas binding of the initiator Met-tRNA$_f^{Met}$ was not (146).

Several groups have reported the isolation of proteins that bind to cap structures. Extracts of the brine shrimp *Artemis salina* contain a ribosome-associated protein that binds to an isolated cap structure, $m^7GpppGpC$ (147). This binding protein was not identical with any of the purified initiation factors. However, studies with reticulocyte ribosomes suggested that a protein capable of recognizing cap structures corresponds to a known initiation factor (148). The binding of purified reticulocyte initiation factor IF-M3 to capped mRNA was inhibited by 7-methylguanylic acid and thus IF-M3 was postulated to function in recognizing the cap portion of mRNA.

The degree to which cap structures stimulate the translation of an mRNA sequence depends on a variety of factors. These results imply that caps are not an absolute requirement for translation. The translation of two mRNAs from brome mosaic virus, both of which were capped, was examined in wheat germ extracts before and after removal of terminal m^7G by periodate (149). Although presence of the cap enhanced translation and the formation of ribosome-mRNA complexes, RNA from which the cap had been removed still functioned to a lesser degree in both systems. The greatest stimulation by the cap was noted at low or limiting concentrations of mRNA, with only minimal enhancement of translation at high mRNA levels. This is consistent with a model in which the cap provides an increment of increased stability in the recognition of mRNA sequences by ribosomes; the increment may be more or less significant when monitored by in vitro measurements.

Other aspects of mRNA structure at or near the 5'-terminus also seem to influence translation. Synthetic RNA polymers which contain cap structures have been synthesized with polynucleotide phosphorylase (150). The primer-dependent form of this enzyme was used to extend caps and produce polymers with various nucleotide compositions. These polymers were used in ribosome binding studies to assess the effect of nucleotide composition in sequences adjacent to the cap. Capped polymers rich in adenosine and uridine bound to 80 S ribosomes, whereas other synthetic polymers of different composition showed little or no binding (150). Thus, it seems likely that the interaction of a specific mRNA with a ribosome is very complex, involving the m^7G portion of the cap structure, possibly the adjacent 2'-O-methylnucleosides, A,U-rich sequences, and perhaps the distance between the cap structure and the AUG initiating codon. In this context, the cap can best be pictured as a facilitator rather than an absolute requirement for translation. The degree to which the 5'-terminal cap enhances translation then depends on the specific mRNA in question and its peculiar nucleotide sequence near the cap as well as on the system in which it is being translated. In the future, it will be of interest to determine the relative effect of cap structures on a family of specific mRNAs, such as those obtained from brome mosaic virus, in an attempt to correlate translational stimulation with distance between cap and initiator AUG, as well as with other structural features of each mRNA. The effect of cap structure on biological function may be subtle, as measured by in vitro assays, but nonetheless of crucial importance to the cell as a post-transcriptional regulatory mechanism. It is also possible that present in vitro conditions do not adequately reflect in vivo events, and under certain conditions cap structures may be an absolute requirement for translation within the cell.

The stimulation of mRNA translation by caps led several investigators to search for the possible role of this modification in biological systems, such as early embryonic development and viral infection, which require modulation of genetic expression at the translational level. Characterization of mRNA obtained from embryonic brine shrimp (80) and sea urchin eggs (151), both of which represent developmental stages in which mRNA is not translated, revealed that the mRNA was nonetheless capped. In a study involving tobacco hornworm oocytes, however, the mRNA was found to contain a cap structure with a terminal unmodified guanosine residue rather than the usual m^7G (79). It was suggested that the reason for mRNA quiescencè in the oocytes of this insect is the lack of methylation in the terminal guanosine residue and that after maturation and fertilization the formation of m^7G renders the mRNA biologically active.

Cap structures have also been considered to function in other systems which display a differential translation of mRNAs. Polio virus RNA is not capped (110, 111), yet the cellular mRNA of the infected host seems to be capped and methylated to the same degree as uninfected cellular mRNA (152). This difference in methylation between host and viral RNA has been suggested as a variation in mRNA structure that could be exploited by the virus in infected cells to translate preferentially its RNA rather than host mRNA.

In vitro methylation of reovirus RNA was inhibited in extracts that were prepared from interferon-treated cells (153). However, the mechanism of interferon inhibition and whether it differentially affects viral and host mRNA methylation are not known. It is also unknown whether interferon affects the methylation of viral RNA in vivo.

Other Possible Roles of Capping and Methylation   The effect of capping and methylation on translation of mRNA is readily demonstrable; it is, therefore, tempting to assume it is the only function of these post-transcriptional modifications. It is possible, however, that methylation may be involved in mRNA processing (73) in some as yet undetermined manner, such as it participates in rRNA processing (4). Methylation enhances RNA stability, both in model systems with synthetic RNAs (154) and in in vivo studies on rRNA (155). As noted earlier, one of the major unanswered questions about RNA metabolism in animal cells concerns the mechanism whereby a small number of mRNA sequences possess or acquire nuclear or cytoplasmic stability and resist rapid degradation. Possibly, nuclear processing involves an enzyme possessing activities for both cleavage and capping of nuclear sequences. The development of in vitro systems with the use of isolated nuclei and in vivo systems involving a single mRNA species should considerably increase understanding of these processes.

REFERENCES

1. Darnell, J. E., Jelinek, W. R., and Molloy, G. R. (1973). Science 181:1215.
2. Brawerman, G. (1974). Annu. Rev. Biochem. 43:621.
3. Greenberg, J. R. (1975). J. Cell Biol. 64:269.
4. Perry, R. P. (1976). Annu. Rev. Biochem. 45:605.
5. Lewin, B. (1975). Cell 4:11.
6. Scherrer, K., and Marcaud, L. (1968). J. Cell. Physiol. 72:181.
7. McKnight, G. S., and Schimke, R. T. (1974). Proc. Natl. Acad. Sci. USA 71:4327.
8. MacNaughton, M., Freeman, K. B., and Bishop, J. O. (1974). Cell 1:117.
9. Imaizumi, T., Diggelmann, H., and Scherrer, K. (1973). Proc. Natl. Acad. Sci. USA 70:1122.
10. Bachenheimer, S., and Darnell, J. E. (1975). Proc. Natl. Acad. Sci. USA 72:4445.
11. Lizardi, P. M. (1976). Cell 7:239.

12. Lizardi, P. M. (1976). Prog. Nucleic Acid Res. Mol. Biol. 19:301.
13. Dunn, J. J., and Studier, F. W. (1975). Brookhaven Symp. Biol. 26:267.
14. Bajszar, G., Samarina, O. P., and Georgiev, G. P. (1974). Mol. Biol. Rep. 1:305.
15. Schmincke, C. D., Herrmann, K., and Hausen, P. (1976). Proc. Natl. Acad. Sci. USA 73:1994.
16. Schibler, U., and Perry, R. P. (1976). Cell 9:121.
17. Brandhorst, B. P., and McConkey, E. H. (1974). J. Mol. Biol. 85:451.
18. Aronson, A., and Wilt, F. H. (1969). Proc. Natl. Acad. Sci. USA 62:186.
19. Levy, B., and McCarthy, B. J. (1976). Biochemistry 15:2415.
20. Price, R. P., Ransom, L., and Penman, S. (1974). Cell 2:253.
21. Lengyel, J., and Penman, S. (1975). Cell 5:281.
22. Herman, R. C., Williams, J. G., and Penman, S. (1976). Cell 7:429.
23. Egyhazi, E. (1976). Cell 7:507.
24. Singer, R. H., and Penman, S. (1973). J. Mol. Biol. 78:321.
25. Perry, R. P., and Kelley, D. E. (1973). J. Mol. Biol. 79:681.
26. Puckett, L., Chambers, S., and Darnell, J. E. (1975). Proc. Natl. Acad. Sci. USA 72:389.
27. Berger, S. L., and Cooper, H. L. (1975). Proc. Natl. Acad. Sci. USA 72:3873.
28. Rottman, F., Gillam, S., and Smith, M., unpublished observations.
29. Milcarek, C., Price, R. P., and Penman, S. (1974). Cell 3:1.
30. Houdebine, L. M. (1976). FEBS Lett. 66:110.
31. Hirsch, M., and Penman, S. (1973). J. Mol. Biol. 80:379.
32. Gorski, J., Morrison, M. R., Merkel, C. G., and Lingrel, J. B. (1974). J. Mol. Biol. 86:363.
33. Sheiness, D., and Darnell, J. E. (1973). Nature (New Biol.) 241:265.
34. Nemer, M., Graham, M., and Dubroff, L. M. (1974). J. Mol. Biol. 89:435.
35. Fromson, D., and Verma, D. P. S. (1976). Proc. Natl. Acad. Sci. USA 73:148.
36. MacLeod, M. C. (1975). Biochemistry 14:4011.
37. Bard, E., Efron, D., Marcus, A., and Perry, R. P. (1974). Cell 1:101.
38. Munoz, R. F., and Darnell, J. E. (1974). Cell 2:247.
39. Soreq, H., Nudel, U., Salomon, R., Revel, M., and Littauer, U. Z. (1974). J. Mol. Biol. 88:233.
40. Huez, G., Marbaix, G., Hubert, E., Leclercq, M., Nudel, U., Soreq, H., Salomon, R., Lebleu, B., Revel, M., and Littauer, U. Z. (1974). Proc. Natl. Acad. Sci. USA 71:3143.
41. Marbaix, G., Huez, G., Burny, A., Cleuter, Y., Hubert, E., Leclercq, M., Chantrenne, H., Soreq, H., Nudel, U., and Littauer, U. (1975). Proc. Natl. Acad. Sci. USA 72:3065.
42. Huez, G., Marbaix, G., Hubert, E., Cleuter, Y., Leclercq, M., Chantrenne, H., DeVos, R., Soreq, H., Nudel, U., and Littauer, U. (1975). Eur. J. Biochem. 59:589.
43. Jeffrey, W. R., and Brawerman, G. (1975). Biochemistry 14:3445.
44. Heywood, S. M., and Kennedy, D. S. (1976). Biochemistry 15:3314.
45. Blobel, G. (1973). Proc. Natl. Acad. Sci. USA 70:924.
46. Jelinek, W., Adesnik, M., Salditt, M., Sheiness, D., Wall, R., Molloy, G., Philipson, L., and Darnell, J. E. (1973). J. Mol. Biol. 75:515.
47. Perry, R. P., Kelley, D. E., and LaTorre, J. (1974). J. Mol. Biol. 82:315.
48. Getz, M. J., Birnie, G. D., Young, B. D., MacPhail, E., and Paul, J. (1975). Cell 4:121.
49. Ryffel, G. U. (1976). Eur. J. Biochem. 62:417.

50. Edmonds, M., and Abrams, R. (1960). J. Biol. Chem. 235:1142.
51. Jelinek, W. R. (1974). Cell 2:197.
52. DePomerai, D. I., and Butterworth, P. H. (1975). Eur. J. Biochem. 58:185.
53. Diez, J., and Brawerman, G. (1974). Proc. Natl. Acad. Sci. USA 71:4091.
54. Brandhorst, B. P., and McConkey, E. H. (1975). Proc. Natl. Acad. Sci. USA 72:3580.
55. Mendecki, J., Lee, S. Y., and Brawerman, G. (1972). Biochemistry 11:792.
56. Birnboim, H. C., Mitchel, R. E. J., and Straus, N. A. (1973). Proc. Natl. Acad. Sci. USA 70:2189.
57. Harbers, K., and Spencer, J. H. (1974). Biochemistry 13:1094.
58. Nakazato, H., Edmonds, M., and Kopp, D. W. (1974). Proc. Natl. Acad. Sci. USA 71:200.
59. Edmonds, M. A., Nakazato, H., Korwek, E. L., and Venkatesan, S. (1976). Prog. Nucleic Acid Res. Mol. Biol. 19:99.
60. Molloy, G. R., and Darnell, J. E. (1973). Biochemistry 12:2324.
61. Nichols, J. L., and Eiden, J. J. (1974). Biochemistry 13:4629.
62. Proudfoot, N. J., Cheng, C. C., and Brownlee, G. G. (1976). Prog. Nucleic Acid Res. Mol. Biol. 19:123.
63. Salser, W., Browne, J., Clarke, P., Heindell, H., Higuchi, R., Paddock, G., Roberts, J., Studnicka, G., and Zakar, P. (1976). Prog. Nucleic Acid Res. Mol. Biol. 19:177.
64. Proudfoot, N. J., and Brownlee, G. G. (1976). Nature (Lond.) 263:211.
65. Moore, P. B. (1966). J. Mol. Biol. 18:38.
66. Starr, J. L., and Sells, B. H. (1969). Physiol. Rev. 49:623.
67. Perry, R. P., and Kelley, D. E. (1974). Cell 1:37.
68. Desrosiers, R., Friderici, K., and Rottman, F. (1974). Proc. Natl. Acad. Sci. USA 71:3971.
69. Furuichi, Y. (1974). Nucleic Acids Res. 1:809.
70. Shatkin, A. J. (1974). Proc. Natl. Acad. Sci. USA 71:3204.
71. Wei, C. M., and Moss, B. (1974). Proc. Natl. Acad. Sci. USA 71:3014.
72. Ro-Choi, T. S., Choi, Y. C., Henning, D., McCloskey, J., and Busch, H. (1975). J. Biol. Chem. 250:3921.
73. Rottman, F., Shatkin, A. J., and Perry, R. (1974). Cell 3:197.
74. Sripati, C. E., Groner, Y., and Warner, J. R. (1976). J. Biol. Chem. 251:2898.
75. DeKloet, S. R., and Andrean, A. G. (1976). Biochim. Biophys. Acta 425:401.
76. Dottin, R. P., Weiner, A. M., and Lodish, H. F. (1976). Cell 8:233.
77. Yang, N. S., Manning, F. R., and Gage, L. P. (1976). Cell 7:339.
78. Faust, M., Millward, S., and Fromson, D. (1975). J. Cell Biol. 67:114.
79. Kastern, W. H., and Berry, S. J. (1976). Biochem. Biophys. Res. Commun. 71:37.
80. Muthukrishnan, S., Filipowicz, W., Sierra, J. M., Both, G. W., Shatkin, A. J., and Ochoa, S. (1975). J. Biol. Chem. 250:9336.
81. Perry, R. P., and Sherrer, K. (1975). FEBS Lett. 57:73.
82. Muthukrishnan, S., Both, G. W., Furuichi, Y., and Shatkin, A. J. (1975). Nature (Lond.) 255:33.
83. Ouellette, A. J., Frederick, D., and Malt, R. A. (1975). Biochemistry 14:4361.
84. Perry, R. P., Kelley, D. E., Friderici, K. H., and Rottman, F. M. (1975). Cell 6:13.
85. Cory, S., and Adams, J. M. (1975). J. Mol. Biol. 99:519.
86. Dubin, D. T., and Taylor, R. H. (1975). Nucl. Acids Res. 2:1653.
87. Desrosiers, R., Friderici, K., and Rottman, F. (1975). Biochemistry 14:4367.
88. Lavi, S., and Shatkin, A. J. (1975). Proc. Natl. Acad. Sci. USA 72:2012.
89. Wei, C. M., Gershowitz, A., and Moss, B. (1976). Biochemistry 15:397.

90. Furuichi, Y., Morgan, M., Shatkin, A. J., Jelinek, W., Salditt-Georgeiff, N., and Darnell, J. E. (1975). Proc. Natl. Acad. Sci. USA 72:1904.
91. Dasgupta, R., Harada, F., and Kaesberg, P. (1976). J. Virol. 18:260.
92. Zimmern, D. (1975). Nucleic Acids Res. 2:1189.
93. Keith, J., and Fraenkel-Conrat, H. (1975). FEBS Lett. 57:31.
94. Pinck, L. (1975). FEBS Lett. 59:24.
95. Keith, J., and Fraenkel-Conrat, H. (1975). Proc. Natl. Acad. Sci. USA 72:3347.
96. Stoltzfus, C. M., and Dimock, K. (1976). J. Virol. 18:586.
97. Colonno, R. J., and Stone, H. O. (1976). Nature (Lond.) 261:611.
98. Furuichi, Y., and Miura, K. (1975). Nature (Lond.) 253:374.
99. Wei, C. M., and Moss, B. (1975). Proc. Natl. Acad. Sci. USA 72:318.
100. Sommer, S., Salditt-Georgieff, M., Bachenheimer, S., Darnell, J. E., Furuichi, Y., Morgan, M., and Shatkin, A. J. (1976). Nucleic Acids Res. 3:749.
101. Moss, B., and Koczot, F. (1976). J. Virol. 17:385.
102. Desrosiers, R. C., Sen. G. C., and Lengyel, P. (1976). Biochem. Biophys. Res. Commun. 73:32.
103. Moyer, S. A., Abraham, G., Adler, R., and Banerjee, A. K. (1975). Cell 5:59.
104. Rose, J. K. (1975). J. Biol. Chem. 250:8098.
105. Krug, R. M., Morgan, M. A., and Shatkin, A. J. (1976). J. Virol. 20:45.
106. Bondurant, M., Hashimoto, S., and Green, M. (1976). J. Virol. 19:998.
107. Dubin, D. T., and Stollar, V. (1975). Biochem. Biophys. Res. Commun. 66:1373.
108. Hefti, E., Bishop, D. H. L., Dubin, D. T., and Stollar, V. (1976). J. Virol. 17:149.
109. Thomason, A. R., Brian, D. A., Velicer, L. F., and Rottman, F. M. (1976). J. Virol. 20:123.
110. Hewlett, M. J., Rose, J. K., and Baltimore, D. (1976). Proc. Natl. Acad. Sci. USA 73:327.
111. Nomoto, A., Lee, Y. F., and Wimmer, E. (1976). Proc. Natl. Acad. Sci. USA 73:375.
112. Frisby, D., Eaton, M., and Fellner, P. (1976). Nucleic Acids Res. 3:2771.
113. Nuss, D. L., and Koch, G. (1976). J. Mol. Biol. 102:601.
114. Lesnaw, J. A., and Reichmann, M. E. (1970). Proc. Natl. Acad. Sci. USA 66:140.
115. Perry, R. P., Kelley, D. E., Friderici, K., and Rottman, F. (1975). Cell 4:387.
116. Furuichi, Y., Muthukrishnan, S., and Shatkin, A. J. (1975). Proc. Natl. Acad. Sci. USA 72:742.
117. Adams, J. M., and Cory, S. (1975). Nature (Lond.) 255:28.
118. Furuichi, Y., Shatkin, A. J., Stavnezer, E., and Bishop, J. M. (1975). Nature (Lond.) 257:618.
119. Wei, C. M., Gershowitz, A., and Moss, B. (1975). Cell 4:379.
120. Friderici, K., Kaehler, M., and Rottman, F. (1976). Biochemistry 15:5234.
121. Wei, C. M., Gershowitz, A., and Moss, B. (1975). Nature (Lond.) 257:251.
122. Moyer, S. A., and Banerjee, A. K. (1976). Virology 70:339.
123. Furuichi, Y., Shatkin, A. J., Stavnezer, E., and Bishop, J. M. (1975). Nature (Lond.) 257:618.
124. Perry, R. P., and Kelley, D. E. (1976). Cell 8:433.
125. Rottman, F., Desrosiers, R., and Friderici, K. (1976). Prog. Nucleic Acid Res. Mol. Biol. 19:21.
126. Dimock, K., and Stolzfus, C. M. (1977). Biochemistry. 16:471.
127. Rottman, F., and Friderici, K., unpublished observations.
128. Steitz, J. A., and Jakes, K. (1975). Proc. Natl. Acad. Sci. USA 72:4734.
129. Salditt-Georgieff, M., Jelinek, W., Darnell, J. E., Furuichi, Y., Morgan, M., and Shatkin, A. (1976). Cell 7:227.
130. Rhodes, D. P., Moyer, S. A., and Banerjee, A. K. (1974). Cell 3:327.

131. Moss, B., Martin, S. A., Ensinger, M. J., Boone, R. F., and Wei, C. M. (1976). Prog. Nucleic Acid Res. Mol. Biol. 19:63.
132. Furuichi, Y., Muthukrishnan, S., Tomasz, J., and Shatkin, A. J. (1976). J. Biol. Chem. 251:5043.
133. Colonno, R. J., Abraham, G., and Banerjee, A. K. (1976). Prog. Nucleic Acid Res. Mol. Biol. 19:83.
134. Martin, S. A., Paoletti, E., and Moss, B. (1975). J. Biol. Chem. 250:9322.
135. Martin, S. A., and Moss, B. (1975). J. Biol. Chem. 250:9330.
136. Groner, Y., and Hurwitz, J. (1975). Proc. Natl. Acad. Sci. USA 72:2930.
137. Winicov, I., and Perry, R. P. (1976). Biochemistry 15:5039.
138. Nuss, D. L., Furuichi, Y., Koch, G., and Shatkin, A. J. (1975). Cell 6:21.
139. Shinshi, H., Miwa, M., Sugimura, T., Shimotohno, K., and Miura, K. (1976). FEBS Lett. 65:254.
140. Both, G. W., Banerjee, A. K., and Shatkin, A. J. (1975). Proc. Natl. Acad. Sci. USA 72:1189.
141. Rose, J. K., and Lodish, H. F. (1976). Nature (Lond.) 262:32.
142. Kemper, B. (1976). Nature (Lond.) 262:321.
143. Both, G. W., Furuichi, Y., Muthukrishnan, S., and Shatkin, A. J. (1975). Cell 6:185.
144. Dasgupta, R., Shih, D. S., Saris, C., and Kaesberg, P. (1975). Nature (Lond.) 256:624.
145. Kozak, M., and Shatkin, A. J. (1976). J. Biol. Chem. 251:4259.
146. Hickey, E. D., Weber, L. A., and Baglioni, C. (1975). Proc. Natl. Acad. Sci. USA 73:19.
147. Filipowicz, W., Furuichi, Y., Sierra, J. M., Muthukrishnan, S., Shatkin, A. J., and Ochoa, S. (1976). Proc. Natl. Acad. Sci. USA 73:1559.
148. Shafritz, D. A., Weinstein, J. A., Safer, B., Merrick, W. C., Weber, L. A., Hickey, E. D., and Baglioni, C. (1976). Nature (Lond.) 261:291.
149. Shih, D. S., Dasgupta, R., and Kaesberg, P. (1976). J. Virol. 19:637.
150. Both, G. W., Furuichi, Y., Muthukrishnan, S., and Shatkin, A. J. (1976). J. Mol. Biol. 104:637.
151. Hickey, E. D., Weber, L. A., and Baglioni, C. (1976). Nature (Lond.) 261:71.
152. Munoz, R. F., and Darnell, J. E. (1976). J. Virol. 129:719.
153. Sen, G. C., Lebleu, B., Brown, G. E., Rebello, M. A., Furuichi, Y., Morgan, M., Shatkin, A. J., and Lengyel, P. (1975). Biochem. Biophys. Res. Commun. 65:427.
154. Stuart, S. E., and Rottman, F. M. (1973). Biochem. Biophys. Res. Commun. 55:1001.
155. Liau, M. C., Hunt, M. E., and Hurlbert, R. B. (1976). Biochemistry 15:3158.

International Review of Biochemistry
Biochemistry of Nucleic Acids II, Volume 17
Edited by B. F. C. Clark
Copyright 1978   University Park Press   Baltimore

4
Structure and Function of Viral Nucleic Acids

W. MIN JOU and W. FIERS
Laboratory of Molecular Biology, State University, Ghent, Belgium

Research in the authors' laboratory was supported by grants from the "Fonds voor Fundamen-
teel Wetenschappelijk Onderzoek" and the "Kankerfonds" of the "Algemene Spaar-en Lijfren-
tekas."

75

In 1969, Sanger and his colleagues reported the first extensive nucleotide sequence of part of a viral genome, and in fact of any large nucleic acid, obtained by adapting their fast sequencing microtechniques with ^{32}P-labeled RNA to the problems raised by large, viral RNAs (1). The sequence was derived for part of the coat protein gene of the RNA bacteriophage R17. (The term "bacteriophage" and its abbreviation "phage" are used interchangeably.) Since then, nucleotide sequence determination of large RNA molecules of all types has progressed rapidly, especially for RNAs of viral origin, but also for ribosomal RNA and to a lesser extent for messenger RNAs (see Chapter 3). Continued effort in the determination of phage RNA structure has shown in molecular detail what polypeptide chain initiation and termination sites look like and how successive code words specify a polypeptide. It has proved the existence of extensive untranslated regions and has provided a basis for the construction of secondary structure models. The complete primary structure of MS2 RNA, as well as good models of its secondary folding (2–4) (for part of the structure see Figure 1), and extensive segments of the sequence of the closely related R17 (23.9% of the total) and f2 RNAs (11.5%) are now known. Large sections of Qβ-RNA (belonging to a serologically different phage group) have also been determined. Because of the inherent difficulties involved, the sequence information obtained so far from plant and animal RNA viruses is much more modest, but interesting results have nevertheless been obtained regarding the genetic organization of several important virus groups by using RNA sequencing technology.

    Until recently, the primary structure determination of DNA presented quite different problems, and, in fact, DNA sequencing has often been reduced to RNA sequencing by the use of in vitro transcription with RNA polymerase. However, the combination of the following relatively recent developments has made DNA sequencing a convenient and extremely powerful technique: 1) the use of a wide variety of restriction enzymes for dissection of DNA molecules, providing defined pieces amenable to molecular analysis; 2) fast methods for direct DNA sequencing (see also Chapter 5). For these reasons, the situation has

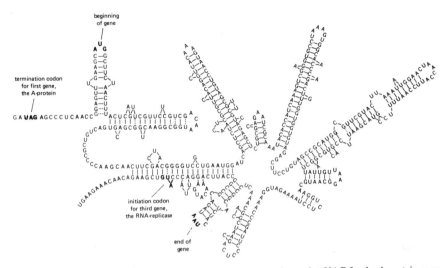

Figure 1. Structure of the MS2 coat protein gene. The termination codon UAG for the A protein gene is followed by a short untranslated region. The gene coding for the coat protein starts with AUG and ends with the double stop signal UAA·UAG. The initiation codon AUG for the third gene, the viral-coded replicase subunit, is located 36 nucleotides beyond the end of the coat gene. The secondary structure model (the "flower" model) is based on experimental observations and on estimates of thermodynamic stability. Putative base pairs are shown by *dashes* rather than the standard dot representation. Reproduced from Min Jou et al. (2) with permission of Macmillan Journals, London.

become inverted and instead of avoiding DNA sequencing it may be advantageous to convert RNAs into DNAs (by using reverse transcriptase or DNA polymerase I) to study their structure. In addition, the double-stranded DNA-RNA hybrid or a complete DNA copy can be introduced into a suitable vector (either of viral origin or a plasmid) and amplified. With the use of restriction enzymes, the amplified DNA can then be dissected into fragments small enough for direct DNA sequencing. These detours for RNA structure determination have only just begun to be exploited, but many results are to be expected in the near future, not only for viral RNA but especially for all large RNAs which are only obtained in low yields (e.g., messenger RNAs) or which are difficult to label adequately in vivo.

This chapter describes the structural information that has been accumulated for the different viral nucleic acids and delineates the different strategies that have been followed to tackle the problems. Sequencing techniques themselves are beyond the scope of this chapter, but excellent reviews of both RNA (5, 6) and DNA (7) sequencing methods (see also Chapter 5 of this volume) are available. In general, nucleic acid, labeled either in vivo or in vitro with ^{32}P, has been used.

SEQUENCE STRATEGY

To determine the structure of a nucleic acid (either RNA or DNA), a piece not exceeding 250 nucleotides in length, in a pure and highly radioactive form, is needed. No viral nucleic acid genome fulfills these criteria; even the RNA of satellite tobacco necrosis virus (dependent on the presence of a helper virus for its replication) contains approximately 1,200 nucleotides, and viroid RNAs have not yet been obtained at high specific activity. The smallest independent RNA viruses consist of around 3,500 nucleotides (group I RNA phages), and the smallest independent DNA viruses of about 5,400 nucleotides (the single-stranded DNA phages) or nucleotide pairs (the animal viruses SV40 and polyoma). The first problem then is to reduce the nucleic acid into subsets of smaller fragments. Alternatively, one or a limited number of pieces of interest can be selected. The more general approaches are discussed first.

Limited Hydrolysis

While studying the tRNA^{Ala} molecule (the first RNA molecule ever sequenced), it was observed that one can obtain reproducible fragments by partial ribonuclease T_1 digestion, presumably because of the highly ordered secondary and tertiary structure of the RNA (8). Rather unexpectedly, limited digests of macromolecular RNA, both from ribosomal (9, 10) or from phage origin (1, 11), also gave a discrete number of fragments as revealed by polyacrylamide gel electrophoresis. These observations indicate that these RNAs also have at least a partially ordered, specific conformation in solution. Obviously, by selecting hydrolysis conditions which predominantly create fragments of 25–200 chain length and by finding ways to fractionate such products, a source of RNA pieces that can be used as a starting point becomes available to study the structure of RNAs of this size. In vivo homogeneously ^{32}P-labeled RNA has chiefly been used for this approach. Because among RNA viral genomes only phage RNA could easily be labeled to a sufficient extent at that time, the system was developed further, mainly by using RNA from the phages R17 or MS2.

To obtain optimum resolution in the 25–200-nucleotide size range, the partially digested RNA was routinely electrophoresed through 12% polyacrylamide gel slabs at neutral pH (1, 12). The gel patterns contained some 35–45 bands, each consisting of several fragments (or complexes) of approximately the same length. One particular disadvantage of this approach is that there is no a priori selection of fragments from any particular region of the molecule. Therefore, each gel band must be screened for the presence of a particular region based on a specific property within the region of interest, e.g., the presence of a unique T_1-oligonucleotide like the 5'- or 3'-terminus or nucleotide sequences corresponding to the known amino acid sequence of the coat protein.

Next, the fragment of interest must be purified from other fragments which are of roughly the same size in the gel band. When the chain length does not exceed 70 nucleotides, the fragment can be subjected to a two-dimensional system by using homochromatography in the second dimension (13). To over-come the limitation in chain length, a two-dimensional polyacrylamide gel elec-trophoresis system has been developed (14). In the first dimension, elec-trophoresis is carried out at pH 3.5 in the presence of urea so that the denatured polynucleotides are separated according to size and base composition. The sec-ond dimension is run at pH 8.0 in a less porous gel; under these conditions, the polynucleotides refold into a secondary and spatial structure so that the fraction-ation is now size- and conformation-dependent. An example of such a separation is shown in Figure 2. Because of the high versatility (e.g., adaptation of the gel concentration in both dimensions according to the problem at hand), this method has been used successfully for the further resolution of all the bands in the primary digests of MS2 RNA (2–4). In this way, a very large number of frag-ments from all over the molecule have been obtained in pure form. Sequencing of pure fragments, obtained either by homochromatography or by two-dimensional gel electrophoresis, was done essentially according to the now classic techniques of Sanger and his colleagues (5, 6). The great number of fragments have proved to be of crucial importance in obtaining the necessary overlaps which allow the construction of long nucleotide sequences. However, partial hydrolysis has only a limited reproducibility, probably related at least in part to the insufficient intactness of the RNA (possibly due to radiolysis) and to the somewhat varying tertiary structure in each preparation. In practice, these characteristics result in a varying yield and purity of RNA fragments between different digests, thereby considerably complicating the problem. The problem has at least partly been offset by the development of faster, more sensitive, and more informative methods for the sequence analysis of oligonucleotides (16–19) and for finger-printing ribonuclease digests of RNA fragments (20). Partial enzymic digestion is achieved preferably by means of a base-specific enzyme such as ribonuclease T_1, which cleaves only after guanosine residues; indeed, this makes subsequent sequencing somewhat simpler.

Careful analysis of partial ribonuclease T_1 digests revealed that there are sites in the RNA molecule which are extremely sensitive to the enzyme, even with the use of milder digestion conditions, so that it most likely would be impossible to find an overlap at these points (2). But, with the use of an enzyme with a different base specificity the required overlaps should be obtained. Under appropriate conditions, ϵ-carboxymethyllysine-41-pancreatic ribonuclease (CM-RNase) has a very high preference for YpA (pyrimidine nucleoside-phosphate-adenosine) bonds (21). The complementary information obtained from the study of the set of fragments created by this latter enzyme allowed the complete elucidation of the MS2 RNA structure (2–4).

In addition, partial digests prepared with RNase IV, an enzyme isolated from *Escherichia coli* MRE-600, have been used for the isolation of regions from phage R17 RNA for sequencing purposes (22–24). T4 endonuclease IV, an enzyme discovered in T4-infected *E. coli* cells, has a high preference for single-stranded DNA and yields oligonucleotides with 5′-terminal cytidine residues.

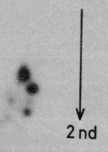

1 st ←

2 nd ↓

xylene
cyanol FF

bromophenol blue

Specific fragments have thus been obtained from partial digests of DNA from the single-stranded DNA phages (25), some of which have been sequenced (26-28).

Hydrolysis with Specific Enzymes

Complete hydrolysis of an RNA fragment with ribonuclease T_1 (guanosine-specific) or pancreatic ribonuclease (specific for pyrimidine nucleoside residues) and two-dimensional separation of the resulting products are routinely the first steps in the structure determination (5, 6). Because of the typical pattern obtained, such a fingerprint is also highly characteristic for any given RNA. To obtain a pattern diagnostic of large RNAs (such as viral RNAs), a satisfactory fractionation of the longest unique ribonuclease T_1 oligonucleotides is required. Jeppesen (29) adapted the two-dimensional technique by using homochromatography in the second dimension to study complete ribonuclease T_1 digests of the RNA from bacteriophage R17. A recent modification of this technique (20), used to study MS2 RNA sequences, allows the deduction of the composition of most oligonucleotides from their position; small, compact spots are obtained. This procedure offers both a high sensitivity and an excellent resolution. Another technique, originally designed to study products from partial digests (see under "Limited Hydrolysis"), involves two-dimensional fractionation on polyacrylamide gels (14). It leads to very nice separations of total ribonuclease T_1 digests of even complex viral RNAs (30). Such systems are being widely used now to study the relationship between different viruses, between RNA molecules contained within one virus, between viral mRNAs appearing in the cell and the genomic RNA, etc. (see below).

Direct structural studies on large sections of DNA have only become possible since the discovery of the type II restriction enzymes (31). Because these enzymes have quite complex, well defined recognition sequences (32) only a limited number of specific breaks are introduced, as evidenced by the reproducible and simple band pattern obtained after gel electrophoresis, e.g., with SV40 DNA as a substrate (33). Unlike partial ribonuclease hydrolysis producing RNA fragments which often vary in yield and are far from molar, a restriction enzyme digest (as it is based on the primary nucleotide sequence itself) provides, in principle, a reproducible set of fragments in molar yield. The great potential of this approach has stimulated the search for more restriction enzymes from other sources recognizing different but equally well defined sequences. At the moment, more than 80 type II restriction enzymes, with at least 30 different spe-

Figure 2. Two-dimensional separation of RNA fragments. ^{32}P-labeled MS2 RNA was partially digested with ribonuclease T_1 and fractionated by electrophoresis on a 6% polyacrylamide gel (cf. De Wachter et al. (15)). The material present in one particular band (γ5) was eluted, concentrated, and further resolved by two-dimensional electrophoresis. The first dimension was on an 8% polyacrylamide gel in 0.025 M citric acid-6 M urea; the second dimension was on a 16% polyacrylamide gel in 0.04 M Tris-acetate, pH 8.0 (Min Jou and Fiers, unpublished observations, conditions according to ref. 14).

cificities, have been described (34) (restriction enzymes can be detected by their selective endonuclease activity on foreign, as opposed to host, DNA). Therefore, by successive digestion with restriction enzymes having different specificities, the whole DNA genome of a virus can be reduced to a reproducible set of fragments amenable to nucleotide sequencing.

The fractionation of DNA restriction fragments is routinely carried out on one-dimensional polyacrylamide (33) or Agarose (35) gels. Even with restriction digests from the smallest (viral) genomes available, such as SV40 (33, 36, 37) or the icosahedral (38, 39) or filamentous (40) DNA phages, fragments with the same or very similar mobility are often observed. In an attempt to extend the fractionation potential to more complex restriction fragment mixtures, a two-dimensional gel system has now been developed (41). The first dimension is run in an acrylamide-agarose matrix polymerized without crosslinker, whereas the second dimension consists of an acrylamide-bisacrylamide gel (20:1) run at 50°C. The method is especially suitable for the analysis of fragments in the size range of 50–1,000 nucleotide pairs. In digests of phage λ-DNA or adenovirus DNA (M.W. 32×10^6 and 22×10^6, respectively), with restriction enzymes recognizing a 4-nucleotide sequence and thus giving rise to a large number of fragments, more than 100 well resolved spots could be discriminated, and more than 200 in partial digests of these DNAs (Figure 3).

Different methods have been used to arrange the fragments from a particular restriction digest in a physical map of the viral DNA genome. As an example, the restriction map of SV40, showing the cleavage sites of some 20 different restriction enzymes, is shown in Figure 4. Some of these methods are general, but in other approaches specific characteristics of the system under study are essential (34). An extremely simple and rapid method of general applicability has recently been published (48). The DNA is labeled at the 5'-termini with the use of polynucleotide kinase and γ-[^{32}P]ATP. A single restriction nuclease cleavage is introduced, and the two fragments, each containing a single label, are separated by gel electrophoresis. Each fragment is then partially digested with the restriction nuclease under study, and the fragments are separated according to size by gel electrophoresis. Only the fragments containing the labeled 5'-end are detected by radioautography, and the successive cleavage sites can be directly located relative to the 5'-end by measuring the length of each partial product (by comparison with a set of DNA molecular weight markers). When different restriction enzyme digests are run in parallel, the relative order of the restriction sites can be read directly from the gel.

The concept of this method for restriction mapping is basically similar to that of the DNA sequencing method developed by Maxam and Gilbert (49). In fact, the difference in this latter method is that the cleavages are in principle base specific and are introduced by four different chemical reactions. The products are then separated according to chain length on a denaturing, high percentage polyacrylamide gel. The radioautograph gives the distance of each nucleotide from the 5'-end; in other words, the nucleotide sequence is read directly from the

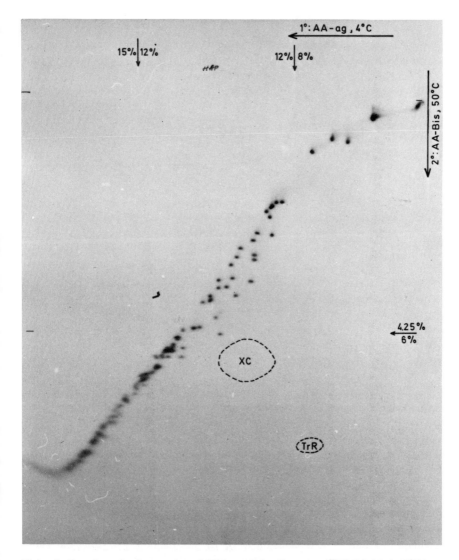

Figure 3. Two-dimensional separation of DNA restriction fragments. ^{32}P-labeled λvir DNA was digested with Hap II (a restriction enzyme from *Haemophilus aphrophilus*) and subjected to two-dimensional gel electrophoresis. Multiple step gels were used in both directions: First dimension: 8% acrylamide-1% Agarose, 12% acrylamide-1% Agarose, 15% acrylamide-1% Agarose; 370 V, 18 hr. Second dimension: 4.25% acrylamide-bisacrylamide (20:1), 6% acrylamide-bisacrylamide (20:1); 100 V, 18 hr. The positions of the dye markers xylene cyanol FF(XC) and trypan red (TrR) are marked on the figure. Reproduced from Derynck and Fiers (41) with permission of Academic Press, London.

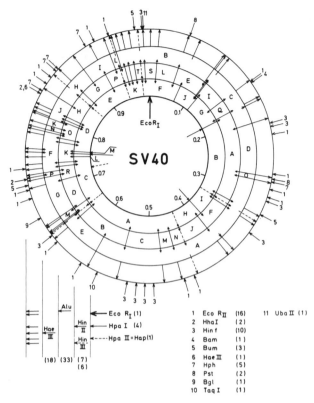

Figure 4. Cleavage map of SV40 DNA. The single cleavage site of Eco R_I is taken as a reference, and map distances are given as fractional lengths of the SV40 genome in a clockwise direction from the Eco R_I site. The number of cleavage sites are given in parentheses for each enzyme. *Inner circle,* Eco R_I (42, 43), Hpa I (35, 44), and Hpa II (35, 45) (or Hap) sites (the latter in *dashes*). *First ring,* Hind II (*straight lines*) and Hind III (*dashed lines*) sites (45, 295). *Second ring,* Alu sites (37) (Hind III sites are also cleaved by Alu). *Third ring,* Hae III sites (36, 45, 47). *Fourth ring,* fragments produced by combined action of Eco R_1, Hind II + III, Hap, Alu, and Hae III. *Bottom right,* cleavage sites by various other enzymes listed. Relevant references are listed by R. Roberts (34).

four lanes on the film. Similar gel fractionations according to size differences down to the nucleotide level are used in the "plus and minus" sequencing technique of Sanger and Coulson (50). In this method, enzymic extensions starting from a primer (either synthetic or a strand from a restriction fragment) are manipulated so that four series of ^{32}P-labeled products are obtained, each containing intermediates specifically terminated at each of the four nucleotides. By using a long and a short fractionation of the same reactions, up to 126 nucleotides could be read in one experiment (51). An additional advantage of these methods is the use of in vitro ^{32}P-labeling procedures so that the methods can be applied to material which is difficult to label adequately in vivo. A complication of the latter method, however, is the requirement of a single-stranded template which is not readily available in most cases. With the help of these extremely powerful methods (especially the Maxam and Gilbert method, already widely used because it is more generally applicable), knowledge of nucleotide sequences will undoubtedly expand considerably in the near future. Knowledge will not remain limited to DNA from directly available sources such as viral or plasmid DNA; in principle, any piece of DNA (by using the technology of genetic engineering) or RNA (after reverse transcription into DNA) can be brought within the scope of these methods. A detailed account of some of these recently developed DNA sequencing techniques may be found in Chapter 5 of this book.

Sequences of Specific Binding Sites

In this technique, advantage is taken of the ability of a specific region of a nucleic acid to form a stable complex with another macromolecule, thus rendering this region resistant to digestion by a nuclease while the bulk of the uncovered nucleic acid is removed by hydrolysis. The piece of nucleic acid thus obtained is usually well defined and suitable for sequencing. Obviously, this method is limited to special regions of a nucleic acid molecule. The three ribosome binding sites of R17 RNA (52) and the $Q\beta$ coat protein ribosome binding site (53) were the first regions determined by this approach starting from in vivo ^{32}P-labeled RNA. When DNA was the primary nucleic acid source, both in vivo labeled DNA and in vitro ^{32}P-labeled RNA transcripts were used to determine the structure (e.g., binding of respressors to operators, of RNA polymerase to promoters). In the future, such regions will most likely be studied with the use of the new DNA sequencing techniques described briefly in the previous section (49, 50).

The nucleotide sequences of ribosome binding sites have now been obtained for many more viral and bacterial systems. In addition, the interaction sites of phage RNA with different proteins playing a role in the life cycle of RNA phages have been determined. From the comparison of the nucleotide sequence of regions from different systems which have apparently identical functions, similar features probably related to this function can be discerned.

Transcription Approaches

Often it is more convenient to study a transcript covering only a limited part of the viral genome rather than the viral genome itself. In its most simple form, the natural, specific initiation and termination processes are used, e.g., when a small virus-induced RNA molecule can be isolated from virus-infected cells labeled with [^{32}P]phosphate (54, 55). Alternatively, such a functional transcription system can sometimes be reproduced in vitro to prepare labeled transcripts. Either discrete and relatively simple end products are isolated (56, 57), or limited synchronous products are generated by controlling reaction conditions such as time, temperature, and the composition of the nucleoside triphosphate mixture (58). The in vitro system offers an additional advantage: nearest neighbor labeling data are obtained which greatly facilitate sequence reconstruction, which involves running four different reactions, each containing a different labeled α-[^{32}P]nucleoside triphosphate (59). Even in a heterologous system in which SV40 DNA is used together with *E. coli* RNA polymerase, the reaction starts from one major specific site, indicating the presence of a single region which is similar to an *E. coli* promoter (60).

Although it offers the possibility of RNA sequencing (until recently considered a technical advantage), transcription of a DNA fragment with *E. coli* RNA polymerase in vitro often poses serious problems. Indeed, when natural initiation and termination signals are not present, a very heterogeneous population of molecules is obtained. For the study of the *lac* operator, use was made of the fact that transcription can be forced to start from an oligonucleotide primer added to the reaction mixture (61).

An alternative approach involves a sequential DNA-RNA hybridization procedure, making use of specialized transducing phages which carry genetic deletions on either side of the *lac* control region and carry the target DNA in opposite orientations. In this way, sufficiently pure RNA transcripts are obtained from both strands for detailed sequence analysis (62).

Synthesis of complementary DNA starting from an internal natural primer is the normal reaction mechanism for reverse transcriptase, an enzyme present in RNA tumor viruses (see below). In order to ensure that synthesis starts from the 3'-end of any RNA, an oligo(dT) primer can be used, forming a complex with the 3'-terminal poly(A) tail (already present (63) or first enzymically added (64)). As an alternative, the reverse transcriptase activity of *E. coli* DNA polymerase I in the presence of manganese can be used (65). The importance of these approaches has been stressed already (see above).

For many virus classes, including RNA tumor viruses, isolation of the self-copying enzyme system is rather simple because it is part of the virion (see below). This feature facilitates the study of the homologous reaction because incubation of disrupted virions with triphosphates is sufficient to start the reaction.

RNA VIRUSES

RNA Bacteriophages

The male-specific RNA coliphages are among the simplest viruses known and can be classified into four serological groups. Most structural work has been done with members of group I (f2, R17, and MS2) and with the group III phage Qβ. The nucleic acid content is about one-third of the particle weight. The RNA consists of about 3,500–4,500 nucleotides (group I and group III phages, respectively). The virion contains 180 coat protein subunits and a single copy of another structural protein, the so-called A protein (A2 in Qβ phage). Qβ particles contain an additional protein, called A1, which is a read-through product of the coat protein. The only other phage-coded polypeptide is part of the viral-specific RNA replicase complex.

As mentioned above, the first extensive viral nucleic acid sequence published was derived from phage R17 (1). Because these viral RNAs can be labeled in vivo to a high specific activity, considerable effort has been devoted to this system, resulting in the primary structure elucidation of many regions of these molecules. Now, the complete nucleotide sequence of bacteriophage MS2 RNA has been established (2–4). During the course of these studies, several techniques which have proved to be of more general applicability have been pioneered or refined (see under "Sequence Strategy").

Gene Order   The structural data of group I phages are discussed by reference to the schematic drawing of phage MS2 RNA in Figure 5A. The molecule contains 3,569 nucleotides and is slightly enriched in G and C, viz. C 26.1; U 24.5; A 23.4; G 26.0. The corresponding molecular weight is 1.23×10^6 (sodium salt). A similar map of the Qβ genome is presented in Figure 5B.

For R17 RNA, the gene order 5' terminus, A protein, coat, replicase, 3' terminus has been determined by locating a series of marker oligonucleotides from the ribosome binding sites of the three genes (52) and from the coat gene (1) in a one-third and a two-third fragment generated by limited ribonuclease IV cleavage (67). The same gene order was independently found by studies on the in vitro translation of nascent strands derived from replicative intermediates (68). The genetic map of Qβ has been constructed by locating the three ribosomal binding sites (53, 69–71) relative to the 5'- and 3'-ends, making use of in vitro partial transcripts (57, 72, 73). Because the coat protein and the A1 protein share the same amino terminal sequence (74–76), it was suggested that the latter is a read-through product of the former, presumably at a UGA termination codon which is known to be leaky. Indeed, synthesis of A1 protein is specifically enhanced with the use of a UGA suppressor strain (75). By comparing amino acid sequences of the A1 protein with the previously determined structure of the coat protein, it has now been directly proved that the A1 protein is a natural

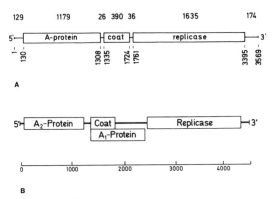

Figure 5. Genetic map of (A) the MS2 and (B) the Qβ genome. A, the length of the different regions, expressed in number of nucleotides, is shown on *top*. B, the nucleotides in the viral RNA are numbered from the 5'-end to the 3'-end (*bottom*). A, reproduced from Fiers et al. (4) with permission of Macmillan Journals, London, B, adapted from Weissmann (66).

read-through product at a single UGA termination site for the coat protein (77). Because the three possible reading phases of translation are blocked in the region preceding the initiation site of the replicase gene (71), a read-through inside this gene is not possible and, therefore, the information for the region of the A1 protein beyond the coat (approximately 600 nucleotides) must be located between the coat and replicase genes. This accounts for the larger genome size of Qβ relative to the group I phages (78).

5′- and 3′-Terminal Regions   Determination of the 5′-terminal nucleotide sequences of phages MS2 (15), R17 (22), and Qβ (58) up to the A protein ribosome binding site revealed the presence of long untranslated regions at the 5′ terminus of both group I (129 nucleotides) and group III phages (62 nucleotides). A similar situation exists at the 3′-terminus, where MS2 has 171 nucleotides not participating in protein synthesis (79) and Qβ a minimum of 58 (73). Although the exact role of these sequences is unknown, the complete conservation of these regions within the group I phages (80) (at least so far as they have been determined) illustrates their essential biological function.

Ribosome Binding Sites   The isolation of regions of RNA by ribosome protection (first from RNA phages and later also from mRNAs both from DNA phages and from *E. coli*) results in a piece of about 35 nucleotides which is suitable for sequencing. This may provide information on at least some of the features which enable the ribosome to select these specific regions and to disregard others also containing (noninitiating) AUG or GUG codons. The three ribosomal binding sites from the phages R17 (52) and Qβ (53, 69–71) and the coat protein initiation site from f2 phage (81) have been determined with this approach. The corresponding regions from phage MS2 have also been determined from partial ribonuclease digests (2, 3).

In all of these sites, except one, protein synthesis is initiated with the AUG codon. The initiation region of the A protein in MS2 RNA, however, contains a GUG codon (82). This finding agrees with previous in vitro observations pointing to the possible initiator role of the GUG codon (83, 84). The only nucleotide difference with the R17 binding site is the G↔A interchange in the initiator codon itself, yet the level of gene expression is approximately the same. Out of 23 prokaryotic initiator sites determined to this time (85), there is only one other natural gene starting with the GUG codon, namely the gene coding for the repressor of the lactose operon (86).

Initially it was rather disappointing that comparison of different initiator nucleotide sequences showed so little sequence homology. Shine and Dalgarno (87) were the first to suggest that a polypyrimidine-rich sequence near the 3′-end of the 16 S ribosomal RNA from *E. coli* interacts directly by complementary base pairing with a polypurine stretch found in all initiator regions some 4–14 nucleotides before the initiation codon. All the initiator sites known can be written with between three and nine such contiguous bases paired with the 16 S 3′-terminal region (85). The theory also explains the finding of Steitz (88), who showed by rebinding the isolated R17 initiation sites to ribosomes that the A protein region is intrinsically the best for binding. The mRNA-rRNA complex implicated by the hypothesis of Shine and Dalgarno was subsequently isolated from protein synthesis initiation complexes (89).

The dependence on the ribosomal protein S1 and the initiation factors IF2 and IF3 seems to be inversely correlated with the stability of the mRNA-rRNA complex, suggestive of a helper function for the latter proteins either in the formation or in the stabilization of the interaction (90). When 70 S ribosomes bind to intact bacteriophage R17 or Qβ-RNA, only the coat protein binding region is effectively recognized (52, 53). As R17 RNA is more severely degraded, recognition of the coat gene becomes less effective (88), whereas the A protein and replicase initiator sites become unmasked and available for ribosome binding (52, 70, 71). On intact phage RNA, the A protein and replicase genes are thus clearly negatively regulated by the RNA secondary and tertiary structure. The Qβ A protein binding site is recognized on short, in vitro, synthesized pieces up to 500 nucleotides in length, but ribosomes fail to detect the same region in larger fragments (69). Presumably, this site is buried either by a long range secondary structure interaction or by tertiary structure (long range refers only to the distance along the primary sequence between the two base-pairing segments). Such a long range secondary structure interaction seems to be present in the MS2 A protein gene (3); the latter structural model was built partly on the basis of experimental evidence and according to thermodynamic stability rules (91, 92), but without taking into account the biological implications. Nevertheless, this large range interaction explains why the A protein gene can only be expressed on RNA chains being synthesized, as has been proposed previously (93). According to the secondary structure model of the MS2 coat protein gene (2), a similar long range interaction exists between part of the coat protein gene and the initiation site for the polymerase, which explains the inaccessibility of the latter site in native RNA and also the fact that at least part of the coat gene has to be translated in order to allow translation of the replicase gene (93) (the polarity effect).

Termination Sites for Translation   The signals for peptide chain termination as determined on group I phage RNA are: UAG (A protein gene of MS2 (94) and R17 (23)), the double stop UAA·UAG (coat gene of R17 (95), MS2 (2), and f2 (96)), and UAG (MS2 replicase gene (79)). Under amber suppressor conditions, a read-through product of both the A protein (97) (in the case of MS2) and the virus-coded replicase subunit (98) (in MS2, R17, and f2 phage) has been identified. Inspection of the MS2 nucleotide sequence in this respect identifies the codons UGA (94) (A protein) and UAA (79) (replicase) as terminators in these events. Although the termination codons of the three genes and even of the read-through product of the replicase gene are located at the top of a hairpin structure, it remains unclear whether this position has a functional significance. At least, termination can take place when the stop codon is not in such a position, as exemplified by the A protein read-through product (94). In the Qβ coat gene, there is good evidence (see above) that a single UGA triplet serves as the termination signal. Because the read-through protein A1 (Figure 5*B*) is a component necessary for reconstitution of infectious Qβ particles (99), a certain level of read-through at this UGA must be maintained. Although it is well known that the UGA stop signal is somewhat leaky, this situation again poses the problem of contextual effects on the efficiency of suppression.

In group I phages, and also in the Qβ region A2 coat, the stop signal of one gene is not immediately followed by the start of the next gene. Successive genes are separated by an intercistronic divide of 26 (MS2 (94) and R17 (23) A protein and coat genes), 36 (R17 (95) and MS2 (2) coat and replicase genes), and a minimum of 23 nucleotides (100) (A2 and coat genes of Qβ). Of course these regions are involved in ribosome binding (discussed above), which may be their principal raison d'être. Such a strict physical separation of translation units is not a general phenomenon. In the tryptophan operon, the *trp*B gene ends with a single UGA codon which overlaps with the *trp*A initiation codon, viz. U̲G̲A̲U̲G̲ (101). The nucleotide sequence preceding this region thus fulfills two functions: encoding the COOH-terminal sequence of the β subunit of the tryptophan synthetase and serving as ribosomal binding site for the *trp*A gene (which codes for the α subunit). A similar situation has now been found at the junction of the D and J genes in the single-stranded DNA phage φX174, the sequence constituting the overlapping stop and start signals which in this case U̲A̲A̲U̲G̲ (51, see also Chapter 5). Even more spectacular in this system is the finding that gene E is within the same segment of DNA as gene D but is read in another reading frame (51). The chain terminator codons are UGA (E gene) and UAA (D gene). The φX gene F also ends with UGA (102). So far, it seems that the stop codons UAA, UAG, and UGA are used with roughly equal frequency in the *E. coli* system and its infecting phages.

Protein Binding Sites   Because of their small size, the relative ease of handling, and the availability of appropriate virus and host mutants, intensive study has led to a detailed understanding of most aspects of phage development. The subject has been reviewed recently (66, 93, 103). Because the genetic information of these phages specifies only three (or four) proteins, it was not unexpected to find several host proteins playing a role in phage development. Purified Qβ-replicase consists of four subunits, one of which is the product of the phage replicase gene (subunit II); the others are the bacterial protein ribosomal protein S1 (subunit I) and the protein elongation factors EF-Tu and EF-Ts (subunits III and IV). In addition, another bacterial protein, host factor (HF), is required for copying the parental plus strand. Viral proteins were also found to serve at least dual functions. Besides its structural role in the virion, coat protein also acts as a translational repressor of the replicase gene, and the replicase complex also serves to block protein synthesis early in infection. The role of these proteins in different stages of phage development, already evident from genetic, physiological, or biochemical data (see refs 66, 73, 93, 103), must be explained on the basis of specific nucleic acid-protein interactions. Characterization of the RNA fragment involved in the specific binding and its localization in the phage genome often have led to a detailed molecular understanding of the biological phenomenon. Such studies with purified replicase could only be carried out with the Qβ-enzyme complex because the replicase of group I phages has proved difficult to isolate because of its instability (103).

It is well known that late in infection replicase synthesis is greatly reduced when a sufficient amount of coat protein is present. A region of 59 nucleotide residues from R17 RNA (104), corresponding to this coat binding site, has been isolated and sequenced. The region comprises the terminator hairpin loop of the coat protein gene, the intercistronic region, and the ribosomal binding site for the replicase gene. It was subsequently shown that the latter is sufficient for the binding (105). The translational repression of the replicase gene is thus exerted by the direct interaction of the coat protein with the replicase ribosomal binding site. A similar interaction has now also been demonstrated in the Qβ system (106).

Conversely, translational repression of the coat cistron by Qβ-replicase also plays an important role in infection. In order to convert the phage RNA from its messenger role (its first function because of the need for viral-specific replicase subunit) into a template for its own replication, the RNA needs to be freed of ribosomes. This is realized by binding of Qβ-replicase to the protein synthesis initiator region of the coat gene (100), as demonstrated by the isolation of an RNA piece protected from nuclease digestion. Because this is the only site independently available to ribosomes on intact RNA (see under "Ribosome Binding Sites"), an RNA-Qβ-replicase complex free of ribosomes is finally obtained. A second site, thought to be essential for the recognition of Qβ-RNA as a template for the replicase, has also been detected. It consists of three nonoverlapping fragments with chain lengths of 21, 60, and 164 nucleotides, whose sequences have been determined (107). Although the three segments have been mapped in the neighborhood of the replicase ribosome binding site, the fragments do not overlap with the known sequences at this site (108). Electron microscopic visualization of binding complexes shows that one replicase molecule indeed attaches to two sites on Qβ-RNA (109), most likely corresponding to the two sites detected previously. A direct interaction of Qβ-replicase with the 3'-end of Qβ-RNA, where RNA synthesis starts, could not be established, although isolated ribosomal protein S1, one of the four constituents of the Qβ-replicase molecule, does bind close to the the 3'-end (110). The interaction with the Qβ -enzyme complex may be too weak to isolate. But also, host factor, a specific cellular protein only required for the initiation of minus strand synthesis on plus strands, has been shown to interact with an adenylate-rich region close to the 3'-terminus of Qβ-RNA (110). Simultaneous binding of host factor at another more internal site (probably in the neighborhood of the second replicase binding site) perhaps could bring the bound replicase molecule to the 3'-end for initiation of replication.

Use of Code Words   Initially, because the amino acid sequences of the R17, f2, and MS2 coat proteins were already known, nucleic acid fragments were identified as belonging to the coat gene because their sequence information matched with a segment from the protein (1, 96, 111, 112). When the whole structure of the MS2 coat gene became known, an almost complete agreement between nucleotide and amino acid sequence was found (2). The only exceptions were at positions 11, 12, and 17, where the nucleotide sequence predicted aspartic acid, asparagine, and aspartic acid, respectively, instead of asparagine, aspartic acid, and asparagine as determined by using the protein. Independent amino acid sequence studies have confirmed this view for MS2 (113) and also for R17 and f2 (Chapter 3 in ref. 103). The agreement between the nucleotide sequence and amino acid sequence confirms in a direct way the genetic code word dictionary determined by the groups headed by Nirenberg, Ochoa, and Khorana from in vitro studies.

Around the same time that the nucleotide sequence corresponding to the MS2 A protein was determined, most of the peptide sequences were also elucidated (3) in an independent effort. The primary structure of the replicase protein subunit was derived entirely from the nucleic acid sequence. The following discussion on the use and frequency of code words is based on the completed nucleotide sequence of MS2 RNA. Sequences in translated regions from other phages are limited, and for the group I phages largely identical (80). Reference to this work is made whenever it is appropriate. Although all 61 sense code words are used to specify the replicase, and all but one specify the A protein gene (the UGU codon for cysteine is absent, but because there are only 3 cysteine residues in the protein, this is considered to be attributable to chance), only 49 codons are used to code for the coat protein. For a few amino acids, the choice between degenerate code words is definitely nonrandom, especially when the code words in the whole genome are considered.

On what basis is the choice made between degenerate codons for a given amino acid? The answer to this question may be sought in several ways. Secondary structure could be one possible constraint on third letters. The percentage of paired bases in the secondary structure models (2–4) is 65–70%. Taking into account long range interactions (such as those proposed in Figure 3 of ref. 4), this percentage is close to the value of $73 \pm 5\%$ measured for the complete viral RNA (114). The secondary structure of randomly synthesized nucleotide sequences is estimated at 50–60% base pairing (115, 116), but such random structures are less stable than natural RNAs (117). Therefore, phage RNAs, like many other natural RNAs, are built so that the already considerable potential of base pairing present in a statistical RNA is slightly enhanced both quantitatively and qualitatively. This enhancement could easily be achieved by the appropriate choice of third letters and by bringing complementary strands into proper register (1, 111). No statistically convincing evidence in favor of such mechanisms is found when the secondary structure models of MS2 RNA are analyzed in this respect. The authors believe that relatively few selected third letters are sufficient to bring about the specific secondary and tertiary structure; therefore, statistical methods may be inadequate to reveal them.

For some amino acids, the choice of the third letter may be directly related to translation in the host cell. It is relevant to compare the code words used in the three different genes. Because the mature virion contains 180 copies of coat protein, only a single A protein molecule, and since the replicase has a catalytic function (and is not incorporated into the virus particle), it is clear that three differently regulated gene products are required and in fact are made in very different amounts. Several control mechanisms (at the level of initiation of protein synthesis and of translational repression) (see under "Ribosome Binding Sites" and "Protein Binding Sites") are known to be operative (66, 93), but this does not rule out a possible additional control during elongation (modulation control). According to this theory (118, 119), certain critical triplets would slow down the speed of translation and thus limit the amount of protein made to satisfy the needs of the system. That this can indeed occur has been demonstrated in an *E. coli* cell-free system with synthetic polynucleotides dependent on the availability of certain tRNAs (120). It is also a well established fact that the level of different isoaccepting tRNAs varies widely in *E. coli*. Although there is only 1 amino acid (histidine) absent, 12 codons are not used to specify the MS2 coat protein, the major viral product made upon infection. Most of these missing code words are undoubtedly not involved in modulation, either because they are used in MS2 coat mutants (121) or have been found in R17 or f2 RNA fragments corresponding to the coat protein (1, 95, 96, 112). The isoleucine codon AUA and the arginine codons AGA/AGG, however, are very good candidates as modulating codons. Such a role would agree with the observations that the specific isoaccepting tRNAs corresponding to the codons AUA (122, 123) and AGA/AGG (122) are present in very low amounts in *E. coli*. Other code words are preferentially used over the entire genome. This is clearly the case for the glycine codon GGU (glycine is coded for by the 4-fold degenerate codon (GGN) (N represents any of the nucleoside letters) and for the tyrosine codon UAC as compared to UAU.

RNA Plant Viruses

In the past, structural information on RNA plant viruses was rather limited. This was at least in part attributable to the unfavorable starting conditions available to the plant molecular virologist. A specific radioactivity of 0.05–0.075 μCi/μg (124) in a macromolecular RNA sample is definitely low to start extensive sequencing. However, recently conditions have been worked out which allow to obtain more than 10-fold higher specific radioactivities (125), almost comparable to those achieved with RNA phages. But copying into DNA is a new and perhaps more favorable way of studying large RNAs (e.g., from plant viruses) in the near future. All plant viral RNAs studied with respect to structure are single-stranded, the sense strand being incorporated into the virus (positive strand viruses). They belong to the tobamovirus, tymovirus, and bromovirus groups. Because these viral RNAs can accept an amino acid at the 3'-end upon incubation with a specific aminoacyl-tRNA synthetase, the hypothesis was formulated that the 3'-ends of these RNAs have at least certain features in common with the cloverleaf structure of a tRNA. The question arose whether this specific feature had a function in the life cycle of the virus. Therefore, because of this special property, and also because fragments derived from the 3'-terminus can be readily detected in a partial ribonuclease digest by their unique chemical properties, substantial effort has gone into the study of these 3'-termini.

After detection of a 3'-terminal fragment in a polyacrylamide band pattern of a partial ribonuclease T_1 digest of in vivo-labeled viral RNA, the fragment was purified further on a second higher percentage polyacrylamide gel and sequenced according to classic methods (5, 6). In this way, the 3'-terminal sequences of the RNA of tobacco mosaic virus (TMV) (126), green tomato atypical mosaic virus (GTAMV, a particular strain of TMV) (127), eggplant mosaic virus (EMV) (128), turnip yellow mosaic virus (TYMV) (129, 130), and RNA 4 from brome-grass mosaic virus (BMV) (131) have been determined. No abnormal bases were encountered within these structures, which all have the 3'-terminal trinu-cleotide· · · pCpCpA$_{OH}$ specific for tRNA.

The sequence of the 71-nucleotide TMV 3'-OH fragment (126) contains 3 or 4 potential nonsense codons: 2 successive UAAs, 1 UAG, and perhaps 1 UGA. Because the sequence does not overlap the coat protein gene which is now known to be at the 3'-terminus (see below), this whole region is extracistronic. The 74-nucleotide sequence at the 3'-end of GTAMV RNA (127), which, like TMV RNA accepts histidine, has several aspects in common with the TMV sequence, especially the last 12 nucleotides and a region (–23 to –35) which is also homologous to the *Salmonella typhimurium* tRNA^{His} TψC loop. But the two sequences differ considerably at other positions, in agreement with a distant relationship between both strains.

In addition, the sequence of the first 59 nucleotides from the 3'-OH end of eggplant mosaic virus RNA (a tymovirus) has been reported (128). For this RNA, as well as for the tobamoviruses TMV and GTAMV, it was impossible to fold the sequence into a cloverleaf structure characteristic of tRNAs. However, if the recently determined sequence of 159 nucleotides from the other tymovirus TYMV (129, 130) is examined, an answer to this problem might be found. Indeed, the last 103 nucleotides can be folded in several ways into secondary structures which resemble certain tRNA-like features. Most interestingly, the anticodon loop has the classic seven-base loop with a CAC-valine anticodon in the middle, but the G-T-ψ-C sequence (or its unmodified equivalent) is absent from the structure. It, therefore, seems that the 3'-end of TYMV RNA might be folded into a cloverleaf structure resembling a tRNA; a tertiary structure folding on this basis seems possible. It should be noted that the TYMV 3'-terminal fragment studied by RajBhandary and his colleagues (130) was obtained by RNase P cleavage, a tRNA processing enzyme. The structure at the end of EMV RNA shares much sequence homology with TYMV RNA, but insertions and deletions as well as substitutions occur in the sequence, suggesting a rather distant relationship. The EMV RNA has a potential CAC-valine anticodon in the same region as the TYMV RNA, but has only been sequenced a few residues beyond this point. It is quite possible that the EMV RNA has been split by ribonuclease T_1 in the easily accessible anticodon region. The presumed tRNA-like nicked complex would then dissociate upon isolation by polyacrylamide gel electrophoresis in the presence of urea (128), and only the 3'-linked polynucleotide half of the complex would be detected. A similar argument can be developed for the 3'-termini of TMV (126) and GTAMV RNA (127), which have also been isolated under denaturing conditions and could have been nicked in the D loop.

Kaesberg and his collaborators (132) have studied the 3'-ends of the 4 RNAs from the multicomponent bromegrass mosaic virus, all of which can accept tyrosine. By partial digestion with ribonuclease T_1, a structure of 159 nucleotides was obtained for the 4 RNAs. Not unexpectedly, no differences could be observed between the RNase T_1 products of RNAs 3 and 4. There is evidence that RNA 4 originates from RNA 3 by processing (133), but the fingerprints of RNAs 2 and 1 were also very similar and only 1 and 2 nucleotide substitutions, respectively, were revealed as compared to RNAs 3 and 4, indicative of a strong conservation of this region. However, although a piece of similar length was obtained when looking at the 3'-ends of the distantly related bromovirus cowpea chlorotic mottle virus (CCMV) RNAs (also accepting tyrosine), many sequence differences were found. Therefore, apart from the aminoacylation property, additional selection pressures ensure a rigid conservation of this region in the components of a multipartite genome. Again, the nucleotide sequence of the 3'-terminal region of BMV RNA 4 (131) can be folded into a secondary structure which has several features in common with a tRNA cloverleaf structure: the molecule folds back on its own, giving rise to an amino acid acceptor stem, and the anticodon loop contains a tyrosine anticodon.

Another interesting aspect of the TYMV 3'-terminal fragment (129, 130) is the presence of the information for the COOH-terminal region of the coat protein (134), followed by a single UAA termination signal and a 108-nucleotide untranslated stretch (presumably folded in a tRNA-like structure as discussed above). This places the coat gene at the 3'-terminus of the viral genome. But, although infectivity is exclusively confined to the viral RNA of 2.0×10^6, this infectious RNA does not promote the synthesis of coat protein. Instead, efficient production of coat protein was observed only with a small RNA of 3×10^5 daltons, also isolated from the virion (135). Both RNAs are capped by 7-methylguanosine triphosphate, but can be discriminated on the basis of the next nucleotide, which is guanosine for the large and adenosine for the small RNA. The large RNA has a closed (covered) coat cistron at its 3'-terminal end, whereas the small RNA corresponds in genetic information to the 3' end of the large TYMV RNA, but has a functional initiation site for this coat protein.

The coat protein gene of TMV is within 1,000 nucleotides of the 3'-end of the RNA in comparisons of RNase T_1 products of intact RNA and of a piece derived from the 3'-end (136), with nucleotide sequences known to be derived from the coat protein gene (124, 137) or from the 3'-terminal region (126). However, TMV RNA also is not an efficient template for synthesis of coat protein and, in fact, coat protein is synthesized on a short mRNA of 750 nucleotides appearing in the cytoplasm after infection. This small RNA corresponds effectively to the 3'-terminal region of the genome (136).

A third analogous situation is found in the BMV multicomponent viral RNA. BMV RNA 4 is wholly contained within the larger RNA 3 (133). BMV RNA 4 is not required for infectivity although it is an efficient messenger for coat protein synthesis, whereas RNA 3 makes another protein but little coat (138). Therefore, it seems that in this case also only the initiation site closest to the 5'-terminus can function properly. The internal site remains silent (hidden) until it becomes functional by formation of a shorter RNA derived from the 3'-end, thus making the initiation site 5'-proximal. This creates a very efficient ribosome binding site which is then used for the synthesis of structural proteins (cf. TYMV, TMV, BMV, Semliki forest virus, Polyoma, SV40). The role of RNA secondary structure and the presence of a 5'-(capped) end in this selective process remains to be elucidated.

Kaesberg and collaborators (139) have determined the ribosome binding site of BMV RNA 4 by sequencing the ribosome-protected RNA fragment. Because the initiating AUG codon is only 10 nucleotides from the 5'-terminus, the isolated fragment also contains the 7-methylguanosine triphosphate. Blocked 5'-termini have now been found in all four BMV (140) and CCMV (141) RNA species, in alfalfa mosaic virus RNAs (142), in the large and small TYMV RNAs (135), in TMV RNA (143, 144), and in cucumber green mottle mosaic virus RNA (145). As is the case with BMV RNA 4, the first base in the chain (linked to 7-methylguanosine) turned out to be unmodified wherever it was determined for these plant viral RNAs (135, 139, 140, 142–144), unlike most eukaryotic messengers. To our knowledge the only other messengers reported so far having an unmodified nucleotide in that position are Sindbis-specific 26 S mRNA (146), Sindbis viral 42 S RNA (147), and Newcastle disease virus mRNA (148). According to earlier investigations (before the existence of capped structures became known), satellite tobacco necrosis virus RNA would be heterogeneous at the 5'-end and start with ppApGpU· · · and pppApGpU· · · (149), which explains why in vitro translation is not inhibited by the cap analogue 7-methylguanosine-5'-phosphate (150). The eight positions between the 5'-terminal unmodified guanosine residue and the initiator AUG are all occupied by adenosine and uridine residues in BMV RNA 4. Besides the cap structure, adenosine and uridine nucleotides are known to be important for stable ribosome binding (151). A four-base complement with the 18 S RNA 3'-terminal sequence (152) of the eukaryotic ribosome is present, in agreement with a proposed direct interaction of mRNA with 18 S ribosomal RNA (87, 153). The binding site contains 53 nucleotides and cannot form a stable hairpin structure as evaluated by the thermodynamic rules available (91, 92). The nucleotide sequence specifies the first 14 amino acids following the AUG initiation codon, as determined on the BMV coat protein (154). Because BMV RNA 4 has an estimated chain length of 900 nucleotides and because roughly 600 bases are required to specify the protein, about one-third of the molecule at the 3′-terminus is left untranslated. Of these 300 nucleotides, the 159-nucleotide 3′-OH region (131), which is largely conserved between the 4 BMV RNAs (132), has been sequenced (discussed above).

Incubation of 25 S tobacco mosaic virus protein double disks with a partial T_1 ribonuclease digest of in vivo ^{32}P-labeled TMV RNA under reconstitution conditions results in the specific protection of only a few RNA segments (124, 137). Because of the low amount of radioactivity available, only partial sequences could be deduced (this is also the case for the 3′-ends discussed above). But the sequences could be related to amino acids 52–95 and 95–129 of the coat protein (the only TMV gene product whose primary amino acid sequence is known (155)), allowing the derivation of an unambiguous sequence. The main RNA product corresponds to amino acids 95–129 and is recovered both as one large intact hairpin and as a nicked structure, whereas the region corresponding to amino acids 52–95 is recovered in much lower quantity. The latter segment can be written as three smaller hairpins with a nick in the middle one. It is not clear whether the different recoveries reflect a difference in affinity for the protein disk or a different yield of the fragments in the partial digest. Considering the secondary structures (137), it is tempting to believe that a different yield is at least partly responsible for the observed difference. It is probably worth noting that both protected fragments have more than 50% adenosine residues among the 15 3′-terminal nucleotides, including an identical A-A-U-A-A-U stretch; this is reminiscent of the specific interaction between Qβ host factor and regions in Qβ-RNA and MS2 R17 RNA which contain a A-A-U-A-A-A sequence (see under "Protein Binding Sites") and also of the common A-A-U-A-A-A sequence at the 3′-end of eukaryotic mRNAs (see next section).

RNA Animal Viruses

From the great diversity of RNA viruses which multiply in animal cells, the genome of several groups is now being studied at the molecular level. Of course, the interest in these classes of viruses stems mainly from their being infectious and often hazardous to man or to domestic animals. The use of RNA sequencing techniques has been restricted largely to the study of the genetic organization within these viruses because it is not easy to achieve sufficient radioactivity for extensive sequencing studies in the viral RNA by in vivo labeling. However, the study of DNA copies made in vitro, using either reverse transcriptase or *E. coli* DNA polymerase I, has already been initiated (see below) and will undoubtedly lead to important results in the near future.

Picornaviridae and Togaviridae   The genome of these classes of viruses consists of one single-stranded RNA which is infectious, thus demonstrating the messenger capacity of the viral RNA. The picornaviruses are the smallest RNA animal viruses (pico-RNA viruses). Their genome consists of about 7,500 nucleotides. The RNA genome of the togaviruses (previously named arboviruses) is roughly twice as large.

Picornaviruses contain a poly(A) tract which is rather heterogeneous in length (polio (156, 157), Mengo (158, 159), encephalomyocarditis (160, 161), foot and mouth disease (162), human rhino (163), Columbia SK (164)). Its length has been estimated to be from 15 to 90 nucleotide residues. The RNA of seven members of the enterovirus (including polio 1 and 2), three members of the cardiovirus (encephalomyocarditis (EMC), Maus-Elberfeld, Mengo), and three members of the foot and mouth disease virus subgroups have recently been fingerprinted (165, 166) with the use of slight modifications of a two-dimensional polyacrylamide gel system (14). This allows the simultaneous detection of the poly(A) tract, which has been estimated to be between 50 and 100 nucleotides. The presence of poly(U) at the 5'-terminus in polio minus strands (167) is interesting. When the poly(A) tail of the plus strand is largely removed, the infectivity of the RNA is drastically reduced (168). A similar observation has been reported for EMC RNA (161). Furthermore, the protein synthesizing capacity of polio RNA in an in vitro system is not impaired by this treatment (169). Conceivably, the RNA replication is blocked because the plus strand without poly(A) is unable to act as a template for the 5'-terminus of the minus strand.

Encephalomyocarditis virus has a large poly(C) tract within the RNA (170), which seems to be the case for all members of the cardioviruses (EMC, Mengo, Maus-Elberfeld) and foot-and-mouth disease viruses (four different strains) but not for the picornaviruses belonging to the enterovirus (polio, Echo-type 12, Cocksackie-Faulkner, Cocksackie 8068, bovine enterovirus, swine vesicular disease virus) and human rhinovirus subgroups (types IA, 2, and 14) (171). Members of the fifth genus, equine rhinoviruses, have not been investigated. The absence or presence of poly(C) has now been confirmed with the use of a two-dimensional polyacrylamide gel system (165). Thus, the occurrence of poly(C) regions in the picornaviruses seems to be genus-specific. The lengths of the poly(C) tracts differ considerably (from 100 up to more than 200), even among members of the same genus and for subtypes of the same serotype. For EMC RNA, the poly(C) sequence has been located near the 5'-end of the molecule (172), most likely to the left of the single initiation site of translation because no polyproline sequences have been reported in EMC-specific proteins. Ribonuclease T_1 digests of picornaviruses are currently being investigated by using two-dimensional gel electrophoresis (14), resulting in a fingerprint highly diagnostic for each of the RNA genomes studied (and, in fact, for any large RNA molecule) because of the many unique longer ribonuclease T_1 products resolved (including poly(A) and poly(C) tracts) (165, 166). For example, the fingerprints of three cardiovirus genomes (EMC, Maus-Elberfeld, and Mengo) had most of the spots in common, in agreement with their very close serological relationships, yet they could readily be distinguished from each other.

With the use of avian myeloblastosis reverse transcriptase and oligo(dT) as a primer in the absence of deoxyguanosine triphosphate, a limited DNA transcript of EMC RNA is made (173). A 26-nucleotide sequence has been determined and on this basis the corresponding region preceding the 3'-terminal poly(A) can be deduced. It contains the nucleotide arrangement A-A-U-A-A-A found in roughly the same position in all eukaryotic messenger RNAs sequenced so far (174) (Table 1). This signal may play a role in polyadenylation of completed viral plus strands (and of processed messengers) but this need not necessarily rule out another possible role in processing of precursors of cellular messengers (174).

Polio (175–178) and EMC (179) RNA do not contain a capped structure at the 5'-end, so this feature cannot be regarded as an absolute requirement for translation of a eukaryotic message (see also Chapter 3). On the other hand, Sindbis viral 42 S RNA (147) as well as the Sindbis-specific 26 S messenger RNA (146) contain a 7-methylguanosine at the 5'-end, but the first nucleotide in the chain is unmodified (see under "RNA Plant Viruses").

The 3'-ends of all togavirus RNAs seem to be polyadenylated. It has been reported for Eastern equine encephalitis (157), Sindbis (164, 180–182) and Semliki forest (183) viruses (SFV). As for polio (167), the SFV minus strand contains a poly(U) stretch at its 5'-end (184) which is probably slightly smaller (60–70 nucleotides) than the complementary poly(A) tail (80 nucleotides). It has now been directly shown that the information for the 26 S message is located at the 3'-terminus of the 42 S viral genome (183). This result was obtained by a comparison on two-dimensional polyacrylamide gels of RNase T_1 fingerprints derived from 42 S RNA, 26 S RNA, and RNA fragments containing the 3'-terminal region of the 42 S RNA (generated by mild alkali treatment) (see under "Retroviridae") where this technique has been further exploited. As expected, the 26 S RNA also contains a poly(A) tail. This relationship is much like the one described for plant viruses between BMV RNAs 3 and 4, between TMV viral RNA and the TMV coat protein mRNA, and between TYMV large and small viral RNA (see under "RNA Plant Viruses"). Indeed, with both the 42 S and the 26 S RNA, only a single, but different, polypeptide is synthesized (185, 186). In addition, in this case, a smaller RNA (26 S) is derived from the 3'-end of the larger species during infection, apparently activating the internal silent site which serves for the synthesis of major viral structural proteins (187).

Rhabdoviridae and Paramyxoviridae   Members of these two virus families contain one single-stranded RNA molecule which is not infectious (negative strand viruses); the RNA-dependent RNA polymerase is a structural component of the virion. The molecular weight of vesicular stomatitis virus (VSV) RNA (a rhabdovirus) has been estimated at $3.6–4.0 \times 10^6$ and of the paramyxoviruses Newcastle disease virus and Sendai virus at $5.3–5.6 \times 10^6$. The RNA of these viruses contains no poly(A) (164, 188, 189) and, although it is the antimessenger, it does not contain poly(U) (190) either. Newcastle disease virus mRNA is capped but has an unmodified nucleotide as the first base in the chain

100   Min Jou and Fiers

Table 1. Sequences adjacent to poly(A) for seven eukaryotic mRNAs and EMC viral RNA

RNA	Nucleotide sequence[a]
Rabbit α-globin mRNA	U G G U C U U U G AAUAAA G U C U G A G U G A G U G G C Poly(A)
Human α-globin mRNA	U G G U C U U U G AAUAAA G U C U G A G U G G G C G G C Poly(A)
Rabbit β-globin mRNA	U G G C U AAUAAA G G A A A U U U A U U U U C A U U G C Poly(A)
Human β-globin mRNA	U G C C U AAUAAA A A A C A U U U A U U U U C A U U G C Poly(A)
Mouse immunoglobulin light chain mRNA	A A U A U U C AAUAAA G U G A G U C U U U G C A C U U G Poly(A)
Chicken ovalbumin mRNA	C C U U U A A U C A U AAUAAA A A C A U G U U U A A G C Poly(A)
Rat preproinsulin mRNA	C U C U G C A A U G AAUAAA G C C U U U G A A U G A G C Poly(A)
Encephalomyocarditis viral RNA	C U A G A G U A G U A A AAUAAA U A G A U A G A G Poly(A)

[a] Boxed sequences denote homology. Taken from refs. 173, 174, and 335. Phosphate symbols are omitted for convenience.

(148). It shares this unusual property with plant viral RNAs and with Sindbis 42 S and 26 S RNA (see under "RNA Plant Viruses").

The VSV genome starts with the sequence pppApCpGp· · · (191) (uncapped) and ends with · · · YpGpU$_{OH}$ (192). All mRNAs synthesized in vitro seem to start with the sequence m^{7}GpppAmpApCpApGp (193), which is not complementary to the 3'-end of the genome. Moreover, only the α-phosphate of ATP is retained in the cap (194). There is now evidence that five messenger RNAs, accounting for almost all the coding potential of the genome (195), are synthesized, starting from a single initiation site close to the 3'-terminus of the viral template (196, 197). Recently, a small RNA species of unique length (68 nucleotides) has been detected (198) which has the 5'-terminal sequence ppApCpGp complementary to the 3'-terminus of the VSV RNA. All this evidence would then fit in a model in which transcription starts from the 3'-end with the small RNA and continues with the information for the different messengers until the 5'-end of the viral genome. The messengers would then be produced by processing from this transcript, followed by capping of the 5'-end and polyadenylation of the 3'-end (188, 199), the small starter transcript remaining unchanged. Presumably, under conditions in which processing is inhibited, full transcript plus strands are made which in turn give rise to new minus strand viral genomes.

Orthomyxoviridae   The influenza virion contains little RNA, only about 1%. Polyacrylamide gel electrophoresis has revealed the presence of at least seven RNA pieces (200) in influenza A strains. The large, medium, and small size RNA classes have different nucleotide sequences (201) as revealed by homochromatography fingerprinting, agreeing with genetic and biological observations showing that the viral genome behaves as if it were composed of several individual units. All RNAs present are initiated with pppA· · · (202) and terminated with · · · U$_{OH}$ (200). With the use of a more powerful (denaturing) gel electrophoretic system according to the methods of Joklik (203), two groups have now obtained evidence for the presence of eight RNAs in roughly molar yields in influenza A and B strains (204, 205). This agrees with extensive genetic studies in which eight recombination groups have been detected (206). Influenza C also has a segmented genome with a minimum of four RNA segments (204).

On the basis of the two-dimensional gel electrophoresis fingerprints of their ribonuclease T$_1$ products, it was concluded that the eight RNA species contain distinct nucleotide sequences (205). No poly(A) or poly(C) stretches could be detected. The total molecular weights have been estimated to be 5.3–5.7 × 10^6 for influenza A (204, 205) and 6.4 × 10^6 for influenza B (204). A minimum molecular weight of 4.7 × 10^6 was assigned to the genome of influenza C (204).

Because different influenza A strains yield distinct gel patterns (207), analysis of recombinant strains can be used to identify a particular gene. With this approach, it could be shown that RNA segment 4 (the largest RNA being segment 1) codes for hemagglutinin and 5 or 6, for neuraminidase in influenza

A/HongKong and A/PR8, respectively (208). All eight influenza genes have now been mapped (209).

Reoviridae   The reovirus particle has an RNA content of about 15%. The genome is multipartite and consists of 10 double-stranded RNA segments (210) (total molecular weight 15×10^6) which are probably all essential for infectivity. It is likely that each RNA represents a monocistronic genome segment (211). Each RNA segment is a perfect duplex and can be represented as follows (212–215):

$$m^7GpppGmpC\cdots \qquad \cdots UpC_{OH}$$

$$_{OH}C\ pG\cdots \qquad \cdots UpApGpp$$

One-quarter of the weight of the RNA in the virion is accounted for by small oligonucleotides, 6–20 nucleotides in length, which consist of either short oligo(A) sequences or abortive transcripts (216).

Reovirus messengers start with the identical $m^7GpppGmpC\cdots$ capped structure (217) and are not polyadenylated (218). Incubation of viral cores (containing the RNA-dependent RNA polymerase) under restrictive conditions leads to synthesis of only the smallest mRNA. Its 5'-terminal sequence was determined up to nucleotide residue 25 by pulse labeling (219). No potential initiator AUG or GUG codon was present within this region. Neither is there a region complementary with the 3'-terminus of the eukaryotic 18 S RNA of the ribosome (152), as could be expected according to the Shine and Dalgarno model (87, 153) for the ribosome binding site.

Retroviridae   This family includes the subfamilies oncovirinae (the RNA tumor virus group), lentivirinae (visnavirus, an agent which causes a slow infection in sheep, and related viruses), and spumavirinae (foamy agents) (220). The viral RNA sediments at 70 S, but after denaturation this value is reduced to about 35 S, indicative of a subunit structure of the viral genome. A provocative finding was the observation that virions contain an RNA-dependent DNA polymerase activity (221, 222), suggesting a DNA intermediate in virus growth. This finding, which actually made plausible Temin's provirus theory for oncogenicity, has greatly stimulated research in this field.

From the structural point of view, answers have been sought to the following questions: how many subunits are there in the genome? are the subunits different in the sense that they contain other coding information, or in the more restricted sense of a permuted sequence? what is the gene order? what is the mechanism of reverse transcription? At least partial answers to these questions are available at present.

An important starting point in investigating these problems is a powerful separation system (homochromatography (20, 29) or two-dimensional gel electrophoresis (14)) for the 20–30 large, unique T_1-oligonucleotides which presumably arise from all over the molecule. The genetic complexity of the RNA can then be estimated from the radioactivity present in a unique large T_1-

oligonucleotide as compared to the radioactivity in the total RNA. In order to obtain a reliable estimate, the calculation is based on several oligonucleotides. Nucleotide sequence complexities have been estimated as 3.5×10^6 daltons (Rous sarcoma virus (30, 223), murine leukemia, and visna virus (224)), 3.9×10^6 daltons (reticuloendotheliosis virus (224)), 3.0×10^6 daltons (mouse mammary tumor virus (225)), and 2.3×10^6 daltons (murine sarcoma virus (224)). Therefore, it must be concluded that the viral genome is polyploid. Electron microscopic observations revealed the presence of two 35 S RNA chains in the 70 S RNA complex of RNA tumor viruses (226, 227). In a further step, this set of T_1-oligonucleotides was mapped relative to the 3'-end (160, 189, 228, 229). RNA was partially cleaved by mild alkali treatment and separated according to size. The poly(A)-containing 3'-terminal fraction of each size class was isolated, hydrolyzed with ribonuclease T_1, and fingerprinted as described above. The location of a given oligonucleotide relative to the 3'-terminus is directly deduced from the smallest poly(A)-tagged RNA piece in which it could be detected. The results showed (230, 231) that the marker T_1-oligonucleotides were distributed all over the RNA, each marker occupying a unique position, thus proving that the sequence arrangement is identical in the different subunits.

A further application of this technique to deletion mutants (230, 232) (transformation defectives and defectives in the viral envelope glycoprotein) and to recombinant strains (233–235) led to the gene order 5'-gag-pol-env-sarc-poly(A). The term *"pol"* denotes the gene for the reverse transcriptase, *"env"* for the viral envelope glycoprotein, *"sarc"* for sarcoma formation, and *"gag"* for the internal virion proteins or group-specific antigens. Therefore, a correlation between a sophisticated T_1-oligonucleotide analysis and the genetic elements has provided a genetic map of this important group of viruses. (The relative order of *gag* and *pol* is still tentative.)

Besides the 70 S RNA genome, purified virions contain a mixture of 4 S RNAs, 5 S RNA, and 7 S RNA. Isolated 70 S RNA still contains a mixture of tRNAs which by sequential heating can be liberated from the complex (236). The most tightly bound RNA fraction is a pure tRNA (tRNA^{Trp} in the case of avian sarcoma viruses) which serves as a primer for DNA synthesis. It is the only tryptophan-accepting species that is detected in chicken cells. The nucleotide sequence of this tRNA has been determined (237) as well as the part of the structure interacting with the viral RNA. This interacting region has 70% GC and extends from the 3'-terminus through the acceptor stem into loop IV of the tRNA (238, 239). Interestingly, this tRNA contains the sequence G-ψ-ψ-C instead of G-T-ψ-C, which is found in most tRNAs. The only other tRNA known to share this feature is the tRNA^{Pro} primer found in murine leukemia virus (quoted in ref. 239 as Dahlberg, personal communication). By using the 70 S RNA as template-primer complex and allowing limited synthesis with reverse transcriptase in the absence of deoxycytidine triphosphate, a unique octadeoxynucleotide sequence extending from the 3'-terminus of the tRNA has been determined (240, 241), indicating either several initiation sites identical up to that position or, more likely, a unique site. The sequence (rA)-d(A-A-T-G-A-A-G-C) is identical for three Rous sarcoma strains, an avian leukemia, and an avian myeloblastosis virus.

Weissmann and collaborators (242) have shown that DNA synthesis in disrupted virions starts at a major initiation site less than 200 nucleotides from the 5'-terminus of the RNA. They characterized the ^{32}P-labeled RNA in the DNA-RNA hybrid and located the major protected RNA piece at the 5'-end of the physical map (231). Luckily, the large reference T_1-oligonucleotide used as a marker for the 5'-terminal region in these studies contains the 5'-terminal capped structure itself, previously identified in RNA tumor viruses (243–245). This discrete reverse transcript made in vitro has been sequenced (246, 247) by using the Maxam and Gilbert direct DNA sequencing technique (49). Taken together with previous data (238–241), a 119-nucleotide RNA stretch from the 5'-terminus of the avian sarcoma virus 35 S RNA subunit could be deduced. The observation that the first 20 nucleotides from the 5'-terminus are reiterated at the 3'-terminal region preceding the poly(A) (quoted in ref. 246 as J. Coffin, personal communication) suggests that the reverse transcriptase somehow makes use of this terminal redundancy to move from the 5'- to the 3'- end.

DNA VIRUSES

DNA Bacteriophages

Single-stranded DNA Bacteriophages   Single-stranded DNA phages exist in two varieties: polyhedral (ϕX174, S13) and filamentous (fl, fd, M13). Their genomes contain approximately 5,400 and 6,000 nucleotides, respectively, probably coding for 9 proteins in both cases. Whereas the former class of viruses causes cell lysis, the filamentous phages are released by continuous extrusion through the cell wall, without killing the host cell. In filamentous phages, a simple transcriptional process helps to account for the different amounts of protein made. After the conversion of the incoming DNA strand into a double-stranded replicative intermediate, this molecule is transcribed into several discrete mRNAs which are initiated at different promoter sites (in vitro at least eight) but are all terminated at a unique termination signal (248). This factor ρ-independent transcription termination site is located immediately after gene VIII coding for the coat protein. Because no promoter has been localized in front of gene III (the first gene encountered past the termination site), a read-through mechanism may be responsible for the presence of the corresponding messenger. In ϕX174, RNA synthesis can be initiated in regions preceding genes A, B, and D (249, 250), respectively, and termination (ineffective) can occur after genes E, F, G, and H (250).

The single-stranded DNA phages comprise the group of autonomous DNA viruses with the smallest genome known (their replicative form has almost the same size as the animal viruses SV40 and polyoma); they can easily be obtained in large quantities or in highly radioactively labeled forms. Therefore, they have often been the study material of choice for pioneering new DNA sequencing techniques, especially in Sanger's group. Of course, special attention has been paid to biologically important regions such as the origin of replication, transcription starts, and ribosome binding sites. But, the recently developed fast DNA sequencing techniques (see under "Hydrolysis with Specific Enzymes" Chapter 5) have now led to the almost complete elucidation of the ϕX174 genome (described in Chapter 5) and of large sections of the fd DNA (Takanami and Schaller, unpublished observations). Hence, a detailed consideration of the available data on the life cycle and the molecular biology of these phages does not seem appropriate for the moment because a much more coherent picture may emerge in the near future. Only the main results are briefly summarized below.

Five regions of the fd genome (or the closely related phage fl) have been published so far. Three ribosome binding sites of bacteriophage fl RNA are known (251), corresponding to fl gene V (coding for a DNA binding protein), to fl gene VIII (specifying a precursor of the major coat protein), and the third site, now identified as belonging to gene IV (252) (function unknown). Specific characteristics of ribosome binding regions have been discussed under "Ribosome Binding Sites." The sequence of the promoter for the fd gene VIII and the first 130 nucleotides transcribed from that site has been determined (253). It contains the ribosome binding site for the coat protein in positions 96–120. The identity of this region with the previously determined fl ribosomal binding site (251) illustrates the close relationship between both phages. The sequence of the messenger RNA coding for the coat protein has recently been completed (254). A region around a second strong promoter site (255) extending from 110 nucleotides before the RNA start site (256) and including some 120 nucleotides of the RNA transcript (257) has also been studied. Current ideas on promoter sites are discussed in Chapter 7, which deals with RNA synthesis. The fifth region is an unlocalized sequence of 50 residues obtained by the extension of a chemically synthesized octadeoxynucleotide primer on fl template DNA with E. coli DNA polymerase 1 (258, 259).

With the use of DNA sequencing techniques, the structure of a 48-nucleotide fragment obtained by partial digestion of ϕX174 DNA with T4 endonuclease IV was determined (26, 27). Many of the other fragments produced by this enzyme are derived from the same region of the genome (28). Combining the results on these fragments of DNA sequencing (28), transcription studies (260), amino acid sequencing (261), and the fast plus and minus method (50), a sequence of 281 nucleotides and partial sequences totaling another 263 nucleotides could be reconstructed (262). They were localized in gene F, which codes for the major coat protein. A ribosome-protected fragment corresponding to the gene G, which specifies the spike protein, was also analyzed (263, 264); it was obtained by exploiting the property that ribosomes specifically bind to single-stranded phage DNAs under proper initiation conditions. A chemically synthesized decanucleotide was used as a primer for DNA polymerase to confirm and extend this sequence (265). The same primer has been used again for the development of a simple and fast direct DNA sequencing method (50) (see under "Hydrolysis with Specific Enzymes" and Chapter 5). This segment has been extended, and the complete intercistronic region between gene F and gene G has now been determined (102). Gene F ends with a UGA codon, immediately followed by UAA, the intercistronic region being 111 nucleotides long. A sequence of 195 nucleotides in gene F (266) (coding for the first 65 amino acids) and the entire region comprising the overlapping genes D and E and the beginning of gene J (51) (certain aspects of which have been discussed under "Termination Sites for Translation") have been established. One of the three in vitro mRNA starts, ppp(C)G-A-U-G-C· · · (249), occurs 30 residues to the left of the initiation codon of gene D. Part of the structure of the restriction fragment containing the origin of ϕX174 DNA replication has also been reported (267).

   T-Bacteriophages   Because of the large size of their DNA molecule (11 × 10^7 daltons), T-even phage genomes are not very attractive for structural studies in general. However, one aspect of their structure has been studied intensively, namely the eight tRNAs coded for by the T-even phage DNA (most work has been done with T4). It is not understood why the information for this set of tRNAs is present in T4 DNA, in addition to the tRNA complement of the *E. coli* cell. Apparently, the T4-specified tRNAs are only required in some *E. coli* strains recently isolated from nature (268). In addition, these T4 genes have proved to be a suitable system for studying the precursor-product relationship in tRNAs (see Chapter 2). The structure of a small stable RNA transcript with unknown function, species I RNA, has also been determined for T2, T4, and T6 (269).

   T7-like phages are almost 5 times simpler than the T-even phages, the molecular weight of their DNA being 2.5 × 10^7. The genetic organization is such that the T7 genes are clustered according to their function, with the early genes in the first 15% segment at the left-hand side. The early region constitutes a single operon transcribed by the *E. coli* RNA polymerase starting from three strong promoters at the far left end of the DNA molecule. These primary transcripts with a molecular weight of approximately 2.5 × 10^6 (but slightly differing in length because of the multiple initiation sites) are then processed by *E. coli* ribonuclease III into five distinct early mRNAs. This process has been studied in some detail (270). Late transcription starts from several late promoters recognized by the T7-specific RNA polymerase, a single polypeptide specified by one of the early messengers. These late T7 promoters have not been studied at the nucleotide sequence level so far.

   The isolation of specific T7 early promoter(s) (named A1, A2, and A3, starting from the physical end of the DNA) by protection with *E. coli* RNA polymerase holoenzyme leads to a mixture of the three strong promoters. Pribnow (271, 272) devised a system in which he took advantage of the salt dependence of noninitiated complexes as opposed to the salt stability of a complex at an already initiated site. He used dinucleotide-primed initiation to lock the polymerase at a single specific site, followed by DNase treatment in the presence of high saline solution to isolate the complex. Radioactive RNA copies were obtained by allowing elongation to occur from that site. Alternatively, RNA was synthesized from an upstream promoter and isolated by hybridizction to the second promoter fragment. In this way, the sequence of the promoters A2 and A3 was determined. The specific characteristics of promoter sequences are discussed in Chapter 7.

The ribosome binding site of the 5'-proximal early gene 0.3 has recently been determined (273). It has the characteristics common to all initiation sequences recognized by *E. coli* ribosomes (see under "Ribosome Binding Sites"). This sequence differs clearly from a region isolated by ribosome protection with the use of short early T7 mRNA in vitro transcripts (274), where one might have expected to find the same region. This discrepancy is explained by restriction endonuclease mapping results (quoted in ref. 273 as Studier, personal communication), showing that the strain used by Arrand and Hindley (274) is apparently T3.

Temperate Bacteriophages   Temperate phages, such as bacteriophage λ, either have a lytic life cycle like virulent phages or persist in the cell for many cell generations as an integrated prophage. Lambda virions contain a double-stranded, linear DNA molecule (MW 3.2×10^7) which circularizes through its cohesive ends upon infection. With the use of repair reactions, Wu and Taylor determined the sequence of the 12 nucleotides of each cohesive end (275). Bacteriophage λ and four other distinct lambdoid phages (cf. ϕ 80, 82, 21, 424) were shown to share an identical sequence at their terminal sequences (276–278). Short sequences adjacent to the cohesive ends of several temperate phages have been determined with 3'-terminal labeling procedures (279–282).

A special effort has been made to understand at the molecular level the maintenance of the prophage state regulated by the repressor (C_I gene product). The repressor binds to two operators, called O_L and O_R, and blocks initiation of early transcription at the corresponding promoters P_L and P_R (for leftward and rightward transcription, respectively). Both control regions have now been completely sequenced (283, 284) (Figure 6). Operator and promoter sequences are only described here in relation to the phage genetic organization. A general discussion concerning the specific properties of operators and promoters is found in Chapter 7 on RNA synthesis. The right and left λ operators each contain three repressor binding sites of 17 base pairs long having similar but not identical sequences separated by AT-rich spacer sequences. The binding sites O_L1 and O_R1, adjacent to the controlled genes N and *cro* (also called *tof*), have the highest affinity for repressor oligomers (di- or tetramers). Operator mutants map in the binding sites O_L1, O_L2, O_R1, or O_R2.

The control of the expression of gene N is relatively straightforward: binding of the RNA polymerase to the P_L promoter is blocked by repressor binding to the O_L sites. If the repressor is inactivated, leftward transcription ensues and gene N becomes expressed. As a result, ρ termination becomes antagonized, and the λ

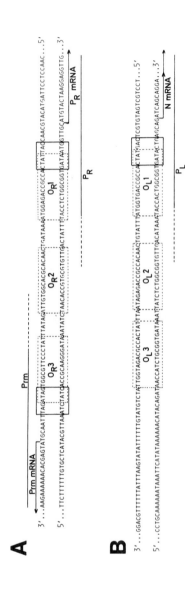

Figure 6. Sequences around two main control regions of phage λ. *A*, Prm- O_R-P_R. *B*, O_L-P_L. The C_I repressor binding sites are indicated by a *dashed rectangle*. For each promoter, the segment of DNA protected from deoxyribonuclease digestion by RNA polymerase is indicated by a *line*. A seven base pair region of approximate homology between different promoters, which is believed to define the binding sequence, is *boxed*. The *arrows* show where transcription is initiated. Adapted from Ptashne et al. (283) and Walz et al. (284).

genes to the left of N also are turned on by transcriptional read-through (i.e., under P_L control).

The situation at the right of the C_I-*rex* region, however, is more complex. There are two promoters, almost adjacent, but oriented in opposite directions (Figure 64). Transcription from P_R is under negative C_I control and regulates rightward transcription (*cro* gene; see above). Just to the left of P_R, however, is Prm (for maintenance of repression), from which the leftward transcription of the C_I-*rex* genes is initiated. If repressor (C_I gene product) is bound to O_R1, transcription from Prm is considerably enhanced, and, therefore, more repressor is made. However, more repressor means occupation of sites O_R2 and O_R3, but binding of repressor to the latter site blocks further transcription from Prm. This control mechanism explains how in the prophage state expression of C_I is completely self-regulated. The presence of these two promoters, P_R and Prm, in the vicinity of O_R leads to a remarkable degree of symmetry in this region around the middle of O_R2.

The Prm mRNA is the only message known which starts immediately with the initiation codon AUG (Figure 64). Therefore, it lacks a polypurine tract which in other prokaryotic messengers precedes the start codon and is thought to interact with the 3'-terminus of the 16 S ribosomal RNA (see under "Ribosome Binding Sites"). Most probably this property limits the expression of the C_I gene in the lysogenic state. Although the P_R mRNA also starts with AUG, AUG is not the start of translation. The AUG initiator codon is situated some 20 nucleotides internally (285) and is indeed preceded by a stretch of 9 bases complementary to the 3'-end of the 16 S ribosomal RNA. This is the longest complementary region observed of this kind. The sequence of the P_L mRNA has been determined up to nucleotide 149, but the exact initiation codon for the gene N protein has not been identified so far (286). Transcription and processing of the major leftward mRNA has been studied both under N^+ and N^- conditions (287).

The structure of a minor leftward 4 S RNA transcript (57) from λcb₂ (occurring both in vitro and in vivo), of an analogous RNA from bacteriophage φ80 (55), and of a minor rightward 6 S RNA from λ (56) have also been reported. It has been proposed that the former 4 S "oop" RNA serves as a primer for λ-DNA replication in the lytic cycle, or, alternatively, depending on some antitermination function, as the leader sequence of a messenger RNA for expression of the C_I gene (288) (establishment of repression). The structure of a 123-base pair Hpa II restriction fragment from this region has been determined (289). It contains the information for the 5'-proximal 65 nucleotides of the oop RNA, which deviates somewhat from the previously reported sequence (57). A further 58 nucleotide pairs contain the λPo promoter, which shows clear homology to the other promoter sequences known (289). The 6 S RNA has been located between the known genetic markers for genes Q and S (290) and is, presumably, together with the gene Q product acting as an antitermination factor (291), implicated in the initiation of transcription of the late genes (rightward late transcription). It is remarkable that these three discrete transcripts, the fd messenger RNA for the major coat protein (254), and the *trp* attenuator sequence (292) have 6–8 uridine residues at the 3′-end, preceded by a GC-rich region (293). These features probably play an important role in termination of transcription. The nucleotide sequences beyond the sites of transcription termination for the 4 S (46, 294) and 6 S RNAs (46, 290) have also been reported.

Animal DNA Viruses

Structural work on animal DNA viruses has been practically restricted to the small viruses of the papova group (mostly SV40) and to the moderately large human adenoviruses. These classes of viruses have received considerable attention not only for practical reasons because of their limited genome size but also intrinsically because of their oncogenic potential.

Papovaviruses Polyomavirus is widespread in mouse populations, whereas the structurally similar SV40 occurs endemically in populations of rhesus monkeys. Papilloma viruses are common in many mammalian species but have not been studied in much molecular detail, in part because of the difficulty of growing these on cells in tissue culture. The polyoma and SV40 virions are icosahedrons, containing a single, circular, supercoiled, double-stranded DNA molecule with a molecular weight of about 3.0×10^6. Both viruses can induce tumors when injected into certain newborn animals and are able to transform cells in vitro. SV40 DNA was the first DNA subjected to restriction mapping (45). The Hind enzymes II + III produce 11 major and 2 minor fragments whose relative positions have been established (295). By now, 100 cleavage sites produced by 18 different restriction nucleases have been mapped on the SV40 genome (Figure 4). This brings the whole genome within the possibilities of direct DNA sequencing (49, 50) (see under "Hydrolysis with Specific Enzymes" and Chapter 5), and it is likely that the complete structure will be elucidated within a reasonably short time. Most results reported so far stem from work initiated at a time when only a few restriction enzymes were available. Generally, in vitro ^{32}P-labeled transcription products of restriction fragments (sometimes made under primer-dependent conditions) and RNA made in vivo and back hybridized to specific fragments have been used for these studies.

From the comparison of the labeling kinetics of Hind II + III restriction fragments during DNA replication (with the use of a pulse label) and consideration of the physical order of the fragments, it followed that the origin of replication is in fragment C near the C-A junction. From there, replication proceeds in both directions at about the same rate and terminates in fragment G (296). When the nucleotide sequence in the region of termination of DNA replication was determined, no specific signal could be discerned (297–299). It was concluded that the termination site is simply the DNA region most distant from the initiation site. Experiments with deletion mutants have confirmed this view (300). Hind G contains a preferred site of initiation for *E. coli* RNA polymerase (a pseudo-promoter), and transcription occurs from the "early strand." This special property has been used, among other techniques, for the sequence determination of this region.

A 28-nucleotide sequence has been determined at the Hind II + III A-C junction, close to the origin of replication, by direct DNA sequencing after 5'-terminal labeling of the A and C fragments (301). The complete nucleotide sequence of the Eco RII-G fragment containing the Hind II + III A-C junction has now also been reported (302). This segment of 316 nucleotides spans the origin of DNA replication and possibly specifies the 5'-ends of both "early" and "late" cytoplasmic 19 S mRNA. By studying early and late RNA, it was found that their 5'-ends are located at least 90 nucleotides from one another in such a way that there is an overlapping region where transcription takes place from both strands (the physical ends of both RNAs have not been accurately mapped so far).

The characterization of deletion mutants in this part of the SV40 genome (303, 304) further defines the origin of replication as within 120 nucleotides of Hind III fragment C starting from the A-C junction (302). This sequence contains several regions with 2-fold rotational symmetry, which perhaps could also exist as double hairpin loop structures. A 27-nucleotide, perfectly symmetrical sequence (palindrome) around the middle of the Eco RII G fragment is especially noteworthy. In addition, several symmetric sequences ("true" palindromes), one of which is 17 nucleotides long, are present. In the late region, outside the origin of replication, a tandem repeat unit of 21 nucleotides is found.

Study of the capacity of different fragments to transform rat embryo cells showed that all the fragments with such activity contained the entire early region (305). The only sequence reported so far which codes for a part of the gene A polypeptide is that of Hind H located near the middle of the early region (306). The fragment contains 275 nucleotides (5.2% of the genome) and has an even lower GC content (34.5%) than total SV40 DNA. The nucleotide sequence can be translated unambiguously into an amino acid sequence (a part of the gene A protein). The polypeptide is high in the sulfur-containing amino acids cysteine and methione and in tyrosine. An earlier version of the sequence based on transcription data (307) showed a systematic error at one end because of artifactual transcription, giving rise to a unique reproducible transcript which contained information derived from both strands. This artifact was detected when direct DNA sequencing methods were applied to that region (49). The structure of fragment Hind I next to Hind H in the direction of transcription (Figure 4) has now also been completed (308).

The two smallest fragments of the Hind II + III digest of SV40 DNA, Hind L and Hind M, are ideally suited for direct DNA sequencing because their length has been estimated at 30 and 20 nucleotides, respectively. Their structure has been determined with use of 5'-^{32}P-labeled fragments either by partial digestion with an exonuclease and analysis by the wandering spot method or by partial chemical degradation with base-specific reagents and chain length-dependent separation on gel (309). Independently, another group reported the same sequence for Hind M with the use of slightly different methods (310). The two fragments are adjacent in the SV40 DNA (Hind C-Hind L-Hind M-Hind D). Because all three possible reading frames are blocked by a stop codon, it must be concluded that the Hind L-Hind M region is untranslated (although synthesis of a polypeptide, presumably dispensable, which ends in Hind M, cannot be excluded entirely). There is recent evidence that synthesis of the three structural proteins starts beyond the Hind M-Hind D junction (311). Also noteworthy is a 13-nucleotide repetition present in both fragments Hind L and Hind M.

A major part of the Hind K fragment derived from the late region Hind C-L-M-D-E-K-F-J-G has been sequenced (307). Because a single reading frame could be selected, the others being blocked by several nonsense codons, an unambiguous amino acid sequence was again deduced. The beginning of the amino acid sequence in Hind K proved to be identical with the seven NH_2-terminal amino acids of the major structural protein, VP1 (312). The NH_2-terminal alanine is immediately preceded by an AUG, where initiation presumably starts. These findings placed the start of the VP1 gene in Hind K close to the boundary with Hind E. Linked transcription-translation experiments with the use of restriction fragments are in agreement with these results (311). They also indicate that the end of the gene is in G. The region immediately preceding the structural gene VP1, located at the 5'-terminus of the Hind K fragment, has also been determined (313). So far, the 5'-end of the late 16 S RNA messenger has not been exactly localized, but it is likely that this region constitutes part of the ribosomal binding site for the VP1 gene. When the putative ribosomal binding sequence is compared with the initiation site of the BMV RNA 4 (the only other eukaryotic ribosome binding site known with certainty), several similarities are apparent. Both sequences are rich in AT, and there is an identical sequence extending over five out of six base pairs. As for the BMV RNA 4 initiator sequence, a complementary region can be constructed between the presumed VP1 ribosome binding site and the 3'-end of the 18 S rRNA. Whether the 3'-end of the 18 S rRNA does indeed serve a function in messenger recognition by sequence complementarity, as appears to happen with the prokaryotic 16 S rRNA (see under "Ribosome Binding Sites"), remains unknown. The nucleotide sequence of the whole area coding for VP1 is nearly completed by now (314). A small part of it around the single SV40 Eco R_I site (map position 0 of the SV40 physical map) has recently been published (315).

Adenoviruses   Human adenoviruses can be divided into three subgroups on the basis of their ability to produce tumors when inoculated into newborn rodents: highly oncogenic (type 12), weakly oncogenic (types 3 and 7), and nononcogenic (types 2 and 5). However, most members of the last group are still able to transform cells in vitro. Virus particles are 600–900 Å in diameter with an icosahedral symmetry. The viral DNA is a linear, double-stranded molecule with a molecular weight of $2.0–2.5 \times 10^7$. The molecule possesses an inverted terminal repetition, as revealed by the formation of single-stranded rings upon denaturation and renaturation.

The first nucleotide sequence coded for by adenovirus DNA was determined in 1970 (54, 316). It is a 5.5 S natural transcript (also called VA RNA) which accumulates in the cytoplasm of adenovirus 2-infected KB cells. Some heterogeneity was observed at both ends. Reinvestigation of this problem revealed the presence of molecules terminating in the homologues series $\cdots$ C-U-C-C-U-$_{OH}$ until $\cdots$ C-U-C-C-U-U-U-U-U-U_{OH} (317). Such a series of uridine residues is also found at transcription termination sites in the *E. coli* system (see under "Temperate Bacteriophages"). A secondary structure model with extensive base pairing can be drawn for the VA RNA molecule. A second, virus-coded, low molecular weight RNA of 5.2 S has now been detected in adenovirus type 2-infected cells (318). Both RNAs map close to one another on the adenovirus DNA (less than 600 base pairs apart), yet they are independently transcribed because both RNA species contain pppG $\cdots$ at their 5'-ends. These RNA species are the only two transcribed by RNA polymerase III on adenovirus DNA (318), whereas both viral strands are transcribed over virtually their entire length by RNA polymerase II (319). The function of the small RNAs remains obscure so far. The study of adenovirus-specific mRNA from infected cells revealed the presence of capped structures at the 5'-end, viz., m^7Gpppm^6-Amp $\cdots$ (80%) and $m^7GpppAmp$ $\cdots$ (20%) (320).

Restriction fragments have been used to map transcripts (e.g., refs. 318 and 319) and to localize the transforming region on the viral genome. It has been shown that the Hsu I-G fragment, located at the extreme left of the map and having a molecular weight of 1.6×10^6, is sufficient to initiate and to maintain transformation (321) in both the cases of adenovirus types 2 and 5. Extensive nucleotide sequence studies in this region are now in progress.

CONCLUSION

Some 10 years ago, the determination of the primary structure of even the smallest viral genome seemed an impossible task. Now, the entire sequence of one viral RNA (MS2 RNA) has been established (2–4), a tentative complete sequence of ϕX174 DNA has been obtained (322) (see Chapter 5), and the primary structure of SV40 DNA will most likely be completed in the near future. In addition, sections of many other viral genomes are known today. Comparison of the structures of genetic control elements such as operators, promoters, and ribosome binding sites (most of them derived . .from viruses but of self-evident inherent general interest) has provided a solid basis for understanding them in molecular terms. The use of restriction enzymes for dissection of DNA molecules (34), in combination with recently developed fast DNA sequencing techniques (49, 50), will undoubtedly lead to a dramatic expansion of structural knowledge about viral and also cellular (see Chapter 6) genomes in the near

future, promising a revolution in insights into the molecular biology of viruses
and their host cells.

NOTE ADDED IN PROOF:

Evidence has been presented that a small protein is covalently attached to the 5'-
terminus of the polio genome and to both 5'-ends of the replicative intermediate
(323, 324), but not to poliovirus polyribosomal RNA (325), pointing to a role of
the protein in RNA replication. A similar protein has been found on Foot and
Mouth disease viral RNA (326).

Quite remarkable are the recent observations that adenovirus late mRNAs
contain at their 5'-ends a leader sequence of about 150 nucleotides which is
transcribed not from DNA sequences immediately preceding the structural gene,
but from sequences which map at relatively distant locations in the viral genome
(327–329). A similar situation has been observed with SV40 late mRNA sequences (330–332). The question now is whether also certain cellular mRNAs
are built up from non-contiguous DNA segments and how such complex splicing
processes (eventually) regulate the expression of genetic information.

The nucleotide sequence of the whole SV40 DNA molecule has recently
been completed (333–334).

ACKNOWLEDGMENTS

We thank our colleagues who contributed valuable information.

REFERENCES

1. Adams, J. M., Jeppesen, P. G. N., Sanger, F., and Barrell, B. G. (1969). Nature
   (Lond.) 223:1009.
2. Min Jou, W., Haegeman, G., Ysebaert, M., and Fiers, W. (1972). Nature (Lond.)
   237:82.
3. Fiers, W., Contreras, R., Duerinck, F., Haegeman, G., Merregaert, J., Min Jou,
   W., Raeymaekers, A., Volckaert, G., Ysebaert, M., Van de Kerckhove, J.,
   Nolf, F., and Van Montagu, M. (1975). Nature (Lond.) 256:273.
4. Fiers, W., Contreras, R., Duerinck, F., Haegeman, G., Iserentant, D., Merregaert,
   J., Min Jou, W., Molemans, F., Raeymaekers, A., Van den Berghe, A., Volckaert, G., and Ysebaert, M. (1976). Nature (Lond.) 260:500.
5. Barrell, B. G. (1971). In G. L. Cantoni and D. R. Davies (eds.), Procedures in
   Nucleic Acid Research, Vol. 2, pp. 751–779. Harper and Row, New York.
6. Brownlee, G. G. (1972). Determination of Sequences in RNA. North Holland
   Publishing Company, Amsterdam.
7. Wu, R., Jay, E., and Roychoudhury, R. (1976). In H. Busch (ed.), Methods in
   Cancer Research, Vol. XII, pp. 87–176. Academic Press, New York.
8. Penswick, J. R., and Holley, R. W. (1965). Proc. Natl. Acad. Sci. USA 53:543.
9. Gould, H. (1966). Biochemistry 5:1103.
10. McPhie, P., Hounsell, J., and Gratzer, W. B. (1966). Biochemistry 6:988.
11. Min Jou, W., Hindley, J., and Fiers, W. (1968). Arch. Int. Physiol. Biochem.
    76:194.
12. De Wachter, R., and Fiers, W. (1971). In L. Grossman and K. Moldave (eds.),
    Methods Enzymol. 21:167.

13. Brownlee, G. G., and Sanger, F. (1969). Eur. J. Biochem. 11:395.
14. De Wachter, R., and Fiers, W. (1972). Anal. Biochem. 49:184.
15. De Wachter, R., Merregaert, J., Vandenberghe, A., Contreras, R., and Fiers, W. (1971). Eur. J. Biochem. 22:400.
16. Volckaert, G., and Fiers, W. (1974). Anal. Biochem. 62:573.
17. Contreras, R., and Fiers, W. (1975). Anal. Biochem. 67:319.
18. Min Jou, W., and Fiers, W. (1976). FEBS Lett. 66:77.
19. Volckaert, G., and Fiers, W. (1977). Anal. Biochem. 83.
20. Volckaert, G., Min Jou, W., and Fiers, W. (1976). Anal. Biochem. 72:433.
21. Contreras, R., and Fiers, W. (1971). FEBS Lett. 16:281.
22. Adams, J. M., Spahr, P. F., and Cory, S. (1972). Biochemistry 11:976.
23. Adams, J. M., Cory, S., and Spahr, P. F. (1972). Eur. J. Biochem. 29:469.
24. Cory, S., Adams, J. M., and Spahr, P. F. (1972). J. Mol. Biol. 63:41.
25. Ling, V. (1971). FEBS Lett. 19:50.
26. Ziff, E. B., Sedat, J. W., and Galibert, F. (1973). Nature (New Biol.) 241:34.
27. Galibert, F., Sedat, J., and Ziff, E. (1974). J. Mol. Biol. 87:377.
28. Sedat, J., Ziff, E. B., and Galibert, F. (1976). J. Mol. Biol. 107:391.
29. Jeppesen, P. G. N. (1971). Biochem. J. 124:357.
30. Billeter, M., Parsons, J. T., and Coffin, J. M. (1974). Proc. Natl. Acad. Sci. USA 71:3560.
31. Smith, H. O., and Wilcox, K. W. (1970). J. Mol. Biol. 51:379.
32. Kelly, T. J., Jr., and Smith, H. O. (1970). J. Mol. Biol. 51:393.
33. Danna, K., and Nathans, D. (1971). Proc. Natl. Acad. Sci. USA 68:2913.
34. Roberts, R. J. (1976). Crit. Rev. Biochem. 4:123.
35. Sharp, P. A., Sugden, B., and Sambrook, J. (1973). Biochemistry 12:3055.
36. Yang, R. C., Van de Voorde, A., and Fiers, W. (1976). Eur. J. Biochem. 61:101.
37. Yang, R. C., Van de Voorde, A., and Fiers, W. (1976). Eur. J. Biochem. 61:119.
38. Lee, A. S., and Sinsheimer, R. L. (1974). Proc. Natl. Acad. Sci. USA 71:2882.
39. Baas, P. D., van Heusden, G. P. H., Vereijken, J. M., Weisbeek, P. J., and Jansz, H. S. (1976). Nucleic Acids Res. 3:8.
40. van den Hondel, C., and Schoenmakers, J. G. G. (1976). J. Virol. 18:1024.
41. Derynck, R., and Fiers, W. (1977). J. Mol. Biol. 110:387.
42. Morrow, J. F., and Berg, P. (1972). Proc. Natl. Acad. Sci. USA 69:3365.
43. Mulder, C., and Delius, H. (1972). Proc. Natl. Acad. Sci. USA 69:3215.
44. Sack, G. H., Jr., and Nathans, D. (1973). Virology 51:517.
45. Dannak, K., Sack, G. H., Jr., and Nathans, D. (1973). J. Mol. Biol. 78:363.
46. Rosenberg, M. De Crombrugghe, B., and Musso, R. (1976). Proc. Natl. Acad. Sci. USA. 73:717.
47. Subramanian, K. N., Pan, J., Zain, S., and Weissman, S. M. (1974). Nucleic Acids Res. 1:727.
48. Smith, H. O., and Birnstiel, M. L. (1976). Nucleic Acids Res. 3:2387.
49. Maxam, A., and Gilbert, W. (1977). Proc. Natl. Acad. Sci. USA 74:560.
50. Sanger, F., and Coulson, A. R. (1975). J. Mol. Biol. 94:441.
51. Barrell, B. G., Air, G. M., and Hutchison, C. A., III. (1976). Nature (Lond.) 264:34.
52. Steitz, J. A. (1969). Nature (Lond.) 224:957.
53. Hindley, J., and Staples, D. H. (1969). Nature (Lond.) 224:964.
54. Ohe, K., and Weissman, S. M. (1970). Science 167:879.
55. Pieczenik, G., Barrell, B. G., and Gefter, M. L. (1972). Arch. Biochem. Biophys. 152:152.
56. Lebowitz, P., Weissman, S. M., and Radding, C. M. (1971). J. Biol. Chem. 246:5120.
57. Dahlberg, J. E., and Blattner, F. R. (1973). Fed. Proc. 32:664.

58. Billeter, M. A., Dahlberg, J. E., Goodman, H. M., Hindley, J., and Weissmann, C. (1969). Nature (Lond.) 224:1083.
59. Bishop, D. H. L., Mills, D. R., and Spiegelman, S. (1968). Biochemistry 7:3744.
60. Dhar, R., Zain, S., Weissman, S. M., Pan, J., and Subramanian, K. (1974). Proc. Natl. Acad. Sci. USA 71:371.
61. Gilbert, W., and Maxam, A. (1973). Proc. Natl. Acad. Sci USA 70:3581.
62. Barnes, W. M., Reznikoff, W. S., Blattner, F. R., Dickson, R. C., and Abelson, J. (1975). J. Biol. Chem. 250:8184.
63. Marotta, C. A., Forget, B. G., Weissman, S. M., Verma, I. M., McCaffrey, R. P., and Baltimore, D. (1974). Proc. Natl. Acad. Sci. USA 71:2300.
64. Gilvarg, C., Bollum, F. J., and Weissmann, C. (1975). Proc. Natl. Acad. Sci. USA 72:428.
65. Proudfoot, N. J., and Brownlee, G. G. (1974). FEBS Lett. 38:179.
66. Weissmann, C. (1974). FEBS Lett. 40:S10.
67. Jeppesen, P. G. N., Argetsinger Steitz, J., Gesteland, R. F., and Spahr, P. F. (1970). Nature (Lond.) 226:230.
68. Konings, R. N. H., Ward, R., Francke, B., and Hofschneider, P. H. (1970). Nature (Lond.) 226:604.
69. Staples, D. H., Hindley, J., Billeter, M. A., and Weissmann, C. (1971). Nature (New Biol.) 234:202.
70. Staples, D. H., and Hindley, J. (1971). Nature (New Biol.) 234:211.
71. Steitz, J. A. (1972). Nature (New Biol.) 236:71.
72. Hindley, J., Staples, D. H., Billeter, M. A., and Weissmann, C. (1970). Proc. Natl. Acad. Sci. USA 67:1180.
73. Weissmann, C., Billeter, M. A., Goodman, HM., Hindley, J., and Weber, H. (1973). Annu. Rev. Biochem. 42:303.
74. Moore, C. M., Farron, F., Bohnert, D., and Weissmann, C. (1971). Nature (New Biol.) 234:204.
75. Weiner, A. M., and Weber, K. (1971). Nature (New Biol.) 234:206.
76. Konigsberg, W., Maita, T., Katze, J., and Weber, K. (1970). Nature (Lond.) 227:271.
77. Weiner, A. M., and Weber, K. (1973). J. Mol. Biol. 80:837.
78. Boedtker, H. (1971). Biochim. Biophys. Acta 240:448.
79. Vandenberghe, A., Min Jou, W., and Fiers, W. (1975). Proc. Natl. Acad. Sci. USA 72:2559.
80. Min Jou, W., and Fiers, W. (1976). J. Mol. Biol. 106:1047.
81. Gupta, S. L., Chen, J., Schaefer, L., Lengyel, P., and Weissman, S. M. (1970). Biochem. Biophys. Res. Commun. 39:883.
82. Volckaert, G., and Fiers, W. (1973). FEBS Lett. 35:91.
83. Clark, B. F. C., and Marcker, K. A. (1966). J. Mol. Biol. 17:394.
84. Ghosh, H. P., Söll, D., and Khorana, H. G. (1967). J. Mol. Biol. 25:275.
85. Steitz, J. A., Sprague, K. U., Steege, D. A., Yuan, R. C., Laughrea, M., and Moore, P. B. (1977). In H. Vogel (ed.), Symposium on Protein-Nucleic Acid Interactions. Academic Press, London.
86. Steege, D. A. In preparation.
87. Shine, J., and Dalgarno, L. (1975). Nature (Lond.) 254:34.
88. Steitz, J. A. (1973). Proc. Natl. Acad. Sci. USA 70:2605.
89. Steitz, J. A., and Jakes, K. (1975). Proc. Natl. Acad. Sci. USA 72:4734.
90. Steitz, J. A., Wahba, A. J., Laughrea, M., and Moore, P. B. (1977). Nucleic Acids Res. 4:1.
91. Tinoco, I., Jr., Borer, P. N., Dengler, B., Levine, M. D., Uhlenbeck, O. C. Crothers, D. M., and Gralla, J. (1973). Nature (New Biol.) 246:40.

92. Borer, P. N., Dengler, B., Tinoco, I., Jr., and Uhlenbeck, O. C. (1974). J. Mol. Biol. 86:843.
93. Kozak, M., and Nathans, D. (1972). Bacteriol. Rev. 36:109.
94. Contreras, R., Ysebaert, M., Min Jou, W., and Fiers, W. (1973). Nature (New Biol.) 241:99.
95. Nichols, J. L. (1970). Nature (Lond.) 225:147.
96. Nichols, J. L., and Robertson, H. D. (1971). Biochim. Biophys. Acta 228:676.
97. Remaut, E., and Fiers, W. (1972). J. Mol. Biol. 71:243.
98. Atkins, J. F., and Gesteland, R. F. (1975). Mol. Gen. Genet. 139:19.
99. Hofstetter, H., Monstein, H. J., and Weissmann, C. (1974). Biochim. Biophys. Acta 374:238.
100. Weber, H., Billeter, M., Kahane, S., Hindley, J., Porter, A., and Weissmann, C. (1972). Nature (New Biol.) 237:166.
101. Platt, T., and Yanofsky, C. (1975). Proc. Natl. Acad. Sci. USA 72:2399.
102. Fiddes, J. C. (1976). J. Mol. Biol. 107:1.
103. Zinder, N. D. (1975). RNA Phages. Cold Spring Harbor Laboratory.
104. Bernardi, A., and Spahr, P. F. (1972). Proc. Natl. Acad. Sci. USA 69:3033.
105. Gralla, J., Steitz, J. A., and Crothers, D. M. (1974). Nature (Lond.) 248:204.
106. Weber, H. (1976). Biochim. Biophys. Acta 418:175.
107. Meyer, F., Weber, H., Vollenweider, H. J., and Weissmann, C. (1975). Experientia 31:743.
108. Porter, A. G., Hindley, J., and Billeter, M. A. (1974). Eur. J. Biochem. 41:413.
109. Vollenweider, H. J., Koller, T. H., Weber, H., and Weissmann, C. (1976). J. Mol. Biol. 101:367.
110. Senear, A. W., and Steitz, J. A. (1976). J. Biol. Chem. 251:1902.
111. Min Jou, W., Haegeman, G., and Fiers, W. (1971). FEBS Lett. 13:105.
112. Jeppesen, P. G. N., Barrell, B. G., Sanger, F., and Coulson, A. R. (1972). Biochem. J. 128:993.
113. Vandekerckhove, J., and Van Montagu, M., personal communication.
114. Boedtker, H. (1967). Biochemistry 6:2718.
115. Gralla, J., and De Lisi, C. (1974). Nature (Lond.) 248:330.
116. Fitch, W. M. (1974). J. Mol. Evol. 3:279.
117. Ricard, B., and Salser, W. (1975). Biochem. Biophys. Res. Commun. 63:548.
118. Ames, B. N., and Hartman, P. E. (1963). Cold Spring Harbor Symp. Quant. Biol. 28:349.
119. Stent, G. S. (1964). Science 144:816.
120. Anderson, W. F. (1969). Proc. Natl. Acad. Sci. USA..62:566.
121. Van Assche, W., Vandekerckhove, J., and Van Montagu, M. (1974). Arch. Int. Physiol. Biochem. 82:1020.
122. Caskey, C. T., Beaudet, A., and Nirenberg, M. (1968). J. Mol. Biol. 37:99.
123. Harada, F., and Nishimura, S. (1974). Biochemistry 13:300.
124. Guilley, H., Jonard, G., Richards, K. E., and Hirth, L. (1975). Eur. J. Biochem. 54:135.
125. Bastin, M., and Kaesberg, P. (1975). J. Gen. Virol. 26:321.
126. Guilley, H., Jonard, G., and Hirth, L. (1975). Proc. Natl. Acad. Sci. USA 72:864.
127. Lamy, D., Jonard, G., Guilley, H., and Hirth, L. (1975). FEBS Lett. 60:202.
128. Briand, J. P., Richards, K. E., Witz, J., and Hirth, L. (1976). Proc. Natl. Acad. Sci. USA 73:737.
129. Briand, J. P., Jonard, G., Guilley, H., Richards, K., and Hirth, L. (1977). Eur. J. Biochem. 72:453.
130. Silberklang, M., Prochiantz, A., Haenni, A.-L., and RajBhandary, U. L. (1977). Eur. J. Biochem. 72:465.

131. Bastin, M., personal communication.
132. Bastin, M., Dasgupta, R., Hall, T. C., and Kaesberg, P. (1976). J. Mol. Biol. 103:737.
133. Shih, D.-S., Lane, L. C., and Kaesberg, P. (1972). J. Mol. Biol. 64:353.
134. Peter, R., Stehelin, D., Reinbolt, J., Collot, D., and Duranton, H. (1972). Virology 49:615.
135. Klein, C., Fritsch, C., Briand, J. P., Richards, K. E., Jonard, G., and Hirth, L. (1976). Nucleic Acids Res. 3:3043.
136. Hunter, T. R., Hunt, T., Knowland, J., and Zimmern, D. (1976). Nature (Lond.) 260:759.
137. Guilley, H., Jonard, G., Richards, K. E., and Hirth, L. (1975). Eur. J. Biochem. 54:145.
138. Shih, D.-S., and Kaesberg, P. (1973). Proc. Natl. Acad. Sci. USA 70:1799.
139. Dasgupta, R., Shih, D.-S., Saris, C., and Kaesberg, P. (1975). Nature (Lond.) 256:624.
140. Dasgupta, R., Harada, F., and Kaesberg, P. (1976). J. Virol. 18:260.
141. Symons, R. H. (1975). Mol. Biol. Reports 2:277.
142. Pinck, L. (1975). FEBS Lett. 59:24.
143. Zimmern, D. (1975). Nucleic Acids Res. 2:1189.
144. Keith, J., and Fraenkel-Conrat, H. (1975). FEBS Lett. 57:31.
145. Kurisu, M., Ohno, T., Okada, Y., and Nozu, Y. (1976). Virology 70:214.
146. Dubin, D. T., and Stollar, V. (1975). Biochem. Biophys. Res. Commun. 66:1373.
147. Hefti, E., Bishop, D. H. L., Dubin, D. T., and Stollar, V. (1976). J. Virol. 17:149.
148. Colonno, R. J., and Stone, H. O. (1976). Nature (Lond.) 261:611.
149. Horst, J., Fraenkel-Conrat, H., and Mandeles, S. (1971). Biochemistry 10:4748.
150. Roman, R., Brooker, J. D., Seal, S. N., and Marcus, A. (1976). Nature (Lond.) 260:359.
151. Both, G. W., Furiuchi, Y., Muthukrishnan, S., and Shatkin, A. J. (1976). J. Mol. Biol. 104:637.
152. Dalgarno, L., and Shine, J. (1973). Nature (New Biol.) 245:261.
153. Shine, J., and Dalgarno, L. (1974). Proc. Natl. Acad. Sci. USA 71:1342.
154. Tremaine, J. H., Ronald, W. P., and Agrawal, H. O. (1974). Proc. Am. Phytopathol. Soc. 1:83.
155. Anderer, F. A., Uhlig, H., Weber, E., and Schramm, G. (1960). Nature (Lond.) 186:922.
156. Yogo, Y., and Wimmer, E. (1972). Proc. Natl. Acad. Sci. USA 68:1877.
157. Armstrong, J. A., Edmonds, M., Nakazato, H., Phillips, B. A., and Vaughan, M. (1972). Science 176:526.
158. Miller, R. L., and Plagemann, P. G. W. (1972). J. Gen. Virol. 17:349.
159. Spector, D. H., and Baltimore, D. (1975). J. Virol. 16:1081.
160. Gillespie, D., Takemoto, K., Robert, M., and Gallo, R. C. (1973). Science 179:1328.
161. Goldstein, N. O., Pardoe, I. E., and Burness, A. T. H. (1976). J. Gen. Virol. 31:271.
162. Chatterjee, N. K., Bachrach, H. L., and Polatnick, J. (1976). Virology 69:369.
163. Nair, C. N., and Owens, M. J. (1974). J. Virol. 13:535.
164. Johnston, R., and Bose, H. R. (1972). Proc. Natl. Acad. Sci. USA 69:1514.
165. Frisby, D. P., Newton, C., Carey, N. H., Fellner, P., Newman, J. F. E., Harris, T. J. R., and Brown, F. (1976). Virology 71:379.
166. Lee, Y. F., and Wimmer, E. (1976). Nucleic Acids Res. 3:1647.
167. Yogo, Y., Teng, M. H., and Wimmer, E. (1974). Biochem. Biophys. Res. Commun. 61:1101.
168. Spector, D. H., and Baltimore, D. (1974). Proc. Natl. Acad. Sci. USA 71:2983.

169. Spector, D. H., Villa-Komaroff, L., and Baltimore, D. (1975). Cell 6:41.
170. Porter, A., Carey, N., and Fellner, P. (1974). Nature (Lond.) 248:675.
171. Brown, F., Newman, J., Stott, J., Porter, A., Frisby, D., Newton, C., Carey, N., and Fellner, P. (1974). Nature (Lond.) 251:342.
172. Chumakov, K. M., and Agol, V. I. (1976). Biochem. Biophys. Res. Commun. 71:551.
173. Merregaert, J., Van Emmelo, J., Devos, R., Porter, A., Gillis, E., and Fiers, W. (1977). Arch. Int. Physiol. Biochim. 85:188.
174. Proudfoot, N. J., and Brownlee, G. G. (1976). Nature (Lond.) 263:211.
175. Wimmer, E. (1972). J. Mol. Biol. 68:537.
176. Nomoto, A., Lee, Y. F., and Wimmer, E. (1976). Proc. Natl. Acad. Sci. USA 73:375.
177. Hewlett, M. J., Rose, J. K., and Baltimore, D. (1976). Proc. Natl. Acad. Sci. USA 73:327.
178. Fernandez-Munoz, R., and Darnell, J. E. (1976). J. Virol. 18:719.
179. Frisby, D., Eaton, M., and Fellner, P. (1976). Nucleic Acids Res. 3:2771.
180. Johnston, R. E., and Bose, H. R. (1972). Biochem. Biophys. Res. Commun. 46:712.
181. Eaton, B. T., and Faulkner, P. (1972). Virology 50:865.
182. Deborde, D. C., and Leibowitz, R. D. (1976). Virology 72:80.
183. Wengler, G., and Wengler, G. (1976). Virology 73:190.
184. Sawicki, D. L., and Gomatos, P. J. (1976). Abstracts of the Annual Meeting of the American Society of Microbiology, p. 250.
185. Cancedda, R., Villa-Komaroff, L., Lodish, H. F., and Schlesinger, M. (1975). Cell 6:215.
186. Glanville, N., Ranki, M., Morser, J., Kääriäinen, L., and Smith, A. E. (1976). Proc. Natl. Acad. Sci. USA 73:3059.
187. Clegg, J. C. S. (1975). Nature (Lond.) 254:454.
188. Ehrenfeld, E., and Summers, D. F. (1972). J. Virol. 10:683.
189. Gillespie, D., Marshall, S., and Gallo, R. C. (1972). Nature (New Biol.) 236:227.
190. Marshall, S., and Gillespie, D. (1972). Nature (New Biol.) 240:43.
191. Hefti, E., and Bishop, D. H. L. (1975). J. Virol. 15:90.
192. Banerjee, A. K., and Rhodes, D. P. (1976). Biochem. Biophys. Res. Commun. 68:1387.
193. Rhodes, D. P., and Banerjee, A. K. (1976). J. Virol. 17:33.
194. Abraham, G., Rhodes, D. P., and Banerjee, A. K. (1975). Cell 5:51.
195. Rose, J. K., and Knipe, D. (1975). J. Virol. 15:994.
196. Ball, L. A., and White, C. N. (1976). Proc. Natl. Acad. Sci. USA 73:442.
197. Abraham, G., and Banerjee, A. K. (1976). Proc. Natl. Acad. Sci. USA 73:1504.
198. Colonno, R. J., and Banerjee, A. K. (1976). Cell 8:197.
199. Ehrenfeld, E. (1974). J. Virol. 13:1055.
200. Lewandowski, L. J., Content, J., and Lepla, S. H. (1971). J. Virol. 8:701.
201. Horst, J., Content, J., Mandeles, S., Fraenkel-Conrat, H., and Duesberg, P. (1972). J. Mol. Biol. 69:209.
202. Young, R. J., and Content, J. (1971). Nature (New Biol.) 230:140.
203. Floyd, R. W., Stone, M. P., and Joklik, W. K. (1974). Anal. Biochem. 59:559.
204. Ritchey, M. B., Palese, P., and Kilbourne, E. D. (1976). J. Virol. 18:2.
205. McGeoch, D., Fellner, P., and Newton, C. (1976). Proc. Natl. Acad. Sci. USA 73:3045.
206. Hirst, G. K. (1973). Virology 55:81.
207. Palese, P., and Schulman, J. L. (1976). J. Virol. 17:876.
208. Palese, P., and Schulman, J. L. (1976). Proc. Natl. Acad. Sci. USA 73:2142.
209. Ritchey, M. B., Palese, P., and Schulman, J. L. (1976). J. Virol. 20:307.

210. Shatkin, A. J., Sipe, J. D., and Loh, P. C. (1968). J. Virol. 2:986.
211. Both, G. W., Lavi, S., and Shatkin, A. J. (1975). Cell 4:173.
212. Muthukrishnan, S., and Shatkin, A. J. (1975). Virology 64:96.
213. Miura, K., Watanabe, K., Sugiura, M., and Shatkin, A. J. (1974). Proc. Natl. Acad. Sci. USA 71:3979.
214. Chow, N.-L., and Shatkin, A. J. (1975). J. Virol. 15:1057.
215. Furiuchi, Y., Muthukrishnan, S., and Shatkin, A. J. (1975). Proc. Natl. Acad. Sci. USA 72:742.
216. Bellamy, A. R., Nichols, J. L., and Joklik, W. K. (1972). Nature (New Biol.) 238:49.
217. Furiuchi, Y., Morgan, M., Muthukrishnan, S., and Shatkin, A. J. (1975). Proc. Natl. Acad. Sci. USA 72:362.
218. Stoltzfus, C. M., Shatkin, A. J., and Banerjee, A. K. (1973). J. Biol. Chem. 248:7993.
219. Nichols, J. L., Hay, A. J., and Joklik, W. K. (1972). Nature (Lond.) 235:105.
220. Fenner, F. (1976). Virology 71:371.
221. Baltimore, D. (1970). Nature (Lond.) 226:1209.
222. Temin, H. M., and Mizutani, S. (1970). Nature (Lond.) 226:1211.
223. Beemon, K., Duesberg, P., and Vogt, P. (1974). Proc. Natl. Acad. Sci. USA 71:4254.
224. Beemon, K. L., Faras, A. J., Haase, A. T., Duesberg, P. H., and Maisel, J. E. (1976). J. Virol. 17:525.
225. Friedrich, R., Morris, V. L., Goodman, H. M., Bishop, J. M., and Varmus, H. E. (1976). Virology 72:330.
226. Mangel, W. F., Delius, H., and Duesberg, P. (1974). Proc. Natl. Acad. Sci. USA 71:4541.
227. Bender, W., and Davidson, N. (1976). Cell 7:595.
228. Lai, M. M. C., and Duesberg, P. H. (1972). Nature (Lond.) 235:383.
229. Green, M., and Cartas, M. (1972). Proc. Natl. Acad. Sci. USA 69:791.
230. Wang, L.-H., Duesberg, P., Beemon, K., and Vogt, P. K. (1975). J. Virol. 16:1051.
231. Coffin, J. M., and Billeter, M. A. (1976). J. Mol. Biol. 100:293.
232. Wang, L.-H., Duesberg, P., Kawai, S., and Hanafusa, H. (1976). Proc. Natl. Acad. Sci. USA 73:447.
233. Joho, R. H., Billeter, M. A., and Weissmann, C. (1975). Proc. Natl. Acad. Sci. USA 72:4772.
234. Wang, L.-H., Duesberg, P., Mellon, P., and Vogt, P. K. (1976). Proc. Natl. Acad. Sci. USA 73:1073.
235. Wang, L.-H., Galehouse, D., Mellon, P., Duesberg, P., Mason, W. S., and Vogt, P. K. (1976). Proc. Natl. Acad. Sci. USA 73:3952.
236. Dahlberg, J. E., Sawyer, R. C., Taylor, J. M., Faras, A. J., Levinson, W. E., Goodman, H. M., and Bishop, J. M. (1974). J. Virol. 13:1126.
237. Harada, F., Sawyer, R. C., and Dahlberg, J. E. (1975). J. Biol. Chem. 250:3487.
238. Eiden, J. J., Quade, K., and Nichols, J. L. (1976). Nature (Lond.) 259:245.
239. Cordell, B., Stavnezer, E., Friedrich, R., Bishop, J. M., and Goodman, H. M. (1976). J. Virol. 19:548.
240. Taylor, J. M., Garfin, D. E., Levinson, W. E., Bishop, J. M., and Goodman, H. M. (1974). Biochemistry 13:3159.
241. Eiden, J. J., Bolognesi, D. P., Langlois, A. J., and Nichols, J. L. (1975). Virology 65:163.
242. Cashion, L. M., Joho, R. H., Planitz, M. A., Billeter, M. A., and Weissmann, C. (1976). Nature (Lond.) 262:186.

243. Keith, J., and Fraenkel-Conrat, H. (1975). Proc. Natl. Acad. Sci. USA 72:3347.
244. Furiuchi, Y., Shatkin, A. J., Stavnezer, E., and Bishop, J. M. (1975). Nature (Lond.) 257:618.
245. Stoltzfus, C. M., and Dimock, K. (1976). J. Virol. 18:586.
246. Shine, J., Czernilofsky, A. P., Friedrich, R., Goodman, H. M., and Bishop, J. M. (1977). Proc. Natl. Acad. Sci. USA. 74:1473.
247. Haseltine, W. A., Maxam, A. M., and Gilbert, W. (1977). Proc. Natl. Acad Sci USA 74:989.
248. Edens, L., Van Wezenbeek, P., Konings, R. N. H., and Schoenmakers, J. G. G. (1976). Eur. J. Biochem. 70:577.
249. Smith, L. H., and Sinsheimer, R. L. (1976). J. Mol. Biol. 103:711.
250. Hayashi, M., Fujimura, F. K., and Hayashi, M. (1976). Proc. Natl. Acad. Sci. USA 73:3519.
251. Pieczenik, G., Model, P., and Robertson, H. D. (1974). J. Mol. Biol. 90:191.
252. Zinder, N. D., personal communication.
253. Takanami, M., Sugimoto, K., Sugisaki, H., and Okamoto, T. (1976). Nature (Lond.) 260:297.
254. Sugimoto, K., Sugisaki, H., Okamoto, T., and Takanami, M. (1977). J. Mol. Biol. 111:487.
255. Schaller, H., Gray, C., and Herrmann, K. (1975). Proc. Natl. Acad. Sci. USA 72:737.
256. Sugimoto, K., Sugisaki, H., Okamoto, T., and Takanami, M. (1975). Nucleic Acids Res. 2:2091.
257. Sugimoto, K., Okamoto, T., Sugisaki, H., and Takanami, M. (1975). Nature (Lond.) 253:410.
258. Sanger, F., Donelson, J. E., Coulson, A. R., Kössel, H., and Fischer, D. (1973). Proc. Natl. Acad. Sci. USA 70:1209.
259. Sanger, F., Donelson, J. E., Coulson, A. R., Kössel, H., and Fischer, D. (1974). J. Mol. Biol. 90:315.
260. Blackburn, E. H. (1976). J. Mol. Biol. 107:417.
261. Air, G. M. (1976). J. Mol. Biol. 107:433.
262. Air, G. M., Blackburn, E. H., Coulson, A. R., Galibert, F., Sanger, F., Sedat, J. W., and Ziff, E. B. (1976). J. Mol. Biol. 107:445.
263. Robertson, H. D., Barrell, B. G., Weith, H. L., and Donelson, J. E. (1973). Nature (New Biol.) 241:38.
264. Barrell, B. G., Weith, H. L., Donelson, J. E., and Robertson, H. D. (1975). J. Mol. Biol. 92:377.
265. Donelson, J. E., Barrell, B. G., Weith, H. L., Kössel, H., and Schott, H. (1975). Eur. J. Biochem. 58:383.
266. Air, G. M., Blackburn, E. H., Sanger, F., and Coulson, A. R. (1975). J. Mol. Biol. 96:703.
267. Van Mansfeld, A. D. M., Vereijken, J. M., and Jansz, H. S. (1976). Nucleic Acids Res. 3:2827.
268. Wilson, J. H. (1973). J. Mol. Biol. 74:753.
269. Paddock, G., and Abelson, J. (1973). Nature (New Biol.) 246:2.
270. Perry, R. P. (1976). Annu. Rev. Biochem. 45:605.
271. Pribnow, D. (1975). Proc. Natl. Acad. Sci. USA 72:784.
272. Pribnow, D. (1975). J. Mol. Biol. 99:419.
273. Steitz, J. A., and Bryan, R. A. (1977). J. Mol. Biol. 114:527.
274. Arrand, J. R., and Hindley, J. (1973). Nature (New Biol.) 244:10.
275. Wu, R., and Taylor, E. (1971). J. Mol. Biol. 57:491.
276. Murray, K., and Murray, N. E. (1973). Nature (New Biol.) 243:134.

277. Padmanabhan, R., and Wu, R. (1972). J. Mol. Biol. 65:447.
278. Padmanabhan, R., Wu, R., and Calendar, R. (1974). J. Biol. Chem. 249:6197.
279. Ghangas, G. S., Jay, E., Bambara, R., and Wu, R. (1973). Biochem. Biophys. Res. Commun. 54:998.
280. Weigel, P., Englund, P. T., Murray, K., and Old, R. W. (1973). Proc. Natl. Acad. Sci. USA 70:1151.
281. Challberg, M. D., Englund, P. T., Isaaksson, A. G., and Murray, K. (1974). Fed. Proc. 33:1424.
282. Bambara, R., and Wu, R. (1975). J. Biol. Chem. 250:4607.
283. Ptashne, M., Backman, K., Humayun, M. Z., Jeffrey, A., Maurer, R., Meyer, B., and Sauer, R. T. (1976). Science 194:156.
284. Walz, A., Pirotta, V., and Ineichen, K. (1976). Nature (Lond.) 262:665.
285. Steege, D. A. (1977). J. Mol. Biol. 114:559.
286. Dahlberg, J. E., and Blattner, F. R. (1975). Nucleic Acids Res. 2:1441.
287. Lozeron, H. A., Dahlberg, J. E., and Szybalski, W. (1976). Virology 71:262.
288. Honigman, A., Hu, S.-L., Chase, R., and Szybalski, W. (1976). Nature (Lond.) 262:112.
289. Scherer, G., Hobom, G. and Kössel, H. (1977). Nature (Lond.) 265:117.
290. Sklar, J., Yot, P., and Weissman, S. M. (1975). Proc. Natl. Acad. Sci. USA 72:1817.
291. Roberts, J. W. (1975). Proc. Natl. Acad. Sci. USA 72:3300.
292. Bertrand, K., Korn, L., Lee, F., Platt, T., Squires, C. L., Squires, C., and Yanofsky, C. (1975). Science 189:22.
293. Gilbert, W. (1976). *In* R. Losick and M. Chamberlin (eds.), RNA Polymerase, pp. 193–205. Cold Spring Harbor Laboratory.
294. Kleid, D., Humayun, Z., Jeffrey, A., and Ptashne, M. (1976). Proc. Natl. Acad. Sci. USA 73:293.
295. Yang, R., Danna, K., Van de Voorde, A., and Fiers, W. (1975). Virology 68:260.
296. Danna, K., and Nathans, D. (1972). Proc. Natl. Acad. Sci. USA 69:3097.
297. Dhar, R., Zain, S., Weissman, S. M., Pan, J., and Subramanian, K. (1974). Proc. Natl. Acad. Sci. USA 71:371.
298. Zain, B. S., Weissman, S. M., Dhar, R., and Pan, J. (1974). Nucleic Acids Res. 1:577.
299. Dhar, R., Weissman, S. M., Zain, B. S., Pan, J., and Lewis, A. M., Jr. (1974). Nucleic Acids Res. 1:595.
300. Lai, C.-J., and Nathans, D. (1975). J. Mol. Biol. 97:113.
301. Jay, E., Roychoudhury, R., and Wu, R. (1976). Biochem. Biophys. Res. Commun. 69:678.
302. Subramanian, K. N., Dhcr, R., and Weissman, S. M. (1977). J. Biol. Chem. 252:355.
303. Lai, C.-J., and Nathans, D. (1974). J. Mol. Biol. 89:179.
304. Shenk, T., Carbon, J., and Berg, P. (1976). J. Virol. 18:664.
305. Abrahams, P. J., Mulder, C., Van de Voorde, A., Warnaar, S. O., and van der Eb, A. J. (1975). J. Virol. 16:818.
306. Volckaert, G., Contreras, R., Soeda, E., Van de Voorde, A., and Fiers, W. J. Mol. Biol. In press.
307. Fiers, W., Rogiers, R., Soeda, E., Van de Voorde, A., Van Heuverswyn, H., Van Herreweghe, J., Volckaert, G., and Yang, R. (1975). Proceedings of the 10th FEBS Meeting, *In* F. Chapeville and M. Grunberg-Manago (eds.), Vol. 39, p. 17. American Elsevier, New York.
308. Van Herreweghe, J., and Fiers, W. In preparation.

309. Ysebaert, M., Thys, F., Van de Voorde, A., and Fiers, W. (1976). Nucleic Acids Res. 3:3409.
310. Tu, C.-P. D., Roychoudhury, R., and Wu, R. (1976). Fed. Proc. 35:1595.
311. Rozenblatt, S., Mulligan, R. C., Gorecki, M., Roberts, B. E., and Rich, A. (1976). Proc. Natl. Acad. Sci. USA 73:2747.
312. Lazarides, E., Files, J. G., and Weber, K. (1974). Virology 60:584.
313. Van de Voorde, A., Contreras, R., Rogiers, R., and Fiers, W. (1976). Cell 9:117.
314. Fiers, W., personal communication.
315. Garfin, D. E., Leong, J.-A. C., and Goodman, H. M. (1976). Biochem. Biophys. Res. Commun. 68:369.
316. Ohe, K., and Weissman, S. M. (1970). J. Biol. Chem. 246:6991.
317. Pan, J., Celma, M., and Weissman, S. M., personal communication.
318. Söderlund, H., Pettersson, U., Venström, B., and Philipson, L. (1976). Cell 7:585.
319. Pettersson, U., Tibbetts, C., and Philipson, L. (1976). J. Mol. Biol. 101:479.
320. Moss, B., and Koczot, F. (1976). J. Virol. 17:385.
321. Graham, F. L., Abrahams, P. J., Mulder, C., Heyneker, H. L., Warnaar, S. O., de Vries, F. A. J., Fiers, W., and Van der Eb, A. J. (1974). Cold Spring Harbor Symp. Quant. Biol. 39:637.
322. Sanger, F., Air, G. M., Barrell, B. G., Brown, N. L., Coulson, A. R., Fiddes, J. C., Hutchison, C. A., III, Slocombe, P. M., and Smith, M. (1977). Nature (Lond.). 265:687.
323. Flanegan, J. B., Pettersson, R. F., Ambros, V., Hewlett, M. J., and Baltimore, D. (1977). Proc. Natl. Acad. Sci. USA 74:961.
324. Nomoto, A., Detzen, B., Pozzatti, R. and Wimmer, E. (1977). Nature (Lond.) 268:208.
325. Pettersson, R. F., Flanegan, J. B., Rose, J. K., and Baltimore, D. (1977). Nature (Lond.) 268:648.
326. Sangar, D. V., Rowlands, D. J., Harris, T. J. R., and Brown, F. (1977). Nature (Lond.) 268:648.
327. Berget, S. M., Moore, C., and Sharp, P. A. (1977). Proc. Natl. Acad. Sci. USA 74:3771.
328. Chow, L. T., Gelinas, R. E., Broker, T. R., and Roberts, R. J. (1977). Cell 12:1.
329. Klessig, D. F. (1977). Cell 12:9.
330. Celma, M. L., Dhar, R., Pan, J., and Weissman, S. M. (1977). Nucleic Acids Res. 4:2549.
331. Aloni, Y., Dhar, R., Laub, O., Horowitz, M., and Khoury, G. (1977). Proc. Natl. Acad. Sci. USA 74:3686.
332. Haegeman, G., and Fiers, W. Nature. Submitted for publication.
333. Fiers, W., Contreras, R., Haegeman, G., Rogiers, R., Van de Voorde, A., Van Heuverswyn, H., Van Herreweghe, J., Volckaert, G., and Ysebaert, M. In preparation.
334. Reddy, V. B., Thimmappaya, B., Dhar, R., Subramanian, K. N., Zain, B. S., Pan, J., Celma, M. L., and Weissman, S. M. In preparation.
335. Ullrich, A., Shine, J., Chirgwin, J., Pictet, R., Tischer, E., Rutter, W. J., and Goodman, H. M. (1977). Science 196:1313.

International Review of Biochemistry
Biochemistry of Nucleic Acids II, Volume 17
Edited by B. F. C. Clark
Copyright 1978   University Park Press   Baltimore

5
Sequence Analysis of Bacteriophage φX174 DNA

B. G. BARRELL

MRC Laboratory of Molecular Biology, Cambridge, England

Sequence analysis of bacteriophage DNA began in this laboratory in 1971. Until then, Sanger and his colleagues had been mainly concerned with the development of simple microtechniques to analyze the base sequence of ^{32}P-labeled RNA (1–3). These techniques were first applied to small RNA molecules such as tRNA and 5 S ribosomal RNA and then to the larger bacteriophage R17 RNA. In the latter case, sequence analysis of the larger ribonuclease T_1 end products and the isolation and sequence analysis of fragments produced by the limited action of ribonuclease T_1 (4) were involved. At the same time, Steitz was able to protect specifically and isolate ribosome binding sites and deduce their base sequences (5). Thus, at that time it seemed feasible to apply a similar approach to the small single-stranded DNA bacteriophages such as f1 and ϕX174. At 5,000–6,000 bases in length, these were nearly twice the size of the bacteriophage RNA. The success of the RNA sequencing methods primarily depended on the base-specific nucleases which had no DNA counterpart. However, in 1969 Sadowski and Hurwitz (6) demonstrated that an enzyme from bacteriophage T4 (endonuclease IV) would cleave single-stranded DNA chains preferentially on the 5′ side of cytosine. In addition, the exonucleases, spleen and venom phosphodiesterases which had been useful in RNA sequencing, had the same specificity with DNA templates. These enzymes, along with the depurination procedure of Burton and Petersen (7), made available the basic prerequisites for DNA sequence analysis. Bretscher (8) had shown that specific binding of ribosomes to viral DNA which had the same sense as the messenger RNA occurred under the proper conditions for initiation of protein synthesis, which suggested that ribosomes would recognize sites on the DNA analogous to real start signals for translation. Thus, the possibility existed that biologically interesting regions of DNA could be isolated in the same manner used by Steitz for RNA.

Initial DNA sequence analysis was performed on the DNAs from both filamentous phage f1 and the isocosahedral phage ϕX174. However, attention focused on the latter because of the well-established genetic map worked out largely by Sinsheimer and his colleagues (9–12) and also because work was in progress on the amino acid sequences of some of the ϕX-coded proteins. The DNA of ϕX174 consists of a single-stranded circle of approximately 5,400

Table 1. Size and function of φX174 proteins

Gene	Protein MW from SDS gels	Protein MW from DNA sequence[a]	Function
A	55,000–67,000	56,000	Replication, specific endonuclease
A*	35,000		Inhibition of double-stranded DNA synthesis
B	19,000–25,000	13,845[b]	Morphogenesis—viral strand synthesis
C	7,000	9,400	Viral strand synthesis
D	14,500	16,811[b]	Viral strand synthesis
E	10,000–17,500	9,940	Lysis
J	5,000	4,097[b]	Structural (spike) protein
F	48,000	46,400	Major coat protein
G	19,000	19,053[b]	Spike protein
H	37,000	35,800	Spike protein

Compiled from data in refs. 14, 15, and 18.

[a] Calculated from the DNA sequence as follows: protein MW = number of nucleotides/3 × 0.00915.

[b] Calculated from the amino acid sequence.

nucleotides. This viral strand (plus strand) has the same "sense" as its mRNA, i.e., mRNA is transcribed from the minus strand. φX174 codes for nine proteins which are listed in Table 1 along with their molecular weights and function. A 10th protein A* is thought to arise from a reinitiation event in the same phase from the 3' half of gene A (13). The nine complementation groups were arranged in the order A, B, C, D, E, J, F, G, H, by Benbow et al. (11), although the assignment of gene J was tentative. For a detailed discussion of φX biology see reviews by Denhardt (14, 15). Much of the early sequence work on φX DNA has been reviewed by Murray in the previous volume (16) and is, therefore, only briefly mentioned here. This chapter concentrates on the rapid plus and minus sequencing method of Sanger and Coulson (17) which has enabled a tentative complete sequence of φX DNA to be deduced (18). This sequence, along with a detailed protocol for the plus and minus method, is contained in the two appendices to this chapter.

For the purpose of this chapter, the more convenient term "gene" is used solely to define a coding or structural region. Thus, the DNA sequence defined by this term includes and extends from the initiation codon ATG up to and including the termination codon.

DIRECT SEQUENCE ANALYSIS OF φX DNA

Sequences of Large Pyrimidine Oligonucleotides

The first nucleotide sequences established in φX were pyrimidine tracts (19, 20) obtained by the Burton and Petersen (7) depurination procedure. With the use of

Table 2. Sequences of large pyrimidine oligonucleotides

Length	Composition	Sequence[a]	Position[b]
10	C_2T_8	C-T-T-T-T-T-T-T-C-T	2,421
10	C_7T_3	C-T-C-C-T-C-T-C-C-C	1,278
9	C_4T_5	T-C-T-T-T-C-T-C-C	1,153
9	C_4T_5	C-T-C-T-T-T-C-T-C	4,072
9	C_5T_4	C-T-T-C-C-T-C-C-T	1,523
9	C_6T_3	T-C-C-T-T-C-C-C-C	2,777
8	C_2T_6	C-T-T-T-T-C-T-T	1,326
8	C_2T_6	T-T-T-T-C-T-T-C	2,550
8	C_3T_5	T-T-T-C-T-C-C-T	2,079
8	C_3T_5	T-C-T-T-C-T-T-C	2,671
8	C_4T_4	C-T-T-C-C-T-T-C $\times$ C-T-C-T-T-T-C-C	Not present[c]
8	C_5T_3	C-C-T-T-T-C-C-C	1,042
7	C_1T_6	C-T-T-T-T-T-T	3,965
7	C_1T_6	T-T-T-C-T-T-T	1,830
7	C_6T_1	C-T-C-C-C-C-C	4,281
7	C_6T_1	C-C-C-T-C-C-C	2,690, 3,924

[a] Compiled by Harbers et al. (21).
[b] The first position of each oligonucleotide in the sequence of Sanger et al. (18) is given (see Appendix I).
[c] The two sequences present in oligonucleotide C_4T_4 are suggested to be any of the possibilities indicated by the crossover position $\times$. None of these possibilities are present in the sequence in Appendix I. However, there is a C-T-T-C-C-T-C-T at position 3,335 and a C-T-C-T-T-C-T-C at position 3,740.

the two-dimensional homochromatography fractionation method, Ling (20) isolated the larger pure pyrimidine oligonucleotides, which were sequenced by using limited digestion with spleen and venom exonucleases; the products were separated two-dimensionally. Then the sequence of each oligonucleotide was deduced by examining the position of each partial digestion product relative to the product one base shorter. If a product loses a cytidine residue on digestion, it tends to migrate slightly faster in the first (electrophoretic) dimension of the fractionation procedure; if it loses a thymidine residue, it migrates considerably more slowly. Thus, the sequence at the 3' half of the oligonucleotide could be deduced from the partial venom phosphodiesterase digestion and that at the 5' half from the partial spleen phosphodiesterase digestion. These two sequences could then be overlapped to give the complete sequence. Recently, Harbers et al. (21) have made a more extensive study of the larger pyrimidine oligonucleotides of ϕX DNA and the closely related S13 DNA with the same approach. They were able to correct two of the sequences deduced by Ling (20). These sequences are shown in Table 2 along with their positions in the complete sequence of Sanger et al. (18) (see Appendix I). Although the closely related sequences of the large pyrimidine oligonucleotides from bacteriophage S13 DNA deduced by Harbers et al. (21) are not included, it is interesting to note that the S13 DNA product C-T-T-C-C-T-C-T-T-C-T, not found in φX DNA, is only 1 residue different from the φX DNA sequence (G)C-T-T-C-C-T-C-T-G-C-T(G) at position 3335, resulting in a change of a GCT codon (alanine) in φX to a TCT codon (serine) in S13 in gene H.

Sequences of Large Purine Oligonucleotides

The hydrazinolysis of DNA to release purine tracts has been technically more difficult to achieve than the depurination process. However, Chadwell (22) improved this procedure. He fractionated the purine tracts of φX DNA produced by treatment with anhydrous hydrazine and piperidine and determined the sequences of the larger products. These are listed in Table 3, showing their position in the φX sequence. Chadwell was unable to find the following shorter purine tracts, A-A-A-G-A, A-G-A-G, and A-G-A-A-A; neither were they found in the sequence deduced by Sanger et al. (18).

Direct Sequencing with Endonuclease IV

Bacteriophage T4 endonuclease IV is the only enzyme which has successfully produced oligonucleotides of a size suitable for DNA sequence analysis. The enzyme was discovered by Sadowski and Hurwitz (6), who demonstrated it to be single strand-specific and found that greater than 95% of the 5'-termini produced by the enzyme were pC-. Use of this enzyme in sequence analysis has been retarded by the difficulties of preparation (23), although a simplified procedure is now available (24).

Table 3. Sequences of large purine oligonucleotides

Length	Composition	Sequence[a]	Position[b]
11	A_7G_4	A-A-A-G-G-A-A-A-G-G-A	3,259
11	A_8G_3	A-A-A-A-A-G-A-A-A-G-G	4,435
11	A_8G_3	A-A-A-A-A-G-A-G-A-G-A	5,247
9	A_4G_5	G-A-G-G-A-G-A-A-G	232
9	A_6G_3	A-G-A-A-A-G-A-G-A	3,407
9	A_7G_2	A-G-A-A-G-A-A-A-A	803
9	A_7G_2	A-A-A-A-A-G-A-G-A	3,440
8	A_6G_2	A-A-A-A-A-G-A-G	3,374
7	A_4G_3	G-G-A-A-A-G-A	3,860
7	A_4G_3	G-A-G-A-A-G-A	4,357
7	A_6G	A-A-A-G-A-A-A	3,667

[a] Deduced by Chadwell (22).
[b] Numbers refer to the first position of the oligonucleotide in the sequence given in Appendix I.

Ling (25) digested in vivo ^{32}P-labeled bacteriophage fd DNA with endonuclease IV and fractionated the products by polyacrylamide gel electrophoresis. Ziff et al. (26) used the single strand specificity of the enzyme and digested ϕX single-stranded DNA under conditions of low temperature and high ionic strength in order to maximize any secondary structure of the DNA. They fractionated a series of discrete products, presumably from regions of low secondary structure. A 48-nucleotide fragment was isolated in this way and further fragmented by endonuclease IV under stronger conditions. Under maximum digestion conditions, however, not all bands of the form 5'-NpC-3' were susceptible to the action of the enzyme. The products were fractionated by two-dimensional homochromatography and were sequenced with the use of digestion with exonucleases and depurination.

Another ^{32}P-labeled fragment of about 50 residues from ϕX was isolated by Robertson et al. (27–29) as a ribosome binding site. ϕX single-stranded viral DNA has the same sense as the messenger RNA, and under suitable conditions a complex with ribosomes is formed. After digestion with pancreatic DNase, several protected fragments could be isolated. The major fragment was sequenced with the use of endonuclease IV and streptococcal nuclease. This sequence contained the deoxyoligonucleotide triplet A-T-G, corresponding to the initiation codon AUG; from the following sequence and the genetic code, an amino acid sequence could be predicted for the NH_2-terminus of a protein inititating at this site. Air and Bridgen (30) studied the sequences of proteins from ϕX and found that this NH_2-terminal sequence occurred in the ϕX spike protein coded by gene G. The function of this piece of DNA and its position on the genetic map were thus defined.

Sedat et al. (31) extended their studies to the higher molecular weight fragments produced by endonuclease IV. All of the fragments studied originated from the same part of ϕX DNA. Concurrent with this work, Air (32) studied the amino acid sequence of the viral coat protein coded by gene F and showed that the DNA in these fragments coded for the amino acid sequence close to the NH_2 terminus of this protein. With these results, the transcription of endonuclease IV bands by Blackburn (33), and the early results of the plus and minus method of Sanger and Coulson (17), the sequence coding for at least the first 100 amino acids of the gene F product was obtained (34).

SEQUENCING ϕX DNA BY COPYING METHODS

Use of RNA Polymerase

Direct sequencing of DNA originally involved the preparation of large amounts of highly labeled [^{32}P]DNA. In addition, no general procedure existed for isolating specific fragments of the genome for sequence analysis. These problems were overcome by copying unlabeled DNA with either RNA polymerase or DNA polymerase by using ^{32}P-labeled triphosphates. In this way, the product can easily be highly labeled for sequence analysis. In addition, when copying starts at a unique site, defined areas can be sequenced specifically by pulse labeling. With this technique Blackburn (33, 35) used RNA polymerase to copy endonuclease IV fragments and restriction enzyme fragments and then used the well established RNA sequencing techniques to define the sequence of the products. In the case of restriction enzyme fragments, RNA polymerase copies both strands of the template. Although this produces a complex fingerprint, the complementary nature of the sequences obtained facilitates the deduction of the sequence by overlapping corresponding sequences from the two strands. With the use of short oligonucleotide primers complementary to the 5' end of one strand, it is possible to copy specifically one strand of the restriction fragment. This approach was used by Gilbert and Maxam (36) in the sequence analysis of the isolated *lac* operator fragment.

Use of DNA Polymerase

Primed synthesis with DNA polymerase was first used by Wu and his colleagues in their studies on the cohesive ends of bacteriophage λ-DNA (37, 38). These cohesive ends are formed by a single-stranded extension of 12 nucleotides from the otherwise double-stranded molecule. This forms a natural template-primer system for DNA polymerase, and by the use of labeled triphosphates the complementary sequences are uniquely labeled and their sequence determined. Other workers (39–42) explored the use of exogenous primers to direct specifically DNA polymerase on various templates. However, the first successful use of such a primer to obtain new sequence information was by Sanger et al. (43). A synthetic octadeoxyribonucleotide which corresponded to the sequence coding for a region of the amino acid sequence of the bacteriophage f1 coat protein was synthesized by Schott et al. (44). Although this particular amino sequence was later found to be in error and the octanucleotide did not prime in this region, it was possible to determine an 80-nucleotide sequence from a binding site elsewhere in the molecule. This sequence was determined by using the ribosubstitution method described below.

Ribosubstitution Method

In 1963 Berg et al. (45) showed that if polymerization with DNA polymerase were carried out in the presence of Mn^{2+} instead of Mg^{2+}, ribonucleotides could be incorporated into the DNA chain. For example, if rCTP were incorporated instead of dCTP with the other three deoxyribotriphosphates present, a DNA chain was produced with all the cytidine nucleotides in the ribo form. These ribose-phosphate bonds could then be cleaved specifically either by alkali or with a suitable ribonuclease. This approach thus overcame the main problem of DNA sequence analysis, the lack of base-specific deoxyribonucleases. The ribosubstitution method was used by Salser et al. (46) with sheared bacteriophage

M13 DNA as a primer-template system, and fingerprint analysis of the riboter-minated products showed no evidence of misincorporation. However, Van de Sande et al. (47) did detect misincorporation of rCTP and rGTP under certain incubation conditions.

The synthesis of a decadeoxyribonucleotide (48) provided the opportunity to check the fidelity of the method and to extend the ribosome binding site sequence of Robertson et al. (27). This site contained the sequence 5'-T-T-T-T-A-T-T-T-C-T-3', and the decamer which was synthesized had the complementary sequence 5'-A-G-A-A-A-T-A-A-A-A-3'. The sequence of the first 36 nucleotides incorporated into the primer could be predicted and thus it could be checked whether DNA polymerase was copying the template faithfully under the incubation conditions used. Both rCTP and rGTP were used in separate incubations in the decamer primed system, so that two overlapping sets of oligonucleotides were produced, from which the complete sequence could be deduced by Donelson et al. (49).

Figure 1 shows an example of the ribosubstitution method. The decamer primer and ϕX viral strand DNA were incubated with DNA polymerase in the presence of rCTP and the three other deoxytriphosphates (one of which was ^{32}P-labeled). The reaction was carried out in such a way that a mixture of newly synthesized DNA varying in length from 20–50 nucleotides was added to the primer. When this was then digested with pancreatic ribonuclease and the products were fractionated two-dimensionally, a number of fragments were obtained, as in Figure 1, *VI*. The fragments could be eluted and their sequences obtained by digestion with spleen and venom exonucleases. In order to deduce the relative order of these fragments, a sample of the incubation mixture was fractionated before hydrolysis (Figure 1, *I*). Products A-E in Figure 1, *I* represent increasing chain lengths of newly synthesized DNA added to the primer. These products were eluted and digested with pancreatic ribonuclease, and the products of each were fractionated two-dimensionally (Figure 1, *II–V*). From the gradually increasing complexity of the fingerprints, the relative order of the digestion products was shown to be $c1$, $c2$, $c3$, $c4$, $c5$. From experiments such as this, it was possible to deduce a sequence of 40 residues extending from the decamer primer and Donelson et al. (49) were able to show that the DNA polymerase was copying faithfully.

The chemical synthesis of primers such as the octanucleotide used in the f1 DNA system and the decanucleotide described above was, of course, time consuming. At that time, the first restriction enzyme maps for ϕX RF DNA had been prepared with the use of the enzymes Hind II and Hae III by Edgell et al. (50), Middleton et al. (51), and Lee and Sinsheimer (52). These double-stranded restriction fragments with unique 5'- and 3'- termini were, therefore, ideal for specific priming. The fragments could be denatured and reannealed in the presence of viral strand ϕX DNA (the plus strand). Because only the minus strand of the restriction fragment hybridizes to the viral strand, the 3'- end of the minus strand is a unique substrate for DNA polymerase in the ribosubstitution method

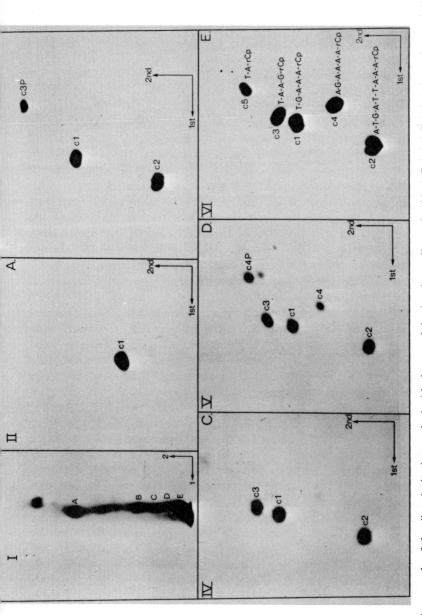

Figure 1. An example of the ribosubstitution method with the use of the decadeoxyoligonucleotide A-G-A-A-A-T-A-A-A-A as primer and φX viral strand DNA as template. The triphosphates rCTP, [α-32P] dATP, dGTP, and dTTP were incorporated by using DNA polymerase I nach Klenow. *Autoradiograph I* shows a two-dimensional fractionation (2, 3) of the first 20—50 nucleotides added to the primer with the use of a 5% "homomix" in the second dimension. *Autoradiographs II-VI* show two-dimensional fractionations of pancreatic ribonuclease digests of the products A—E in autoradiograph I with the use of a 3%, 30-min, hydrolyzed homomix in the second dimension. Reproduced from Donelson et al. (49) with permission of Springer-Verlag.

(see under "Plus and Minus Method"). This method was first used to determine the sequence of a repressor binding site in bacteriophage λ by Maniatis et al. (53); the first large scale sequence analysis of ϕX DNA was started in this way by using all the different fragments from the Hind II and Hae III restriction enzyme digests. However, this approach was soon superseded by the development of the more rapid plus and minus technique of Sanger and Coulson (17) (see under "Plus and Minus Method").

Isolation of Restriction Fragments

The recent discovery of many different restriction enzymes with a large range of different specificities (54) has enabled specific areas of DNA to be located in restriction enzyme fragments. These fragments can then be used for DNA sequence analysis in basically two ways, either as primers for DNA polymerase in the copying methods described in this chapter or as substrates for the chemical modification procedure of Maxam and Gilbert (55). Restriction enzyme fragments are normally purified by electrophoresis of restriction enzyme digests on either polyacrylamide or Agarose gels. If primed synthesis methods for DNA sequencing are used, it is wise to avoid Agarose gels. It is thought that some poison for DNA polymerase is eluted from the Agarose along with the fragments.

A general procedure for the isolation of preparative amounts of fragments is described below. Typically, 200 μg of ϕX RFI DNA is digested at 37°C in 6.6 mM Tris-Cl, pH 7.4, 6.6 mM $MgCl_2$, 6.6 mM 2-mercaptoethanol, with the appropriate quantity of restriction enzyme determined by a trial experiment. When digestion is complete, the sample is extracted twice with phenol, then with ether, and the DNA precipitated with ethanol. The fragments are fractionated on an appropriate gradient polyacrylamide gel as described by Jeppesen (56) except that 40 × 20-cm gels, approximately 3 mm thick, are used.

After electrophoresis, the gel bands can be visualized in two ways, either by staining with 0.2% methylene blue followed by destaining overnight with several changes of water, or with ethidium bromide at a concentration of 1 μg/ml for 1 hr. In the latter case, the bands are visualized in ultraviolet light. Apart from the drawback of excising the bands under ultraviolet light, this method has the important advantage that the DNA can be located at all stages of the extraction procedure. The bands are excised with a scalpel and eluted electrophoretically by a slight modification of the procedure of Galibert et al. (57). After elution, the bands are precipitated with ethanol, and the DNA is spun down at 30,000 rpm for 45 min in an ultracentrifuge. Each fragment is dissolved in 200 μl of H_2O and stored at −20°C.

Small fragments (fewer than 200 base pairs (bp)) can be conveniently eluted from the gel by soaking. The gel is diced and covered with 0.5–1.0 ml of buffer (0.5 M NaCl, 0.1 M Tris-Cl, pH 8.5, 0.005 M EDTA). The tube is sealed with parafilm and left overnight at 37 °C. Alternatively, the extraction can be done for shorter periods and repeated several times. The material is then filtered through glass wool by a low speed centrifuge spin and precipitated with ethanol.

Mapping of Restriction Fragments

The construction of a restriction fragment map of the region of DNA under investigation is normally an essential prerequisite to sequence work. Earlier studies generally employed two methods: either partial digests were produced, followed by complete digestion with the same enzyme (52), or double digests were produced with different combinations of enzymes (58, 59). These methods were laborious, however, and required considerable interpretation. Recently, two rapid mapping techniques have been developed by Jeppesen et al. (60) and by Hutchison (61) and are described briefly below. The first method requires an isolated set of each restriction enzyme fragment from a particular digest for use as primers with ϕX viral strand DNA as a template. Each fragment is separately annealed to the template. A short radioactive pulse of synthesis with DNA polymerase is performed with one ^{32}P-labeled triphosphate; the other three triphosphates are unlabeled. The ^{32}P-labeled triphosphate is then diluted out with the corresponding unlabeled one, and synthesis is continued. This nonradioactive chase must extend at least beyond the next restriction site. The extended DNA is then digested to completion with the restriction enzyme used to produce the primer fragment. The products of each reaction with each fragment are then fractionated by gel electrophoresis. It is convenient to fractionate a marker nick-translated (62) restriction enzyme digest alongside to locate accurately which fragments are labeled in the pulse-chase reactions. Sometimes more than one fragment is labeled, particularly where the adjacent fragment is short. This fragment is normally present in much lower yield and can usefully give additional information. Nick-translation (62) is a convenient method of labeling double-stranded DNA by using DNA polymerase I and the four deoxyribo-triphosphates, one of which is [α-^{32}P]labeled. Commercial preparations of DNA polymerase I seem to contain a DNase activity which introduces nicks into the DNA, and new labeled DNA is synthesized onto the 3' ends. The original DNA is either displaced or degraded by the 5'-exonuclease activity of the DNA polymerase I.

An example of this method is shown in Figure 2. Some Hinf fragments of ϕX DNA are used as primers with a marker digest of nick-translated Hinf fragments for reference. Fragment 1 labels fragment 16a or 16b, fragment 2 labels 14a or 14b, fragment 3 labels 5c, which in turn labels fragment 15. In this way, by using all the fragments as primers, the complete map can be deduced. At the same time, additional information can be obtained by digesting the primed fragments with different restriction enzymes. In this way, the map can be aligned with other known restriction maps of the same DNA.

The second method, that of Hutchison (61), is a two-dimensional hybridization procedure, the principle of which is shown in Figure 3. In the first dimension, an unlabeled restriction enzyme digest is fractionated on a square acrylamide or Agarose slab gel. Following electrophoresis, the gel is stained with ethidium bromide and photographed. The fragments are denatured in the gel by

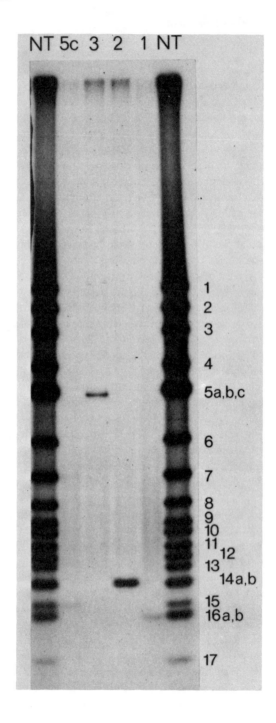

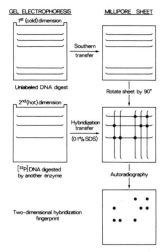

Figure 3. The principle of the two-dimensional restriction fragment mapping technique (61).

soaking in alkali and are then transferred to a sheet of Millipore by a slight modification of the procedure described by Southern (63). In the second dimension, a radioactive preparation of the same DNA digested with a different restriction enzyme is fractionated on a second gel. It is then autoradiographed and denatured in the gel. These radioactive fragments are transferred onto the Millipore sheet containing the unlabeled fragments. The two band patterns should be rotated by 90° with respect to each other and aligned so that each radioactive band intersects each unlabeled band bound to the Millipore sheet. Under the conditions used, each radioactive fragment sticks to the Millipore sheet only at those points where the radioactive band intersects a nonradioactive band containing homologous DNA sequences. The sheet is then autoradiographed, producing a two-dimensional fingerprint of spots where homologous bands intersect. These spots can be aligned with the photograph and autoradiograph of the one-dimensional gels. From the information contained in the fingerprint, the order of restriction sites can be determined with respect to the known map of the second restriction enzyme. Overlaps as small as 25 bp have been observed. Alternatively, instead of using two different restriction enzyme digests, a complete digest and a partial digest of DNA with the same restriction enzyme can be fingerprinted. This technique allows the order of restriction sites within the

Figure 2. An example of restriction fragment mapping by primed synthesis (60) with the use of Hinf I restriction fragments as primers on φX viral strand DNA. The nearest neighbor fragments can be deduced from the Hinf I restriction digest of nick-translated φX RF DNA fractionated alongside. A 4–12% gradient polyacrylamide gel was used. (G. M. Air and B. G. Barrell, unpublished observations.)

partial products to be determined. The method is quick and large numbers of individual restriction fragments do not have to be isolated and redigested.

The two methods described above have been used extensively in the construction of the restriction maps of ϕX DNA (18). The large numbers of fragments have enabled the complete sequence of ϕX DNA to be determined with the use of the plus and minus technique.

Plus and Minus Method

This method is based on the ability of DNA polymerase I to copy the sequence of a single strand of DNA into the complementary sequence by using a region of double-stranded DNA as a primer. In the primed ribosubstitution reaction, when incorporations were performed with one of the triphosphates in low amounts (e.g., when one triphosphate was ^{32}P-labeled), a series of discrete products of different chain lengths could be fractionated on homochromatography or acrylamide gel electrophoresis. These products correspond to positions in the sequence where the radioactive triphosphates were incorporated. For example, if $[\alpha$-^{32}P]dATP were the limiting triphosphate, the reaction slowed up where this was incorporated, producing a series of products ending before adenosine nucleotides. If this could be done for each triphosphate, therefore, it would be possible to read the complete sequence by fractionating these products by polyacrylamide gel electrophoresis. This theory formed the basis of the method, which was first worked out with the use of the decamer (see under "Ribosubstitution Method") as a primer. As restriction fragments became available, however, they were used as primers.

Initial Incorporation   Normally, one strand of a restriction fragment is annealed to a single template strand of DNA in the same manner used in the primed synthesis method for mapping restriction fragments. If the polymerase reaction is then carried out under limiting conditions, it is possible to produce a mixture of copies of every possible chain length between 1 and about 200 nucleotides added onto the primer. The newly synthesized DNA is labeled by substituting an α-^{32}P-labeled triphosphate for one of the four unlabeled triphosphates in the polymerase reaction. The reaction is then terminated by phenol extraction, and excess triphosphates are removed by a Sephadex G-100 column. The mixture which is still hybridized to the template DNA is then divided and treated in different ways.

Minus System   Samples are reincubated with DNA polymerase I lacking the 5'-exonuclease activity (e.g., DNA polymerase I nach Klenow, the mutant enzyme of Heijeneker et al. (64), or T4 DNA polymerase) in the presence of only three triphosphates.

Synthesis then proceeds as far as it can on each chain until it reaches the site where the missing triphosphate ought to be incorporated and then stops. Thus, if dATP is the missing triphosphate (i.e., in the $-$A system), each chain terminates at its 3' end at a position before an adenosine residue. Four separate samples are incubated, each with one of the four triphosphates missing.

Plus System   Samples are incubated in the presence of bacteriophage T4 DNA polymerase and only one triphosphate. This polymerase degrades the newly synthesized DNA from the 3'-end, but "stops" at nucleotides corresponding to the one triphosphate that is present (65). For example, in the +A system only dATP is present and all the chains then terminate at the 3' end with adenosine residues. Four samples are incubated, each with one of the triphosphates present.

Restriction Enzyme Cleavage   If the primer is a strand of a restriction fragment, then the initial extension recreates the restriction site. Prior to the fractionation of the eight plus and minus reactions on acrylamide gels, the primer can be cleaved with the restriction enzyme. Alternatively, the restriction enzyme digest can be performed prior to the plus and minus reactions. In some cases, restriction enzymes do not cleave the plus and minus reactions because of the inhibition of the enzyme by the presence of the single-stranded DNA template. This happens with the enzyme Alu I. An alternative method of cleavage (N. L. Brown, manuscript in preparation) employs the ability of DNA polymerase I to insert ribonucleoside triphosphates in the presence of Mn^{2+} as well as deoxyribonucleotides. In the case of Alu I, the enzyme cleaves the sequence 5'-A- G$\downarrow$C-T-3'; therefore, the first nucleotide to be incorporated with the use of an Alu I fragment as a primer would be dC. Hence, an initial incorporation is performed by using rCTP with DNA polymerase I in the presence of Mn^{2+}. The reaction is continued by adding high amounts of dCTP to flood out the rCTP, the other three deoxytriphosphates, one of which is ^{32}P-labeled, and Mg^{2+} to flood out the Mn^{2+}. Incubation is at 0°C and is performed as for the initial incorporation of the plus and minus procedure. The same procedure is used up to the stage of restriction enzyme cleavage when the samples are cleaved by alkali or pancreatic ribonuclease at the single rC residue prior to electrophoresis.

If the primer is short (less than 75 bp), then the samples can be fractionated uncut with the restriction enzyme. The samples are denatured in formamide containing blue dye markers (bromphenol blue, xylene cyanol FF) and fractionated on denaturing 12% acrylamide gels. The products which have a common 5' end are fractionated according to size, each oligonucleotide being separated from its neighbor which contains one more nucleotide. Samples of the initial incorporation (i.e., those not subjected to the plus and minus reactions, but treated in the same way with the restriction enzyme) containing all possible chain lengths are fractionated alongside as markers. This is termed the zero (0) system. The experimental details for the plus and minus procedures are given in Appendix II.

Deduction of Sequence   Figure 4 shows the principle of the method and the deduction of a hypothetical sequence. It is important to remember that the autoradiograph contains bands, for example in the −A system, that correspond to positions before adenosine residues and, for example, in the +A system that terminate with adenosine residues. Thus, the +A products are one nucleotide larger than the corresponding −A products except if there is more than one consecutive adenosine nucleotide. In this case, the distance between the bands in

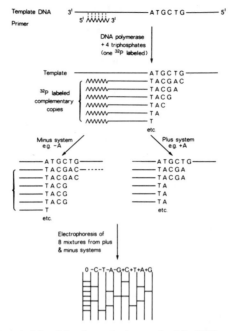

Figure 4. The principle of the plus and minus method for DNA sequencing (17).

the −A and +A systems indicates the number of such consecutive nucleotides. The exact number can be obtained from the marker bands in the zero system alongside. In the example given in Figure 4, the smallest oligonucleotide gives a band in the −T position; this is confirmed by the presence of a band in the +T position in the next largest oligonucleotide. Thus, the bands in the +T and −A positions in this product show that its 3′-terminus is thymidine and the next nucleotide is an adenosine, thus defining the sequence T-A. Similarly, the next largest oligonucleotide defines the sequence A-C, establishing the sequence T-A-C. In the perfect example illustrated, both the plus and minus systems define the complete sequence T-A-C-G-A-C- independently, but in practice it is wise to check the sequence in both systems as described above. This is important because occasionally artifact bands appear and sometimes a band may be missing in one system.

Figure 5 shows a good example of the plus and minus method. The samples have been divided in two and a short and a long fractionation performed as described in Appendix II. An almost unambiguous sequence of at least 126 nucleotides starting 14 nucleotides from the restriction site can be deduced. Generally, the first 15 or so nucleotides are not seen, probably because of the instability of short products on the template in the presence of the DNA polymerase used for the plus and minus reactions. No attempt is made to remove the polymerase before the restriction enzyme digest; these products are probably degraded by the 3'-exonuclease activity of the polymerase. Normally, this region can be sequenced by priming from another adjacent restriction enzyme site, but if this is not possible another method must be used, such as ribosubstitution. If the restriction enzyme fragment is short, the reactions can be fractionated uncleaved and the first 15 residues obtained in this way. If this latter method is used, then the 5'-ends of the primer fragment must be intact and not frayed by any exonuclease present in the enzyme preparation used to produce the fragment. Equally important is the use of a DNA polymerase I, which lacks a 5'-exonuclease activity in the initial incorporation, as well as in the plus and minus incorporations.

In the interpretation of the autoradiographs shown in Figure 5, artifact bands are represented by *broken lines*. Artifact bands are usually in low yield and probably arise from incomplete plus and minus reactions. They are only rarely found in the same positions in both the plus and the minus reactions and can be ignored on these grounds alone. However, there is a variation in the intensity of bands, and some may be missing. For this reason, it is wise to repeat the experiment several times. Only very rarely is an artifact or a missing band consistent from one experiment to another. If a consistent serious anomaly does occur then it should be checked by other methods.

The most serious anomaly that can arise is termed "compressions." They can be easily recognized by a region on the gel where several different bands line up together (Figure 6A). The denaturing gel illustrated was run at low current (10–20 mamp) and did not heat up appreciably. The sequence around the compression region can be easily read with reference to the zero system. Considering only the minus system, the following sequence can be deduced:

- strand 5'-A-A-G-G-G-G-C-C-G-A-A(G,T,C)G-C-A-A-T-T-3'

If the experiment is run with the denaturing gel at high current (40 mamp), the gel shown in Figure 6B is produced, giving the following sequence in this region:

- strand 5'-A-A-G-G-G-G-C-C-G-A-A-G-C-C——T-G-C-A-A-T-T-3'

However, if the sequence in this region is determined on the complementary strand with the gel again run at high current (Figure 6C), the following sequence can be deduced:

+ strand 3'-T-T-C-C——G-G-C-T-T-C-G-G-G-G-A-C-G-T-T-A-A-5'

Compare this with the sequence deduced for the minus strand:

- strand 5'-A-A-G-G-G-G-C-C-G-A-A-G-C-C-——T-G-C-A-A-T-T-3'

+ strand 3'-T-T-C-C-——G-G-C-T-T-C-G-G-G-G-A-C-G-T-T-A-A-5'

The sequence is as follows:

(a)   (b)
- strand 5'-A-A-G-G-G-G-C-C-G-A-A-G-C-C-C-C-T-G-C-A-A-T-T-3'

This particular compression has been independently sequenced with the use of depurination analysis of both strands and occurs at nucleotides 963 to 985 in the

J/F intercistronic region. The compression of bands on the − strand sequence corresponds to region b, which is complementary to region a. It is likely that chains ending in region b "flip back" and base pair with region a and that chains with this secondary structure migrate anomalously. In fact, they migrate as if they were shorter than the corresponding denatured chains. When priming is carried out on the complementary strand, the compression occurs in region a but gives the correct sequence in region b.

Thus, it is important to sequence both strands, especially where a compres-

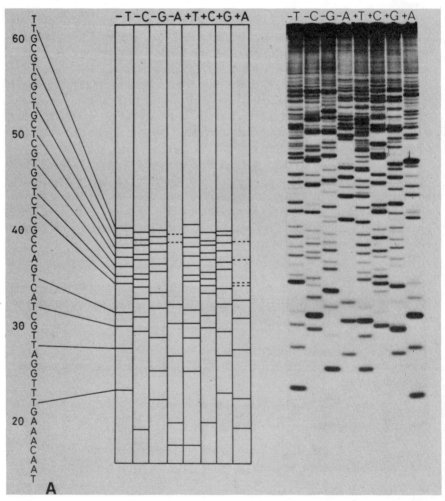

Figure 5. *A*, an autoradiograph of a plus and minus sequencing gel with the use of restriction fragment Taq 4 as a primer on φX − strand template. The products were cleaved by Mbo II. A diagrammatic interpretation of the sequence is shown alongside. Electrophoresis was continued until the bromphenol blue marker reached the bottom of the gel.

sion is recognized, and to check this sequence in a different way, e.g., by depurination as in the case given above. It is unfortunate that even very hot denaturing gels did not completely resolve the sequence on either strand. Gels made to 50% formamide instead of 7 M urea and run at 40 mamp gave the same pattern as shown in Figure 6, *B* and *C*. Higher concentrations of formamide probably would resolve this problem.

In conclusion, the plus and minus procedure has been extensively used in this laboratory to obtain the sequence of φX DNA, which has a chain length of

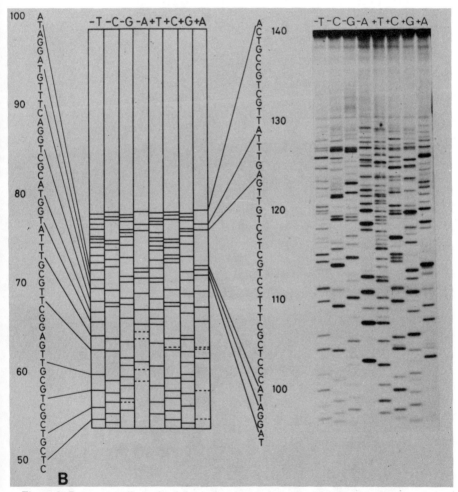

Figure 5. *B*, an autoradiograph of the same reaction described in *A*, but fractionated for a longer time, so that the xylene cyanol marker migrated close to the bottom of the gel. Reproduced from Barrell et al. (66) with permission of MacMillan Journals, London.

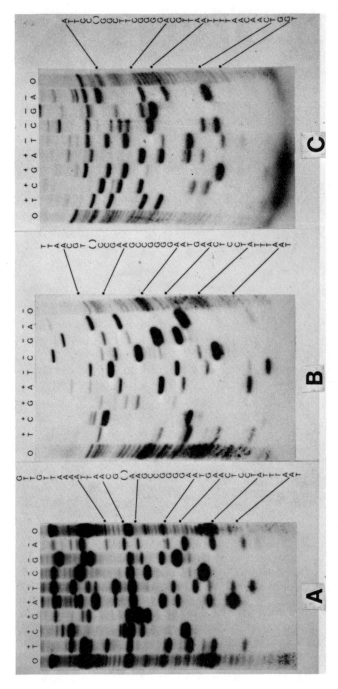

Figure 6. Compressions on plus and minus sequencing gels. *A*, Hha I fragment 10 primed on φX + strand. The gel was run at low current (10–20 mamp) and did not heat up appreciably. O, the compression of bands (shown in the sequence deduced alongside each denaturing gel). *B*, same as in *A*, but this gel was run at 40 mamp. *C*, the same region of DNA shown in *A* and *B* but with Hha I fragment 14 as a primer on φX – strand DNA. See text for the deduction of the sequence in this region. (B. G. Barrell, unpublished observations.)

approximately 5,370 nucleotides. The greatest advantage of the method is its simplicity; experimental time involved is trivial. In addition, it incorporates two systems (the plus system and the minus system), each of which can independently give the sequence and serve as a check on the other. In general, it can be applied to any DNA molecule from which it is possible to separate strands for the template and from which restriction fragments or synthetic oligonucleotides are available as primers. Alternatively, the method can be used on a restriction fragment template which is first degraded by exonuclease III (J. Maat, personal communication). Both 3'-ends of the fragment, which are separated by an asymmetric restriction enzyme cleavage, are then a substrate for the plus and minus method. In another variation (W. Barnes, personal communication), the plus and minus system is replaced by a ribosubstitution technique. The initial incorporation is divided into four samples, each of which is treated with DNA polymerase in the presence of Mn^{2+}; all four deoxyribonucleoside triphosphates with one different ribotriphosphate are in each reaction. The ribotriphosphate is present in greater concentration than the deoxytriphosphates. Because of the less efficient incorporation of the ribotriphosphate, on average each chain contains only one ribonucleotide; ribonucleotides are randomly distributed. The reactions can then be cleaved by base or nucleases at the ribo bond. The four reactions, each terminating at nucleotides corresponding to the input ribotriphosphate, are fractionated by gel electrophoresis as in the plus and minus technique. The main advantage is that a band corresponding to every nucleotide should be seen, as in the chemical modification method of Maxam and Gilbert (55).

Overlapping sequences are built up with the use of the plus and minus methods by priming with all possible adjacent restriction fragments on both strands. Each priming is normally repeated several times to resolve any uncertainties. The inaccuracies, if any, should be recognizable, and other techniques can be employed to resolve these specific problems. The use of the zero system as a reference graticule, coupled with both the plus and minus bands, has resolved the earlier problem of detecting how many nucleotides are present in a run of a particular nucleotide. If there is any doubt, it can be checked by depurination analysis either of nick-translated restriction fragments or of primed synthesis on the appropriate strand. Two-dimensional fingerprints of depurination products are easily done, and the composition of the products can usually be deduced from the graticule (20, 67, 68).

SEQUENCE OF BACTERIOPHAGE ϕX174 DNA

With the development of the plus and minus method and the mapping of an increasing number of restriction enzyme sites on ϕX, a complete sequence for the whole molecule was obtained in about 18 months. The sequence is shown in Appendix I (18). Some of the most interesting aspects of this sequence are described below.

Genes F, G, and H

Early work with the plus and minus system was mainly concerned with extending the sequences obtained by direct methods in the genes coding for the structural proteins, the gene G spike protein, and the gene F coat protein, all of which, along with the gene H spike protein, form the ϕX capsid. Air (32), studying the amino acid sequence of the F protein, showed that the sequences of the related endonuclease IV bands (26, 31, 57) were coded by gene F. In fact, the 5'-end of band 5A coded for the third amino acid of the F protein. Thus, the direct relationship of the sequence with the genetic map (11) and the early restriction enzyme maps (50–52) was known. The sequence was then extended by the plus and minus method with the use of restriction fragment primers mapped in this region of the genome. The sequence near the 5'-end of gene F coding for approximately the first 100 amino acids of the gene F protein has been reported by Air et al. (34). The tentative sequence of the whole gene extends from nucleotide 1,001 to 2,278 in the sequence given in Appendix I.

Similarly, the sequence corresponding to the gene G initiation site (27–29, 69) was extended by a combination of amino acid sequence analysis and the plus and minus method with the use of restriction fragment primers. The complete sequence of gene G was deduced by Air et al. (70).

Third Position Thymidine in Codons   These studies revealed an unexpected phenomenon, the high incidence of thymidine residues in the 3rd position of the DNA codons. Of the 175 codons of gene G, thymidine occurs in the 3rd position of 95 codons (54.3%). Air et al. (34) found a similar preference for 3rd position thymidine (56%) in the sequence at the NH_2-terminus of the gene F product. The DNA of ϕX174 is, in any case, rich in thymidine. The nucleotide composition calculated from the sequence in Appendix I is: A, 23.9%; C, 21.5%; G, 23.3%, and T, 31.2%. These figures are in close agreement with those of Sinsheimer (71) and Seigal and Hayashi (72). However, if it is assumed that the first two positions of each codon are random, i.e. 25% of each base and the high incidence of thymidine in ϕX are only because of selection in the 3rd position, then the frequency of thymidine in this position would be 48%. From the fully confirmed sequences in Appendix I, Sanger et al. (18) calculated that from a total of 1,346 codons 42.9% terminated in thymidine.

The reason for the high thymidine content of ϕX DNA is not known. It also occurs in the closely related bacteriophage S13 DNA and the filamentous phages fd and f1, but it has not been observed in the single-stranded RNA phages (73–75) or in the closely related isometric DNA phage G4 (G. N. Godson, personal communication). The packaging of the DNA in the icosahedral phages and the filamentous phages would be expected to be quite different, and this may rule out the possibility that the high thymidine content of these phages is concerned with packaging. Another possibility, suggested by Denhardt and Marvin (76), is that it is a translation control. However, high 3rd position thymidine content was useful in the early sequence analysis in defining the translational reading frame for unknown amino acid sequences. The gene H initiation codon was originally identified in this way and was later confirmed by amino acid sequence analysis (G. M. Air, unpublished observation). By using the 3rd position thymidine rule early errors in the sequence were corrected and a reading frame for the whole gene was predicted. The five peptides from the H protein since isolated confirm this reading frame. Terminal amino acid determinations with carboxypeptidase confirm the termination codon at position 3,907. The H protein has been implicated as a pilot protein in penetration of the host cell and in the early stages of replication (77–79).

Rare Codons   Although the distribution of codons used by ϕX is biased to those ending in thymidine, some codons seem to be used only rarely. For example, out of a total of 112 arginine codons in the provisional sequence (Appendix I) only 2 are AGG. In addition, out of 126 glycine codons only 4 are GGG. In general, all ϕX codons ending in GG and GA are rare. These may have a modulating role in translation, as suggested by Fiers et al. (75).

Gene J

Benbow et al. (10) suggested that a cistron J, defined by the mutant *am*6 between cistrons E and F, existed and that it coded for the smallest component of the virion, a protein of $\sim$5,000 molecular weight. This protein was sequenced by D. Freymeyer, P. R. Shank, T. Vanaman, C. A. Hutchison, III, and M. H. Edgell (manuscript in preparation), who showed it to be a small basic protein of 37 amino acids. DNA sequence analysis with the use of restriction fragment primers near the 5'-end of gene F established that the protein was coded in this region.

The mutant *am*6 was originally mapped in gene E (10) but was then assigned to gene J (11) because it differed from gene E mutations by producing defective particles. P. J. Weisbeek (personal communication) has now shown that *am*6 is a double amber mutant mapping in fragments F9 (gene A) and Z7 (gene E). The gene E mutation is identical with *am*3 (gene E, see below), and the gene A mutation is a G $\rightarrow$ A change at position 117, resulting in an amber codon that would produce a gene A protein 6 amino acids shorter. In addition, a missense mutation (serine to leucine in gene A) is produced by a C $\rightarrow$ T change at position 57, and another mutation is produced which changes an arginine codon CGA to CGG at position 103 in gene A (B. G. Barrell, unpublished observation). It is not yet known whether there is a mutation in gene J specified by *am*6.

Overlapping Genes D and E

Sequence analysis of the region preceding gene J gave a totally unexpected result (66). This DNA sequence was found to code for the COOH terminus of the gene D protein, the amino acid sequence of which had been established by G. M. Air. Gene E had been mapped between genes D and J by Benbow et al. (11) but was now completely ruled out, especially as the termination codon of gene D overlapped the initiation codon of gene J by one base.

However, in a sense a result such as this was not completely unexpected, because if the φX protein molecular weights were close to correct, then φX DNA would need to be 600–700 nucleotides longer provided that each protein was separately coded.

In order to locate gene E on the restriction fragment map of φX, the marker rescue technique (80, 81) was used. Wild type φX RF DNA was cleaved with the restriction enzyme Hae III, and the purified fragments were annealed to viral strand DNA bearing a gene E mutation. The complex was then used to transfect *Escherichia coli* spheroplasts, and the progeny phage were assayed for wild type plaque-forming particles. Wild type marker can only be rescued from fragments which span the site of the mutation. All gene E mutants tested in this way mapped in Hae III fragment 7. Sequence analysis of the gene D region and of the D protein established that this fragment was wholly contained in gene D; because gene E amber mutants made full length D protein, these amber mutations could not be in the same translational phase as gene D. Although the precise function of the gene D protein is not known, it is made in large amounts and is necessary for the production of viral single-stranded DNA. Gene E is responsible for the lysis of the host cell; this function is not impaired in the gene D amber mutants so far isolated. The complete sequence of the gene D protein and its DNA unequivocally established the reading frame for gene D. The presence of gene E amber mutants in this sequence showed that gene E was being translated at some point within this sequence and that this translation was in a different reading frame from that of gene D.

In order to establish the gene E reading frame, restriction fragment primers such as Hae III fragment 5 were used in the plus and minus method with φX *am3* viral strand as a template, and the determined sequence compared with that of the wild type. A C → T change was observed which corresponded to a G → A change on the viral strand, as shown below:

$$\text{wt } 5'\text{-G-T-G-G-G-A-T-A-3'}$$
$$\text{am3 } 5'\text{-G-T-A-G-G-A-T-A-3'}$$

Thus, a TGG codon (tryptophan) was changed to a TAG amber codon. The phase of this amber codon was displaced one nucleotide to the right of the gene D reading frame. It did not produce a change in the gene D reading frame because one valine codon (GTG) was changed to another valine codon (GTA). Similarly, analysis of an independently isolated gene E amber mutant (*am34*) was shown to have the same nucleotide sequence as *am3*. In addition, the gene E amber mutants, *am27* and N11, were sequenced, and both were shown to be a G → A transition, and again resulted in a TGG codon change to TAG. As with *am3* and *am34*, these codons were displaced one nucleotide to the right of the gene D reading frame and again produced no coding change in gene D because one leucine codon CTG is changed to another, CTA.

By inspection of the nucleotide sequence of the gene D region shown in Figure 7 it was possible to define the gene E reading frame by looking for

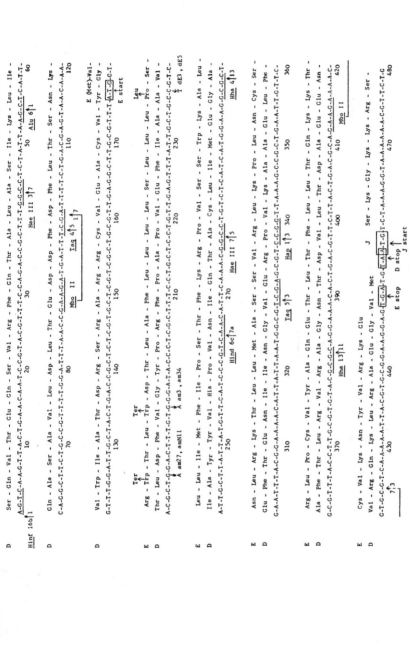

Figure 7. The nucleotide sequence of the gene D region. The amino acid sequences of the D protein and the J protein are aligned with the DNA sequence. The amino acid sequence of the gene E product predicted from the DNA sequence is also shown. Note that the numbering system is different from that used in Appendix I. The amino acid sequence of the first eight amino acids of the D protein has been reported by Farber (82), who found an arginine at residue 7. Reproduced from Barrell et al. (66) with permission of MacMillan Journals, London.

initiation and termination codons in the same phase as the amber codons. Although the gene E product has not yet been isolated, the complete amino acid sequence of this protein can be predicted. Including the NH_2-terminal methionine, the protein consists of 91 amino acids with a molecular weight of 9,940. As seen in Figure 7, the NH_2-terminus of this protein is very rich in hydrophobic residues, as are other proteins which have a membrane-related activity (83, 84). Gene D codons have a high proportion of thymidine residues (40%) in the 3rd position. If the reading frame is displaced one nucleotide to the right, these thymidine residues occur in the second position of the resulting codons, all of which specify hydrophobic amino acids. This second reading frame also discriminates against termination codons because these codons have thymidine in the 1st position. Thus, if a series of mutational events led to the evolution of an initiation site in phase 2 of gene D, a polypeptide possibly would be produced that could form the basis for natural selection of an efficient lysis gene. Although there must have been severe constraints on the evolution of gene E in order to preserve gene D, because of the high incidence of 3rd position thymidine in the gene D codons, these constraints are not as high as would be expected from a more random sequence. Lysozyme activity has not been found in infected cells, and after lysis no apparent morphological change in the cell wall structure exists, except for a few ruptures (85). Probably the gene E protein interferes in some way with cell wall biosynthesis, leading to disruption and lysis.

Overlapping Genes A and B

The definition of the boundaries of genes D, E, J, F, G, and H by sequence analysis has been described above. Although the overlapping of gene E upon gene D alleviated the coding capacity problem, it was still not possible to fit genes A, B, and C into the remaining unassigned DNA. The gene A product, estimated molecular weight 55,000–65,000, would need at least 1,500 of the remaining 1,865 nucleotides. Thus, it was probable that a similar overlap would be found in this region, particularly in the region of genes A and B because gene B mutants had been mapped in gene A (9, 11, 86).

Sequence analysis of the region downstream (i.e., in the 5' to 3' direction) from the 3'-end of gene H revealed a likely candidate for the gene A initiation site at position 3,973. This site contained a sequence capable of forming base pairs with the 3'-end of 16 S ribosomal RNA (87, 88) (see under "Translation"). The sequence after the A-T-G sequence contained codons in this phase which had a high proportion of 3rd position thymidine residues. Van Mansfeld et al. (89) also predicted the gene A reading frame by the sequence analysis of a restriction fragment from gene A. Sequence analysis of the gene A mutants *am*86 and *am*33 confirmed the gene A reading frame (P. M. Slocombe, personal communication). Both mutants produced a C → T change at positions 4,108 and 4,372 which resulted in amber codons in phase with the A-T-G sequence at 3,973. The reading frame continued to the termination codon at position 134. The A protein is thus 512 amino acids in length, with a molecular weight of around 56,000, in good agreement with the SDS gel estimates (Table 1).

The location of the initiation site of the A* protein is not known. This protein arises from a reinitiation event in gene A (13). Its molecular weight is 35,000 so it would need approximately 960 nucleotides to code for it. This places the initiation site at about position 4,550. If the initiation codon is ATG, several starts are possible in the gene A reading frame—at positions 4,522, 4,531, and 4,657.

The first proof that genes A and B overlap on the same DNA sequence was provided by the sequence of a φX mRNA ribosome binding site isolated as a protected fragment by Ravetch et al. (90). The sequences of the ribonuclease T₁ and pancreatic ribonuclease products matched the sequence at positions 5,041 to 5,079 in gene A (91). The ATG initiation codon was in a phase which displaced two nucleotides to the right of the gene A phase. This reading frame continued to the termination codon at position 49 in the total DNA sequence (note that the DNA is a single-stranded circle) and would code for a protein of 120 amino acids with a molecular weight of 13,845. Confirmation that this reading frame belonged to gene B came from sequence analyses by Brown and Smith (92) and Smith et al. (93) of the gene B mutant *am*16, the gene A mutants *am*18 and *am*35, and the revertant of *am*18, *ts*116, which is temperature-sensitive in gene B. The gene B mutant *am*16 was mapped in Alu I fragment 5, Hae III fragment 3, and Hind II fragment 5 by Weisbeek et al. (94). With the use of the plus and minus method, this mutant was shown to be a G → T transversion at position 5,265, as shown in Figure 8. This transversion produces an amber codon in phase with the A-T-G sequence in the proposed gene B initiation site. Unlike the *am*3 (gene E) mutation which did not produce a coding change in gene D, *am*16 would produce a leucine(TTG) to phenylalanine(TTT) change in gene A. Mutant *am*16 is not known to be temperature-sensitive for gene A. Presumably, a leucine to phenylalanine change is acceptable considering that the residues involved are similar.

Figure 9, *A–D*, shows regions of plus and minus gels comparing the wild type (*am*3) sequence with those for *am*35, *am*18, and *ts*116, respectively. These

A, sequence changes produced by the gene B mutant *am*16. **B**, sequence changes produced by the gene A mutants *am*18, *am*35, and the gene B mutant *ts*116.

Figure 8. *A*, sequence changes produced by the gene B mutant *am*16. *B*, sequence changes produced by the gene A mutants *am*18, *am*35, and the gene B mutant *ts*116.

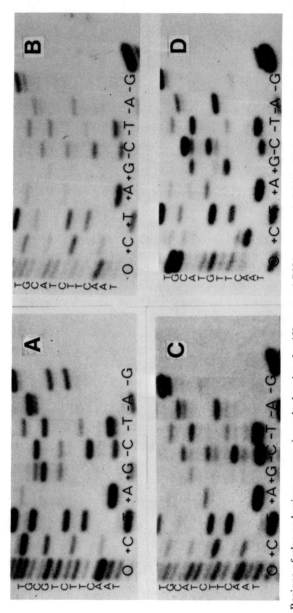

Figure 9. Regions of plus and minus sequencing gels showing the φX − strand DNA sequence of (A) wild type φX, (B) am18, (C) am35, and (D) ts116. Taq 1 fragment 8 was used as a primer on φX viral strand DNA and the products were cleaved with Hinf I. The + strand sequences are shown in Figure 8. Reproduced from Smith et al. (93) with permission of MacMillan Journals, London.

mutants were mapped in Hha I fragment 8b by Baas et al. (59). The sequences deduced in this figure are for the minus strand; Figure 8 shows the plus strand sequences and the effects of the mutations. Both *am*35 and *am*18 produce the same C → T transition at nucleotide 23, giving the expected amber codons in the gene A reading frame. In the gene B reading frame, this transition gives rise to an alanine to valine change. No evidence exists for any impairment in gene B function because of this change. The gene B mutant *ts*116 was isolated as a revertant of *am*18 and gives a G → C change at position 25. Thus, the amber codon TAG reverts to a TAC (tyrosine) codon in the gene A reading frame, not to the original CAG (glutamine). In the gene B reading frame, the mutation results in a GAA (glutamic acid) to CAA (glutamine) codon change which is probably responsible for the *ts* phenotype. However, the glutamine to tyrosine change in gene A in *ts*116 is not known to be *ts*. The results described above establish the reading frames for gene A and gene B in the DNA sequence. Gene B is thus overlapped upon gene A from nucleotides 5,064 to 51 in the circle. Gene A extends beyond the 3'-end of gene B for 85 nucleotides (28 codons).

It is interesting to follow the incidence of 3rd position thymidine through this overlap region. In the D/E overlap, gene D maintained a high content of 3rd position thymidine in the overlap region. As shown in Figure 10, gene B takes over the 3rd position thymidine from gene A, in the overlap region, which raises the possibility that gene A was originally shorter, ending near the 5'-end of gene B and then mutating to read through the gene B region. However, if this is the case, it is not clear why the region of gene A beyond the 3'-end of B also has a preference for 3rd position thymidine in its codons.

Why ϕX has evolved overlapping genes remains a mystery. Constraints on the length of the molecule imposed by packaging, for example, is an obvious possibility. One argument against this is that the closely related phage G4 is some 200 nucleotides longer (G. N. Godson, personal communication). The constraints on the evolution of overlapping genes must be great. The constraints on ϕX are alleviated to some extent because one gene of the pair has a high

Third position Thymidine

Gene A	Gene B	Gene D	Gene E
10		4/11	
12		9	
12		12	
11		9	3/12
7		13	3
7		6	2
12		6	4
8		5	1/19
1	7/19		
1	7		
2	7		
3	7		
5	8		
6	8		
11			
2/9			

Figure 10. The incidence of 3rd position thymidine in each 20 codons through the overlapping genes A/B and D/E. At gene ends, a figure such as 7/19 means 7 cases out of only 19 codons.

incidence of 3rd position thymidine residues in its codons. Therefore, genes D and B automatically select against termination codons in their 2nd position reading frames, i.e., the reading frame is displaced one nucleotide to the right. This, of course, applies to all φX genes with high 3rd position thymidine, and extra open reading frames exist all over the genome. The significance of these and allied potential initiation sites is discussed under "Translation."

One obvious disadvantage in the D/E region is that both proteins are made from the same messenger RNA. Therefore, initiation of synthesis of the E protein may be obstructed by ribosomes synthesizing the D protein. However, if the gene E product is required only in small amounts, the overlap may serve as a convenient control. In the gene B product, this "control" is overcome by the presence of a promoter at position 4,888. Thus, the A and B proteins are translated from different mRNAs (see under "Transcription").

The existence of two overlapping genes in φX raises the possibility that others may exist in different organisms. If they are to be found, other isometric bacteriophages with possible similar constraints on their size seem to be the obvious candidates. *E. coli* has overlapping termination and initiation codons, as found in the D–J region (see under "Boundaries Between Genes") in the *tryp* operon (95), and longer overlaps must be a possibility. In eukaryotes, however, the large amount of DNA present would seem to make evolution of overlapping genes unnecessary.

Gene C

The boundary of gene C, which mapped between genes B and D, was the last to be characterized. The gene C mutant *och*6 was mapped between the A18/6 site and the F9/13 site (59) (170–204, Appendix I). Sequence analysis of this mutant showed a C → T change at nucleotide 196, producing an expected TAA codon from CAA (glutamine) (B. G. Barrell, unpublished results). Examination of the sequence in this reading frame established that the ATG codon at position 133 was the only possible initiation codon for gene C. However, because of the uncertainty in the sequence at nucleotides 314–319, it was impossible to be certain of the location of the gene C termination codon. Recently, J. Sims and D. Dressler (personal communication) have confirmed the sequence in this region (nucleotides 263–375). The first termination codon in phase with the *och*6 TAA codon is at position 391. Therefore, the gene C coding region extends from the end of gene A, with the ATG codon overlapping the gene A termination codon, through the gene D promoter (see below under "Transcription") and terminating at and overlapping with the gene D initiation codon. Gene C thus has 86 codons, which would produce a protein of molecular weight 9,400. No marked preference for 3rd position thymidine exists in the gene C codons (32.5%).

Boundaries Between Genes

From a chain of 5,375 nucleotides, only 215 can be termed noncoding (excluding termination codons), i.e., 4% of the genome. Noncoding nucleotides are con-

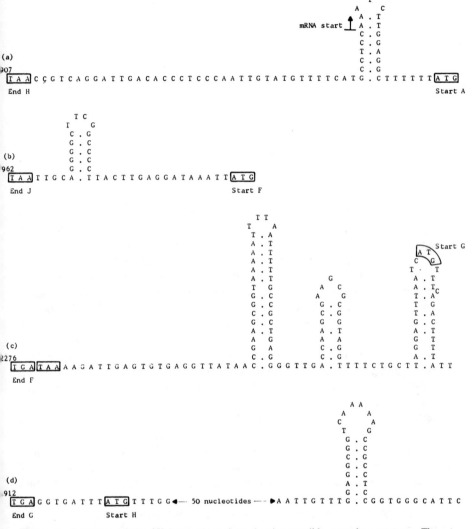

Figure 11. Sequences of the φX intergenic regions showing possible secondary structure. The calculated free energies of these hairpin loops are given in the text. The two loops near the start of gene G are not expected to be stable.

tained in the H–A (63 nucleotides), J–F (36 nucleotides), F–G (108 nucleotides), and G–H (8 nucleotides) regions shown in Figure 11. The H–A noncoding region contains the "gene A promoter;" however, the "gene B promoter" and the "gene D promoter" are contained in the coding regions for gene A and gene C, respectively (see under "Transcription"). The role of the hairpin loop followed by T₆A in this region is discussed also in the same section. The J–F and the F–G

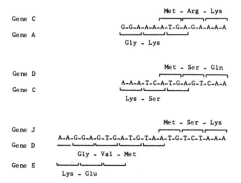

Figure 12. Overlapping termination and initiation codons in ϕX.

noncoding regions also contain regions of secondary structure. The size of the J–F region is more consistent with those found in bacteriophage RNA (96), whereas the F–G region sequenced by Fiddes (68) is comparable in size to the sequence between the *lac i* and *z* genes (97) (122 nucleotides). This latter sequence contains the binding sites for three proteins—the catabolite activator protein, RNA polymerase, and the *lac* repressor. Therefore, from size considerations alone, it seems that the F–G region has an important function. Besides the hyphenated 2-fold symmetry shown base paired in Figure 11*C*, the same sequence has a nonrandom distribution of G/C and A/T sequences which consists of 17 adjacent A/T nucleotides in the center of the symmetry, flanked by 2 G/C-rich regions. There is no evidence that this region contains a promoter site although it may contain a transcription termination site (see under "Transcription"). The origin of replication has been located in gene A (see under "Origin of Replication"). The gene H protein has been shown to bind to ϕX viral strand DNA (78, 79), so this region may be a likely candidate.

Platt and Yanofsky (95) found that in the *E. coli tryp* operon the termination codons of the B gene overlapped by one nucleotide the initiation codon of the A gene. Besides the completely overlapped genes D/E and A/B, ϕX contains three examples of overlapping initiation and termination codons of genes A and C, C and D, and D and J (Figure 12). In these cases, the sequence complementary to the 3′-end of 16 S rRNA involved in initiation of translation (see under "Translation") is contained in a coding region. The fact that four examples have been found so far implies that this may be a more general phenomenon and that they may be involved in some kind of translational control. Figure 13 summarizes the positions of all the known ϕX genes aligned with the Hae III restriction map.

Transcription

Chen et al. (98) located three binding sites for RNA polymerase on restriction fragments of ϕX. The binding sites in Hind II fragments 4 and 6b (6c in Appendix I) correspond to promoters at the start of genes A and D, respectively. The third site is probably internal in gene G and does not correspond to a real

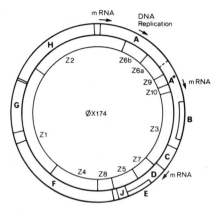

Figure 13. The new genetic map of bacteriophage φX174, aligned with the Hae III restriction fragment map.

promoter site because it is not in association with any in vitro mRNA starts. A promoter before gene A was proposed because polarity in genes F, G, and H does not extend into gene A (11, 99, 100). Smith and Sinsheimer (101) isolated in vitro mRNA synthesized from φX RF DNA and detected 18 different species of mRNA in varying yields. Some of the more prominent species represented copies of greater than the unit length of φX DNA. These observations are in general agreement with the results of Hayashi and Hayashi (99) and Clements and Sinsheimer (102) for in vivo φX mRNA. The nucleotide sequences at the 5'-ends of these mRNAs are pppA-A-A-U-C-U-U-G-G, pppA-U-C-G-C, and pppG-A-U-G (103, 104). They were mapped relative to restriction fragments (105–107) and corresponded with the starts of genes A, B, and D. From the DNA sequence of φX, Sanger et al. (18) located precisely the gene A and gene D promoters, but two locations were possible for the gene B promoter (Table 4). Recently, J. Sims and D. Dressler (personal communication) also determined the sequence of the gene B promoter region and identified the A-T-G-C- sequence at position 4,888 as the mRNA start for this promoter. Studies on other promoter sites (108, 109) have shown that a common sequence, 5'-T-A-T-R-A-T-R-3', is located just before the mRNA start. The sequences corresponding to this region in φX are *boxed* in Table 4. The gene D and gene B promoters are also coding sequences and may not be expected to possess a perfect common sequence, but this does not apply to the gene A promoter. Although elements of symmetry and common sequences can be seen in these promoters, it is difficult at this time to say what signal in the DNA sequence determines a promoter site or the efficiency with which it initiates transcription. Interestingly, the possible gene G RNA polymerase binding site contains the sequence T-A-T-G-A-T-G at positions 2,705–2,711.

The identification of transcription termination sites in φX is more difficult. The sequence at the 3'-end of certain RNA species synthesized by *E. coli* RNA

Table 4. Promoter sequences in φX174 DNA

Promoter	Sequences[a]
A	A-G-G-A-T-T-G-A-C-A-C-C-C-T-C-C-C-A-A-T-T-G-T-A-T-G-T [T-T-T-C-A-T-G] $\overset{3954}{C\text{-}C\text{-}C\text{-}T\text{-}C\text{-}C\text{-}A\text{-}A\text{-}A\text{-}T\text{-}C\text{-}T}$ — — ↑ 18 nucleotides to A protein start
D	R7b/R6c G-T-T-G-A-C-A-T-T-T-A-A-A-A-G-A-G-C-G-T-G-G-A-T-T-A-C [T-A-T-C-T-G-A-] $\overset{358}{G\text{-}T\text{-}C\text{-}C\text{-}G\text{-}A\text{-}T\text{-}G\text{-}C\text{-}T}$ ↑ 32 nucleotides to D protein start
B?	R3/R8 C-A-G-G-T-A-G-G-C-G-T-T-G-A-C-C-C-T-A-A-T-T-T-G-G-T-C-G [T-C-G-G-G-T-A-] $\overset{4832}{C\text{-}G\text{-}C\text{-}A\text{-}A\text{-}T\text{-}C\text{-}G\text{-}C\text{-}C\text{-}}$ ↑ 232 nucleotides to B protein start
B?	A-G-C-T-T-G-C-A-A-A-A-T-A-C-G-T-G-G-C-C-T-T-A-T-G-G-T [T-A-C-A-G-T-A-] $\overset{4888}{T\text{-}G\text{-}C\text{-}C\text{-}C\text{-}A\text{-}T\text{-}C\text{-}G\text{-}C\text{-}A}$ ↑ 176 nucleotides to B protein start

[a] mRNA initiation sequences (103, 104) are *underlined*. *Boxed regions* indicate sequences they may correspond to the TATRATR sequence found in other promoters (108), taking into account the distance from the mRNA starts. The second B promoter sequence has recently been identified as the correct one (J. Sims, personal communication).

polymerase is $U_{5-6}A_{OH}$ (96). Rosenberg et al. (110) compared the sequence distal to the 3'-end of the bacteriophage λ 4 S and 6 S transcripts and found the sequence T-T-T-T-T-T-A-T in common. This sequence precedes the gene A initiation codon at position 3,973 in ϕX. A ρ-independent termination site in this region has been proposed (106, 107, 111). Surprisingly, this sequence occurs after the gene A promoter mRNA start. It could produce a small RNA species (20 nucleotides), but this has not been detected. Possibly, the T_6A acts as an "attenuator" similar to that proposed for the *tryp* operon (112). Hairpin loop structures before the termination site have recently been implicated in termination (M. Rosenberg, unpublished observation; Sugimoto et al. (113)). Figure 11 shows that such a loop is present in this position in ϕX. mRNAs initiated at the gene B and gene D promoters would contain this loop and terminate at this point. Initiation at the gene A promoter would be from within this loop and would thus lack this structure at the 5'-end but would contain it at the 3'-end. This site would thus act as a central terminator in ϕX although Smith et al. (114) locate the major termination site near the start of gene C. Other termination sites have been mapped near the beginning of genes F, G, and H (106, 107, 111). Although there are hairpin loops in these regions (see Figure 11), the termination sites have not been located in the DNA sequence.

Translation

Shine and Dalgarno (87) proposed, from an examination of known ribosome binding site sequences, that the region preceding the initiation codon was characteristically complementary to the 3'-end of 16 S ribosomal RNA. This varies from three to eight base pairs. Steitz and Jakes (88) demonstrated a complex between the bacteriophage R17 gene A initiation site and the 3'-end of 16 S rRNA. The sequences complementary to this 3'-sequence are *boxed* in Table 5 (see also Chapter 4). Ravetch et al. (90) isolated protected fragments by binding ribosomes in vitro to ϕX mRNA. However, as in the case of bacteriophage f1, this method did not give all the possible binding sites. Fragments derived from the initiation sites for only genes B, J, F, and G were found. Robertson et al. (27) showed that protection of the ϕX viral strand DNA with ribosomes against DNase digestion yielded only one major site, that for gene G. This was found only in low yield as a protected RNA fragment by Ravetch et al. (90). Examination of the sequences of these sites showed no correlation between those fragments protected. In addition, it was difficult to find any correlation between the degree of complementarity with 16 S rRNA and the amounts of protein synthesized either in vivo (115, 116) or in vitro (117).

Examination of the sequences preceding all 136 ATG triplets in ϕX reveals that many have complementary sequences to 16 S rRNA besides the known initiation sites shown in Table 5. Thirteen of these (shown in Table 6) potentially could synthesize polypeptides longer than 20 amino acids out of phase in other coding regions. The largest of these potential initiation sites is in gene H and could code for a protein of 200 amino acids. It is difficult to see why at least

Table 5. Initiation sequences of φX174 coded proteins

Initiation site	Sequences[a]
A	C-A-A-A-T-C-T-T- [G-G-A-G-G] -C-T-T-T-T-T-A-T-G-G-T-T-C-G-T-T-C-T-T-A-T
B	A-A-A-G-G-T-C-T- [A-G-G-A-G] -C-T-A-A-A-G-A-A-T-G-G-A-A-C-A-A-C-T-C-A-C-T
C	A-A-G-T-G-G-A-C-T-G-C-T-G-G-C- [G-G-A] -A-A-A-T-G-A-G-A-A-A-A-T-T-C-G-A-C
D	C-C-A-C-T-A-A-T- [A-G-G-T] -A-A-G-A-A-A-T-C-A-T-G-A-G-T-C-A-A-G-T-T-A-C-T
	Ser Gln Val Thr
E	C-T-G-C-G-T-T- [G-A-G-G] -C-T-T-G-C-G-T-T-T-A-T-G-G-T-A-C-G-C-T-G-G-A-C-T

J   C-G-T-G-C-G-G- │A-A-G-G-A-G│ -T-G-A-T-G-T-A-A-T-G-T-C-T-A-A-A-G-G-T-A-A-A
                                                    Ser   Lys   Gly   Lys

F   C-C-C-T-T-A-C-T-T-G- │A-G-G-A│ -T-A-A-A-T-T-A-T-G-T-C-T-A-A-T-A-T-T-C-A-A
                                                    Ser   Asn   Ile   Gln

G   T-T-C-T-G-C-T-T- │A-G-G-A-G│ -T-T-T-A-A-T-C-A-T-G-T-T-T-C-A-G-A-C-T-T-T-T
                                              Met   Phe   Gln   Thr   Phe

H   C-C-A-C-T ⌐T-A-A-G⌐ T- │G-A-G-G-T-G-A│ -T-T-T-A-T-G-T-T-T-G-G-T-G-C-T-A-T-T
                                                        Met   Phe   Gly   Ala   Ile

16 S RNA
3' end   HOA-U-U-C-C-U-C-C-A-C-U-A-G

ᵃ Where a protein start has been independently confirmed by protein sequencing data, the amino acid sequences are indicated. The other initiation regions were identified as described in the text. Sequences complementary to the 3' end of 16 S ribosomal RNA (87, 88) are boxed; broken lines indicate further complementarity if some nucleotides are looped out or not matched. Ribosome binding to mRNA has been demonstrated in these regions for genes J, F, G, and B (90).

Table 6. Possible extra initiation sites in φX174 DNA

Sites[a]	Position[b]	Sequence[c]	Length of protein[d]
C1	51	C-T-T-T-C- G-G-A -T-A-T-T-C-T-G-A-T-G	56
F1	1,038	G-C-C- G-A-G -C-G-T-A-T-G-C-C-G-C-A-T-G	52
F2	1,272	A-T-G- A-A-G-G-A -T-G-G-T-G-T-T-A-A-T-G	39
F3	1,452	A-C-C- G-A-G-G -C-T-A-A-C-C-C-T-A-A-T-G	34
F4	1,689	G-G-A-G-G-T -A-A-A-A-C-C-T-C-A-T-A-T-G	29
F5	2,064	G-G-T-C-A-G-T- G-G-T -A-T-C-G-T-T-A-T-G	68
G1	2,535	G-A-T-A-G-T-T-T-G-A-C- G-G-T -T-A-A-T-G	77

H1a	3,068	A-C-T-G-T- [A-G-G] -C-A-T-G-G-G-T-G-A-T-G	202
H1b	3,101	G-C-C-A-T-T-C- [A-A-G-G] -C-T-C-T-T-A-T-G	191
H2	3,776	G-C-A- [A-A-G-G-A] -T-A-T-T-T-C-T-A-A-T-G	48
A1a	4,418	C-T-T- [A-A-G-G-A] -T-A-T-T-C-G-C-G-A-T-G	34
A1b	4,454	A-A-G-A-A- [A-G-G-T] -A-T-T-A-A-G-G-A-T-G	22
A2	4,610	C-G-A-T-T-A- [G-A-G-G] -C-G-T-T-T-A-T-G	78

16 S RNA
3' end   HOA-U-U-C-C-U-C-C-A-C-U-A-G

[a] Named according to the known gene in which they occur and the order in which they appear in that gene; e.g., F4 is the fourth possible out-of-phase initiation site of gene F. *Bracketed names* are in the same reading frame.

[b] Numbers refer to those in Appendix I and to the A of the ATG codon.

[c] Sequences complementary to the 3' end of the 16 S ribosomal RNA are *boxed.*

[d] Length of the polypeptide that could be produced starting at the ATG codon. Only those sites that could specify a polypeptide of greater than 20 amino acids have been considered.

some of these sequences are not initiation sites and not coding for extra φX proteins. Other genes have been postulated (11) (gene I), and additional functions such as a methylase activity have been ascribed to φX (118). It will be interesting to see whether the sequences of any of these additional possibly overlapping genes are conserved in related bacteriophage.

Origin of Replication

The gene A product is a site-specific endonuclease which introduces a nick into the viral strand in the gene A coding region in restriction fragment Hae III 6b (119, 120). This product is probably the origin of viral strand replication. During the stage of progeny viral strand synthesis, Johnson and Sinsheimer (121) iso-lated RF II molecules which contained a specific gap in the viral strand in this region. Gap-filling reactions with DNA polymerase and labeled triphosphates have also located the gap in restriction fragment Hind II 3 (122). Depurination of the filled-in gap showed the presence of the characteristic product C_6T. The sequence -C-T-C-C-C-C-C- occurs at position 4,285, and the location of the origin in this region agrees precisely with the results of Baas et al. (120). Consideration of the nucleotide sequence in this region shows no outstanding feature that can be predicted to be the site of the specific nick. However, deter-mination of the actual site of cleavage by the A protein should be possible since its recent isolation (119, 123). Friedman and Razin (124) suggested that the single 5-methylcytosine in the φX viral strand acts as a signal for cleavage by the A protein. Lee and Sinsheimer (125) have located the methylated nucleotide in Hind II fragment 6a which is at least 500 nucleotides from the origin. Recently, however, φX DNA has been shown to contain other minor nucleotides (126). Infection of spheroplasts with heteroduplex RF and analysis of the percentage of mixed bursts suggested that the origin of replication lay toward the 3'-end of gene A in fragment Hind II 5 (127, 128). These results depended on the accurate mapping of the gene A mutants employed. However, sequence analysis of these mutants (see under "Overlapping Genes A and B") and the repositioning of am16 in the gene A and B overlap may explain in part this anomalous result. They do not explain the drastic change in the amount of repair of the heterodup-lexes between am18 and am35, both of which have the same sequence change. Possibly, there is another mutation in am35.

Secondary Structure

It is not known whether φX viral DNA has a high degree of secondary structure as found in bacteriophage MS2 RNA (72–74) or in the 16 S ribosomal RNA (129). Sinsheimer (71, 130) showed that heating φX viral DNA to 100°C pro-duced a 25% increase in 260 nm absorption. In addition, nondenatured φX viral DNA appears compact and tangled in the electron microscope (131). However, studies with single strand-specific nucleases showed that only 2% of the viral strand was resistant (132). Depurination analysis of the resistant fractions indi-cated that the major hairpin loop in the gene F–G intergenic region (68) accounted for approximately half of this resistant fraction (see Figure 11). Similar results have been obtained by H. Schaller (personal communication), who also obtained depurination products characteristic of the major loop in the F–G region. Additionally, he found depurination products that could be assigned to the hairpin loop preceding gene A (Figure 11A). These two loops would have a similar stability ($\Delta G = -14$ kcal) according to the rules of Tinoco et al. (133). However, preliminary analysis of the sequence showed that several other loops with similar or higher stability values can be postulated; e.g., in Figure 11B, the J–F loop has a ΔG of -17.2 kcal, whereas, the loop in Figure 11D would have a ΔG of -22.6 kcal, although this sequence is not fully established. T4 endonuclease IV, under limiting conditions of low temperature and high ionic strength, produces a series of well defined fragments from the viral strand (31). These products all originate from the 5'-terminal region of gene F (34), suggesting that this particular region is lacking in secondary structure. If ϕX has a secondary structure, it may be very complex, with long range interactions involving different regions of the circular genome as well as the short range hairpin loops discussed above. A detailed computer analysis of the sequence is at present under way and should, hopefully, provide an answer to this problem.

APPENDIX I

Nucleotide Sequence of Bacteriophage φX174 *am* 3DNA[1] (18)

1. Restriction enzyme cleavage sites are shown above the sequences.

One letter code	Conventional name	Recognition site
A	Alu I	$\downarrow$ A-G-C-T
T	Taq I	$\downarrow$ T-C-G-A
Y	Hap II	$\downarrow$ C-C-G-G
Z	Hae III	$\downarrow$ G-G-C-C
H	Hha I	$\downarrow$ G-C-G-C
R	Hind II	$\downarrow$ G-T-Y-R-A-C
F	Hinf I	$\downarrow$ G-A-N-T-C
M	Mbo II	G-A-A-G-A 8bp$\downarrow$ C-T-T-C-T
Q	Hph I	G-G-T-G-A 8bp$\downarrow$ C-C-A-C-T
P	Pst I	$\downarrow$ C-T-G-C-A-G

Restriction fragments are numbered from the largest, as estimated by gel electrophoresis. Multiple bands are lettered in order of decreasing size.

H15/8b contains two overlapping Hha I sites. M9/2 contains two overlapping Mbo II
$\downarrow$
sites. Hae II (R-G-C-G-C-Y) sites are found at H4/13, H11/14, H14/12, H12/10, H10/7,
$\downarrow$
H7/5, H3/2, and H9b/1. Hpa I (G-T-T-A-A-C) sites are found at R5/7b, R6b/1, and R8/5.

Solid underlines indicate sequences that are fully confirmed; sequences with no underlines probably do not contain more than one mistake for 50 residues. Dashed underlines indicate more uncertain sequences.

---

[1] For further discussion, see "Note Added in Proof," page 175.

```
p1/1          R5/7b                                              F6/9 T7/8
GAGTTTTATC GCTTCCATGA CGCAGAAGTT AACACTTTCG GATATTTCTG ATGAGTCGAA AAATTATCTT GATAAAGCAG GAATTACTAC TGCTTGTTTA CGAATTAAAT CGAAGTGGAC
         10         20         30         40         50         60         70         80         90        100        110        120
                                          A5/18                    F9/13                                               F16a/16b
              T9/10         H8b/4   T10/4    A18/6
TGCTGGCGGA AAATGAGAAA ATTCGACCTA TCCTTGCGCA GCTCGAGAAG CTCTTACTTT GCGACCTTTC GCCATCAACT AACGATTCTG TCAAAAACTG ACGCGTTGGA TGAGGAGAAG
        130        140        150        160        170        180        190        200        210        220        230        240
End A ↑                                                                                                              mRNA start ↑
                                   F13/17                              F17/16a      R7b/6c
TGGCTTAATA TGCTTGGCAC GTTCGTCAAG GACTGGTTTA GATATGAGTC ACATTTTGTT CATGGTAGAG ATTCTCTTGT TGACATTTTA AAAGAGCGTG GATTACTATC TGAGTCCGAT
        250        260        270        280        290        300        310        320        330        340        350        360
                              F16b/1                                              Z3/7              A6/1
                              D start ↑
GCTGTTCAAC CACTAATAGG TAAGAAATCA TGAGTCAAGT TACTGAACAA TCCGTACGTT TCCAGACCGC TTTGGCCTCT ATTAAGCTCA TTCAGGCTTC TGCCGTTTTG GATTTAACCG
        370        380        390        400        410        420        430        440        450        460        470        480
M1/7   T4/5                                                                                            H4/13
AAGATGATTT CGATTTTCTG ACGAGTAACA AAGTTTGGAT TGCTACTGAC CGCTCTCGTG CTCGTCGCTG CGTTGAGGCT TGCGTTTATG GTACGCTGGA CTTTGTAGGA TACCCTCGCT
        490        500        510        520        530        540        550        560        570        580        590        600
                                                                              E start ↑
                                        R6c/7a              Z7/5
T5/3   Y1/3                                          H13/11               M7/3
TTCCTGCTCC TGTTGAGTTT ATTGCTGCCG TCATTGCTTA TTATGTTCAT CCCGTCAACA TTCAAACGGC CTGTCTCATC ATGGAAGGCG CTGAATTTAC GGAAAACATT ATTAATGGCG
        610        620        630        640        650        660        670        680        690        700        710        720
                                                                        M7/3
TCGAGCGTCC GGTTAAAGCC GCTGAATTGT TCGCGTTTAC CTTGCGTGTA CGCGCAGGAA ACACTGACGT TCTTACTGAC GCAGAAGAAA ACGTGCGTCA AAAATTACGT GCGGAAGGAG
        730        740        750        760        770        780        790        800        810        820        830        840
                                                                                                                    End E ↑
                         H11/14                                     H14/12                                R7a/6b
TGATGTAATG TCTAAAGGTA AAAAACGTTC TGGCGCTCGC CCTGGTCGTC CGCAGCCGTT GCGAGGTACT AAAGGCAAGC GTAAAGGCGC TCGTCTTTGG TATGTAGGTG GTCAACAATT
        850        860        870        880        890        900        910        920        930        940        950        960
End D ↑ I I J start
```

Continued

```
                    Z5/8                                          H12/10
       TTAATTGCAG GGGCTTCGGC CCCTTACTTG AGGATAAATT ATGTCTAATA TTCAAACTGG CGCCGAGCGT ATGCCGCATG ACCTTTCCCA TCTTGGCTTC CTTGCTGGTC AGATTGGTCG
1 End J  970        980        990       1000       1010       1020       1030       1040       1050       1060       1070       1080
                                         F start ↑

                              Y3/2              F1/14b     T3/1                     H10/7                  Z8/4            F14b/2
       TCTTATTACC ATTTCAACTA CTCCGGTTAT CGCTGGCGAC TCCTTCGAGA TGGACGCCGT TGGCGCTCTC CGTCTTTCTC CATTGCGTCG TGGCCTTGCT ATTGACTCTA CTGTAGACAT
       1090       1100       1110       1120       1130       1140       1150       1160       1170       1180       1190       1200

                              Q1/3c                                                                       R6b/1
       TTTTACTTTT TATGTCCCTC ATCGTCACGT TTATGGTGAA CAGTGGATTA AGTTCATGAA GGATGGTGTT AATGCCACTC CTCTCCCGAC TGTTAACCAA ACTACTGGTT ATATTGACCA
       1210       1220       1230       1240       1250       1260       1270       1280       1290       1300       1310       1320

                                                                                                    H7/5
       TGCCGCTTTT CTTGGCACGA TTAACCCTGA TACCAATAAA ATCCCTAAGC ATTTGTTTCA GGGTTATTTG ATATCTATAG CGTATTTTAA AGGCGCGTGG ---ATGCCTG ACCGTACCGA
            A1/12c
       1330       1340       1350       1360       1370       1380       1390       1400       1410       1420       1430       1440
                                                                                               A12c/13

                                                                                                                   M3/4
       GGCTAACCCT AATGAGCTTA ATCAAGATGA TGCTCGTTAT GGTTTCCGTT GCTGCCATCT CAAAAACATT TGGACTGCTC CGCTTCCTCC TGAGACTGAG CTTTCTCGCC AAATGACGAC
       1450       1460       1470       1480       1490       1500       1510       1520       1530       1540       1550       1560

                            A13/2                                      H5/9a                                              Z4/1
       TTCTACCACA TCTATTGACA TTATGGGTCT GCAAGCTGCT TATGCTAATT TGCATACTGA CCAAGAACGT GATTACTTCA TGCAGCGTTA CCATGA-GTT ATTTCTTCAT TTGGAGGTAA
       1570       1580       1590       1600       1610       1620       1639       1640       1650       1660       1670       1680

                                                                       H9a/8a                                       F2/11
       AACCTCATAT GACGCTGACA ACCGTCCTTT ACTTGTCATG CGCTCTAATC TCTGGGCATC TGGCTATGAT GTTGATGGAA CTGACCAAAC GTCGTTAGGC CAGTTTTCTG GTCGTGTTCA
       1690       1700       1710       1720       1730       1740       1750       1760       1770       1780       1790       1800

                                                                                                                   Q3c/6F11/7
       ACAGACCTAT AAACATTCTG TGCCGCGTTT CTTTGTTCCT GAGCATGGCA CTATGTTTAC TCTTGCGCTG GTTCGTTTTC CGCCTACTGC GACTAAAGAG ATTCAGTACC TTAACGCTAA
       1810       1820       1830       1840       1850       1860       1870       1880       1890       1900       1910       1920

       AGGTGCTTTG ACTTATACCG ATATTGCTGG CGACCCTGTT TTGTATGGGA ACTTGCCGCC GCGTGAAATT TCTATGAAGG TCTATGGGAT ATGTTTTCCG TTCTGGTGAT TCGTCTAAGA AGTTTAAGAT
       1930       1940       1950       1960       1970       1980       1990       2000       2010       2020       2030       2040
```

H8a/6

Q6/5   Q5/3b

TGCTGAGGGT CAGTGGTATC GTTATGCGCC TTCGTATGTT TCTCCTGCTT ATCACCTTCT TGAAGGCTTC CCATTCATTC AGGAACCGCC TTCTGGTGGT TTGCAAGAAC GCGTACTTAT
2050   2060   2070   2080   2090   2100   2110   2120   2130   2140   2150   2160

F7/5b

TCGCAACCAT GATTATGACC AGTGTTTCAG TCGTTCAGTT GTTGCAGTGG ATAGTCTTAC CTCATGTGAC GTTTATCGCA ATCTGCCGAC CACTCGCGAT TCAATCATGA CTTCGTGATA
2170   2180   2190   2200   2210   2220   2230   2240   2250   2260   2270 End F1 2280

R1/9   H6/3

AAAGATTGAG TGTGAGGTTA TAACCGAAGC GGTAAAAATT TTAATTTTTG CCGCTGAGGG GTTGACCAAG CGAAGCGCGG TAGGTTTTCT GCTTAGGAGT TTAATCATGT TTCAGACTTT
2290   2300   2310   2320   2330   2340   2350   2360   2370   2380   2390   2400
                                                                                                  G start 1

A2/16   A16/15a M4/10   A15a/3   R9/10

TATTTCTCGC CACAATTCAA ACTTTTTTTC TGATAAGCTG GTTCTCACTT CTGTTACTCC AGCTTCTTCG GCACCTGTTT TACAGACACC TAAAGCTACA TCGTCAACGT TATATTTTGA
2410   2420   2430   2440   2450   2460   2470   2480   2490   2500   2510   2520

M10/9   R10/2

TAGTTTGACG GTTAATGCTG GTAATGGTGG TTTTCTTCAT TGCATTCAGA TGGATACATC TGTCAACGCC GCTAATCAGG TTGTTTCAGT TGGTGCTGAT ATTGCTTTTG ATGCCGACCC
2530   2540   2550   2560   2570   2580   2590   2600   2610   2620   2630   2640

F5b/8 M9/2   Y2/5

TAAATTTTTT GCCTGTTTGG TTCGCTTTGA GTCTTCTTCG GTTCCGACTA CCCTCCCGAC TGCCTATGAT GTTTATCCTT TGGATGGTCG CCATGATGGT GGTTATTATA CCGTCAAGGA
2650   2660   2670   2680   2690   2700   2710   2720   2730   2740   2750   2760

Q3b/4   H3/2   F8/4

CTGTGTGACT ATTGACGTCC TTCCCCGTAC GCCGGGCAAT AACGTCTACG TTGGTTTCAT GGTTTGGTCT AACTTTACCG CTACTAAATG CCGCGGATTG GTTTCGCTGA ATCAGGTTAT
2770   2780   2790   2800   2810   2820   2830   2840   2850   2860   2870   2880

Y5/4   Q4/7   Q7/2

TAAAGAGATT ATTTGTCTCC AGCCACTTAA GTGAGGTGAT TTATGTTTGG TGCTATTGCT GGCGGTATTG CTTCTGCTCT TGCTGGTGGC GCCATGTCTA AATTGTTTGG AGGCGGTCAA
2890   2900   2910   2920   2930   2940   2950   2960   2970   2980   2990   3000
                     End G 1   H start 1

AAAGCCGCCT CCGGTGGCAT TCAAGGTGAT GTGCTTGCTA CCGATAACAA TACTGTAGGC ATGGGTGATG CTGGTATTAA ATCTGCCATT CAAGGCTCTA ATGTTCCTAA CCCTGATGAG
3010   3020   3030   3040   3050   3060   3070   3080   3090   3100   3110   3120

Continued

```
Z1/2                        A3/9
GCCGCCCCTA GTTTTGTTTC GTGTGCTATT GCTAAAGCTG GTAAAGGACT TCTTGAAGGT ACGTTGCAGG CTGGCACTTC TGCCGTTTCT GATAAGTTGC TTGATTTGGT TGGACTTGGT
      3130       3140       3150       3160       3170       3180       3190       3200       3210       3220       3230       3240
                                                                                                            R2/6a
GGCAAGTCTG CCGCTGATAA AGGAAAGGAT ACTCGTGATT ATCTTGCTGC TGCATTTCCT GAGCTTAATG CTTGGGAGCG TGCTGGTGCT GATGCTTCCT CTGCTGGTAT GGTTGACGCC
      3250       3260       3270       3280       3290       3300       3310       3320       3330       3340       3350       3360
                                       A9/12d
Y4/1    A12d/7c                                                        F14a/12
GGATTTGAGA ATCAAAAAGA GCTTACTAAA ATGCAACTGG ACAATCAGAA AGAGATTGCC GAGATGCAAA ATGAGACTCA AAAAGAGATT GCTGGCATTC AGTCGGCGAC TTCACGCCAG
      3370       3380       3390       3400       3410       3420       3430       3440       3450       3460       3470       3480
F4/14a                                          F12/10
AATACGAAAG ACCAGGTATA TGCACAAAAT GAGATGCTTG CTTATTC-AC AGAAGGAGTC TACTGCTGCG TTGCGTCTAT TATGGAAAAC ACCAATCTTT CCAAGCAACA GCAGGTTTCC
      3490       3500       3510       3520       3530       3540       3550       3560       3570       3580       3590       3600
                           A7c/8                              R6a/4            F10/15
H2/9b
GAGATTATGC GCCAAATGCT TACTCAAGCT CAAACGGCTG GTCAGTATTT TACCAATGAC CAAATCAAAG AAATGACTCG CAAGGTTAGT GCTGAGGTTG ACTTAGTTCA TCAGCAAACG
      3610       3620       3630       3640       3650       3660       3670       3680       3690       3700       3710       3720
F15/5c     M2/5            H9b/1                                                                        A8/14
CAGAATCAGC GGTATGGCTC TTCTCATATT GGCGCTACTG CAAAGGATAT TTCTAATGTC GTCACTGATG CTGCTTCTGG TGTGGTTGAT ATTTTTCATG GTATTGATAA AGCTGTTGCC
      3730       3740       3750       3760       3770       3780       3790       3800       3810       3820       3830       3840
                    A14/7b
GATACTTGGA ACAATTTCTG AAAGACGCGT AAGCTGATG GTATTGGCTC TAATTTGTCT AGGAAATAAC CGTCAGGATT GACACCCTCC CAATTGTATG TTTTCATGCC TCCAAATCTT
      3850       3860       3870       3880       3890       3900       3910       3920       3930       3940       3950       3960
                                                     End H I                                            mRNA start I
                                                                                                    T1/6
GGAGGCTTTT TTATGGTTCG TTCTTATTAC CCTTCTGAAT GTCACGCTGA TTATTTTGAC TTTGAGCGTA TCGAGGCTCT TAAACCTGCT ATTGAGGCTT GTGGCATTTC TACTCTTTCT
      3970       3980       3990       4000       4010       4020       4030       4040       4050       4060       4070       4080
A start I                                                                          T6/2        Q2/3a       R4/3     Z2/6b
                         A7b/7a      M5/8   F5c/3
CAATCCCCAA TGCTTGGCTT CCATAAGCAG ATGGATAACC GCATCAAGCT CTTGGAAGAG ATTCTGTCTT TTCGTATGCA GGGCGTTGAG TTCGATAATAG TGTTGACGGC
      4090       4100       4110       4120       4130       4140       4150       4160       4170       4180       4190       4200
```

```
CATAAGGCTG CTTCTGACGT TCGTGATGAG TTTGTATCTG TTACTGAGAA GTTAATGGAT GAATTGGCAC AATGCTACAA TGTGCTCCCC CAACTTGATA TTAATAACAC TATAGACCAC
  4210       4220       4230       4240       4250       4260       4270       4280       4290       4300       4310       4320

                                   M8/6                  A7a/4
CGCCCGAAG GGGACGAAAAATGGTTTTTA GAGAACGAGAAGACGGTTAC GCAGTTTTGC AAGCTGGCTG CTGAACGCCC TCTTAAGGAT ATTCCGCATG AGTATAATTA CCCCAAAAAG
  4330       4340       4350       4360       4370       4380       4390       4400       4410       4420       4430       4440

                                   Z6b/6a
AAAGGTATTA AGGATGAGTG TTCAAGATTG CTGGAGGCCT CCACTAAGAT ATCGCGTAGA GGCTTTGCTA TTCAGCGTTT GATGAATGCA ATGCGACAGG CTCATGCTGA TGGTTGGTTT
  4450       4460       4470       4480       4490       4500       4510       4520       4530       4540       4550       4560

F3/5a
ATCGTTTTTG ACACTCTCAC GTTGGCTGAC GACCGATTAG AGGCGTTTTA TGATAATCCC AATGCTTTGC GTGACTATTT TCGTGATATT GGTCGTATGG TTCTTGCTGC CGAGGGTCGC
  4570       4580       4590       4600       4610       4620       4630       4640       4650       4660       4670       4680

                                              A4/11      Z6a/9  M6/1
AAGGCTAATG ATTCACACGC CGACTGCTAT CAGTATTTTT GTGTGCCTGA GTATGGTACA GCTAATGGCC GTCTTCATTT CCATGCGGTG CACTTTATGC GGACACTTCC TACAGGTAGC
  4690       4700       4710       4720       4730       4740       4750       4760       4770       4780       4790       4800

R3/8                                          A11/10     Z9/10
GTTGACCCTA ATTTTGGTCG TCGGGTACGC AATCGCCGCC AGTTAAATAG CTTGCAAAAT ACGTGGCCTT ATGGTTACAG TATGCCCATC GCAGTTCGCT ACACGCAGGA CGCTTTTTCA
  4810       4820       4830       4840       4850       4860       4870       4880       4890       4900       4910       4920

Z10/3            Q3a/1      A10/12b                                  A17/12a    F5a/6             R8/5
CGTTCTGGTT GGTTGTGGCC TGTTGATGCT AAAGGTGAGC CGCTTAAAGC TACCAGTTAT ATGGCTGTTG GTTTCTATGT GGCTAAATAC GTTAACAAAA AGTCAGATAT GGACCTTGCT
  4930       4940       4950       4960       4970       4980       4990       5000       5010       5020       5030       5040

A12b/15b              A15b/17
GCTAAAGGTC TAGGAGCTAA AGAATGGAAC AACTCACTAA AAACCAAGCT GTCGCTACTT CCCAAGAAGC TGTTCAGAAT CAGAATGAGC CGCAACTTCG GGATGAAAAT GCTCACAATG
  5050       5060       5070       5080       5090       5100       5110       5120       5130       5140       5150       5160
           B start 1

H1/15                 A12a/5                            H15/8b     T2/7
ACAAATCTGT CCACGGAGTG CTTAATCCAA CTTACCAAGC TGGGTTACGA CGCGACGCCG TTCAACCAGA TATTGAAGCA GAACGCAAAA AGAGAGATGA GATTGAGGCT GGGAAAAGTT
  5170       5180       5190       5200       5210       5220       5230       5240       5250       5260       5270       5280

                                                                              P1/1
ACTGTAGCCG ACGTTTTGGC GGCGCAACCT GTGACGACAA ATCTGCTCAA ATTTATGCGC GCTTCGATAA AAATGATTGG CGTATCCAAC CTGCA
  5290       5300       5310       5320       5330       5340       5350       5360       5370  5375
```

APPENDIX II
Experimental Procedure for Plus and Minus Method
This procedure uses ϕX DNA as an example.

ANNEALING

Add 10 μl of restriction fragment prepared as described under "Isolation of Restriction Fragments" (approximately 1 pmol), 2.5 μl of + or − strand at an A_{260} of 15 (approximately 0.7 pmol), and 2.5 μl of 0.5 M NaCl, 100 mM Tris-Cl, pH 7.5, 50 mM MgCl$_2$, and 5 mM dithiothreitol. Seal in a capillary tube and place in a boiling water bath for 3 min. Incubate the tube at 67°C for 15–30 min.

INITIAL INCORPORATION

Total volume 25 μl. Dry in vacuo 10–20 μCi of [α-^{32}P]deoxyribonucleoside triphosphate ($\approx$100 Ci/mmol, New England Nuclear) in a siliconized test tube (1/2 × 2 inch). Add:

1. 15 μl of annealed DNA
2. 2.5 μl of each of the other three triphosphates (unlabeled) to a final concentration of 0.05 mM each
3. 1.5 μl of 10 × DNA polymerase mix (0.2 M Tris-Cl, pH 7.5, 0.1 M MgCl$_2$, 10 mM dithiothreitol)
4. 1 μl of 3 units/μl of DNA polymerase I (Boehringer Mannheim GmbH, grade I).

Incubate at 0°C (i.e., in an ice water bath), removing half of the solution at 1 min and the other half at 3 min. Terminate the reaction by adding the aliquots to 25 μl of H$_2$O-25 μl of water-saturated redistilled phenol in a 1-ml "reactivial" (Pierce Chemical Company, Rockfold, Illinois) or "microvial" (Camlab, Cambridge, England). If longer extensions are required, use longer incorporation times.

Mix on a vortex mixer and extract the phenol twice with ether, removing the last traces of phenol by a jet of air. There is no need to separate the phenol and aqueous layers by centrifugation.

REMOVAL OF EXCESS OF TRIPHOSPHATES

Remove the excess of triphosphates before the plus and minus reactions are carried out by passage over a Sephadex G-100 column made from a 1-ml plastic serological pipette (Falcon Plastics, Oxnard, California). Elute with 2.5 mM Tris-Cl, pH 7.9, 0.05 mM EDTA, accomplished within a few minutes. Monitor the column with a minimonitor, g–m meter type 5.10 (Mini-Instruments Ltd., London, England), and collect the fast peak of incorporated material in a few drops (200–300 μl). This is well separated from the slower peak of triphosphates. Take the sample to dryness in vacuo in a siliconized test tube (1/2 × 2). When monitored, the dry sample should give a reading of at least 100 cps through the glass tube.

PLUS AND MINUS REACTIONS

Dissolve sample in 25 μl of H$_2$O. Pipette eight 2-μl aliquots for the plus and minus reactions into the ends of drawn out capillary tubes (3) with the use of graduated 1-5-μl micropettes (Clay Adams, New Jersey). For each plus and minus reaction, mix 2-μl of sample with 2 μl of the appropriate mix and 1 μl of T4 DNA polymerase. Although T4 DNA polymerase is necessary for the plus system, it does not have a 5'-exonuclease activity and so can be used for the minus system as well. Alternatively, for the minus system, DNA polymerase nach Klenow (Boehringer Mannheim) or a mutant form of DNA polymerase (64), both lacking 5'-exonuclease activity, can be used. This is important because the polymerase is still present during digestion with the restriction enzyme and the method relies on the released product having an intact 5' end.

Minus mixes contain 17 mM Tris-Cl, pH 7.5, 17 mM MgCl$_2$, 17 mM dithiothreitol, and three triphosphates at 0.025 mM, whereas plus mixes contain, in the same buffer, a single triphosphate at 0.5 mM. Place the capillary tubes with the reaction mixtures in the tips tip down in a siliconized test tube and incubate at 37°C for 30 min.

RESTRICTION ENZYME CLEAVAGE

Release the newly synthesized DNA from the primer by digestion with the appropriate restriction enzyme. The conditions for digestion vary according to the activity of the particular enzyme preparation and to the amount of contaminating exonuclease. Normally, 1 μl of most restriction enzyme preparations is enough to cleave the small amounts of DNA in each plus or minus reaction. Mix with each plus and minus reaction and continue incubation at 37°C for a further 30 min.

At the same time, cleave two 4-μl aliquots of the initial incorporation with 1 μl of restriction enzyme and 1 μl of 33 mM Tris-Cl, pH 7.5, 33 mM MgCl$_2$, 33 mM dithiothreitol. These reactions, containing every possible chain length, are fractionated alongside the plus and minus reactions as markers. Commonly referred to as a zero system, it is particularly useful in elucidating runs (sequences) of the same nucleotide.

Stop all 10 reactions by the addition to each of 20 μl of deionized formamide containing 10 mM EDTA, 0.3% xylene cyanol FF, and 0.3 % bromphenol blue. Deionize formamide by stirring 100 ml of formamide with 5 g of Amberlite MB-1 (mixed bed ion exchange resin) for 1 hr and then remove the resin by filtration. Place the samples, each in a siliconized test tube, in a boiling water bath for 3 min and load onto the acrylamide gel.

SINGLE SITE RIBOSUBSTITUTION

The annealing procedure is the same as above except that a different buffer is used (2.5 μl of 0.5 M NaCl, 0.01 M dithiothreitol, 0.2 M Tris-Cl, pH 7.5).

After annealing at 67°C, add 2.5 μl of 0.01 M MnCl$_2$, 1.0 μl of 5 mM rCTP, 4.5 μl of H$_2$O, and 2.0 μl of DNA polymerase I. Incubate at 0°C for 10 min.

Then add 25 μl of "Flood mix," i.e., 0.02 M MgCl$_2$, 0.02 M dithiothreitol, 0.25 M dCTP, 10–20 μCi of [α-^{32}P]dNTP (e.g., dA*TP), and 0.1 mM other dNTPs (e.g., dTTP, dGTP).

Incubate at 0°C for 1 min and 3 min (or as required) and continue as above.

Denature the plus and minus reactions and cleave with pancreatic ribonuclease as follows. Seal each sample (5 μl) in a capillary tube and place in a boiling water bath for 3 min; immediately place in iced water. Add 1 μl of 10 mg/ml of pancreatic ribonuclease and incubate at 37°C for 30 min.

Add the dye/EDTA/formamide mixture and denature the samples as described above, prior to electrophoresis.

An alternative method is to cleave with alkali by adding 1 μl of 2 N NaOH to each plus and minus reaction. Seal the samples in a capillary tube and incubate at 100°C for 20 min. Add the dye/EDTA/formamide mixture and immediately apply the sample to the gel.

If the sample cannot be applied immediately, neutralize the alkali with 1 μl of 2 N acetic acid and add the dye mixture. Before running, denature the samples at 100°C for 3 min.

The method (N. L. Brown, manuscript in preparation) has been used for all four triphosphates. rU and rC can be cleaved by pancreatic ribonuclease, rG by T_1 ribonuclease, or alkali for all four.

ACRYLAMIDE GEL ELECTROPHORESIS

It is essential that the oligonucleotides be fractionated according to size and that there are no anomalous migratory effects due to secondary structure. Denature samples before loading as described above and fractionate on denaturing 12% acrylamide, 7 M urea gels. Run the gels at high current, which has the effect of heating the gel in order to keep the oligonucleotides denatured during electrophoresis.

Prepare 12% acrylamide, 7 M urea slab gels according to Maniatis et al. (134) with TBE buffer (0.09 M Tris-borate, pH 8.3, 2.5 mM EDTA (135)). Run gels, approximately 40 × 20 cm and 1.5 mm thick, in a slab gel apparatus (Model RGA 505, Raven Scientific Ltd., Haverhill, England) which provides direct contact between the gel and both buffer reservoirs. In the construction of the gel mold, it is convenient to tape the plates and side spacers with vinyl electrophoresis tape (Universal Scientific, London) thus avoiding the use of grease. Pour the acrylamide solution into the mold, which is held slightly off the horizontal and allowed to polymerize in this position after insertion of the slot former. Slot widths of about 1 cm give good sharp bands for a 25-μl sample. As polymerization takes place, the gel plates become warm. When cool again, remove the slot former carefully, allowing air to enter the slots. Remove the bottom tape and place the gel in the apparatus. Before loading the samples, flush out the slots to remove concentration of urea. Gradually increase the current in the first 10 min to 35–40 mamp.

The blue marker dyes correspond to DNA chain lengths of approximately 15 nucleotides (bromphenol blue) and 55 nucleotides (xylene cyanol FF). A short fractionation separates products in the 15–75-nucleotide range and takes about 3 1/2 hr. Longer runs, where the xylene cyanol FF marker migrates near the bottom of the gel, are required to fractionate oligonucleotides in the 55–150 range. This is normally complete in 9 hr.

After completion of electrophoresis, fix the gel left on one of the glass plates in 10% acetic acid for about 5 min or until the bromphenol blue marker turns green. Wash it in water and wrap in plastic film (Saranwrap, Dow Chemical Company). Autoradiograph the gel.

12% ACRYLAMIDE GEL RECIPES

The following stock solutions are required:

1. 10 × TBE buffer: 108 g of Tris base, 9.3 g of EDTA, and 55 g of boric acid. Make up to 1 l with H_2O, pH 8.3.
2. 1.6% ammonium persulphate made fresh every 2 weeks.
3. 38% acrylamide (BDH, specially purified for electrophoresis), 2% NN′-methylene bisacrylamide, in H_2O, deionized for 45 min by stirring gently with Amberlite MB-1 and then filtered.

For each gel, 100 ml of the following solution is needed: 10 ml of solution 1, 3.2 ml of solution 2, 30 ml of solution 3, and 42 g of urea (ultrapure, Schwarz/Mann), made up to 100 ml with H_2O. Polymerization starts with the addition of 100 μl of TEMED (N,N,N′,N′-tetramethylethylenediamine).

Note added in proof: The following corrections to the φX174 sequence (18) given in Appendix I have been made (F. Sanger, personal communication). The whole sequence can now be considered fully confirmed.

Delete 1298–1300
Insert A between 1390–1391
Delete GCG at 1400–02, replace by ACAAC
Delete 1421–1423
Replace A by G at 1653
T at 1657
Replace A by T at 1687
Replace G by T at 1870
Replace A by C at 2165
Insert C between 2188–2189
Insert C between 2191–2192
Insert A between 2210–2211
Insert C between 2230–2231
Insert AGG between 2216–2217
Delete CCTC at 2220–2223, replace by AATTTA
Insert CG between 2303–2304
Replace C by T at 2413
Replace A by T at 2608
Replace G by A at 2723
Replace C by G at 2794
Replace C by T at 2806
Replace C by T at 2809
Replace GT by TG at 3141–2
Replace T by G at 3150
Delete T at 3526
A at 3528
Insert C between 3547–3548
Insert CGC between 4380–4381
Replace A by T at 4487
Replace T by A at 4490

REFERENCES

1. Sanger, F., Brownlee, G. G., and Barrell, B. G. (1965). J. Mol. Biol. 13:373.
2. Brownlee, G. G., and Sanger, F. (1969). Eur. J. Biochem. 11:395.
3. Barrell, B. G. (1971). In G. L. Cantoni and D. R. Davies (eds.), Procedures in Nucleic Acid Research, Vol. 2, p. 751. New York: Harper and Row.
4. Adams, J. M., Jeppesen, P. G. N., Sanger, F., and Barrell, B. G. (1969). Nature (Lond.) 223:1009.
5. Steitz, J. A. (1969). Nature (Lond.) 224:964.
6. Sadowski, P. D., and Hurwitz, J. (1969). J. Biol. Chem. 244:6192.
7. Burton, K., and Petersen, G. B. (1960). Biochem. J. 75:17.
8. Bretscher, M. S. (1969). Cold Spring Harbor Symp. Quant. Biol. 34:651.
9. Sinsheimer, R. L. (1968). Prog. Nucleic Acid Res. Mol. Biol. 8:115.
10. Benbow, R. M., Hutchison, C. A., III, Fabricant, J. D., and Sinsheimer, R. L. (1971). J. Virol. 7:549.

11. Benbow, R. M., Mayol, R. F., Picchi, J. C., and Sinsheimer, R. L. (1972). J. Virol. 10:99.
12. Benbow, R. M., Zuccarelli, A. J., Davis, G. C., and Sinsheimer, R. L. (1974). J. Virol. 13:898.
13. Linney, E., and Hayashi, M. (1974). Nature (Lond.) 249:345.
14. Denhardt, D. T. (1975). CRC Crit. Rev. Microbiol. 4:161.
15. Denhardt, D. T. Compr. Virol. VII. In press.
16. Murray, K. (1974). *In* K. Burton (ed.), MTP International Review of Science, Biochemistry Series One, Volume 6. Butterworths, London.
17. Sanger, F., and Coulson, A. R. (1975). J. Mol. Biol. 94:441.
18. Sanger, F., Air, G. M., Barrell, B. G., Brown, N. L., Coulson, A. R., Fiddes, J. C., Hutchison, C. A., III, Slocombe, P. M., and Smith, M. (1977). Nature (Lond.) 265:687.
19. Hall, J. B., and Sinsheimer, R. L. (1963). J. Mol. Biol. 6:115.
20. Ling, V. (1972). Proc. Natl. Acad. Sci. USA 69:742.
21. Harbers, B., Delaney, A. D., Harbers, K., and Spencer, J. H. (1976). Biochemistry 15:407.
22. Chadwell, H. A. J. Mol. Biol. In press.
23. Sadowski, P. D., and Bakyta, I. (1972). J. Biol. Chem. 247:405.
24. Bernardi, A., Maat, J., de Waard, A., and Bernardi, G. (1976). Eur. J. Biochem. 66:175.
25. Ling, V. (1971). FEBS Lett. 19:50.
26. Ziff, E. B., Sedat, J. W., and Galibert, F. (1973). Nature (New Biol.) 241:34.
27. Robertson, H. D., Barrell, B. G., Weith, H. L., and Donelson, J. E. (1973). Nature (New Biol.) 241:38.
28. Robertson, H. D. (1975). J. Mol. Biol. 92:363.
29. Barrell, B. G., Weith, H. L., Donelson, J. E., and Robertson, H. D. (1975). J. Mol. Biol. 92:377.
30. Air, G. M., and Bridgen, J. (1973). Nature (New Biol.) 241:40.
31. Sedat, J., Ziff, E. B., and Galibert, F. (1976). J. Mol. Biol. 107:391.
32. Air, G. M. (1976). J. Mol. Biol. 107:433.
33. Blackburn, E. H. (1976). J. Mol. Biol. 107:417.
34. Air, G. M., Blackburn, E. H., Coulson, A. R., Galibert, F., Sanger, F., Sedat, J. W., and Ziff, E. B. (1976). J. Mol. Biol. 107:445.
35. Blackburn, E. H. (1975). J. Mol. Biol. 93:367.
36. Gilbert, W., and Maxam, A. (1973). Proc. Natl. Acad. Sci. USA 70:3581.
37. Wu, R., and Kaiser, A. D. (1968). J. Mol. Biol. 35:523.
38. Wu, R., and Taylor, E. (1971). J. Mol. Biol. 57:491.
39. Wu, R. (1972). Nature (New Biol.) 236:198.
40. Oertel, W., and Schaller, H. (1972). FEBS Lett. 27:316.
41. Padmanabhan, R., Padmanabhan, R., and Wu, R. (1972). Biochem. Biophys. Res. Commun. 48:1295.
42. Goulian, M., Goulian, S. H., Codd, E. E., and Blumenfield, A. Z. (1973). Biochemistry 12:2893.
43. Sanger, F., Donelson, J. E., Coulson, A. R., Kössel, H., and Fischer, D. (1973). Proc. Natl. Acad. Sci. USA 70:1209.
44. Schott, H., Fischer, D., and Kössel, H. (1973). Biochemistry 12:3447.
45. Berg, P., Fancher, H., and Chamberlin, M. (1963). *In* Symposium on Informational Macromolecules, p. 467. Academic Press, New York.
46. Salser, W., Fry, K., Brunk, C., and Poon, R. (1972). Proc. Natl. Acad. Sci. USA 69:238.
47. Van de Sande, J. H., Loewen, P. C., and Khorana, H. G. (1972). J. Biol. Chem. 247:6140.

48. Schott, H. (1974). Makromol. Chem. 175:1683.
49. Donelson, J. E., Barrell, B. G., Weith, H. L., Kössel, H., and Schott, H. (1975). Eur. J. Biochem. 58:383.
50. Edgell, M. H., Hutchison, C. A., III, and Sclair, M. (1972). J. Virol. 9:574.
51. Middleton, J. H., Edgell, M. H., and Hutchison, C. A., III. (1972). J. Virol. 10:42.
52. Lee, A. S., and Sinsheimer, R. L. (1974). Proc. Natl. Acad. Sci. USA 71:2882.
53. Maniatis, T., Ptashne, M., Barrell, B. G., and Donelson, J. (1974). Nature (Lond.) 250:394.
54. Roberts, R. J. (1976). Crit. Rev. Biochem. 4:123.
55. Maxam, A. M., and Gilbert, W. (1977). Proc. Natl. Acad. Sci. USA 74:560.
56. Jeppesen, P. G. N. (1974). Anal. Biochem. 58:195.
57. Galibert, F., Sedat, J., and Ziff, E. B. (1974). J. Mol. Biol. 87:377.
58. Vereijken, J. M., van Mansfeld, A. D. M., Baas, P. D., and Jansz, H. S. (1975). Virology 68:221.
59. Baas, P. D., van Heusden, G. P. H., Vereijken, J. M., Weisbeek, P. J., and Jansz, H. S. (1976). Nucleic Acids Res. 3:1947.
60. Jeppesen, P. G. N., Sanders, L., and Slocombe, P. M. (1976). Nucleic Acids Res. 3:1323.
61. Hutchison, C. A., III. Nucleic Acids Res. In press.
62. Maniatis, T., Jeffery, A., and Klied, D. G. (1975). Proc. Natl. Acad. Sci. USA 72:1184.
63. Southern, E. M. (1975). J. Mol. Biol. 98:503.
64. Heijneker, H. L., Ellens, D. J., Tjeerde, R. H., Glickman, B. W., van Dorp, B., and Pouwels, P. H. (1973). Mol. Gen. Genet. 124:83.
65. Eglund, P. T. (1971). J. Biol. Chem. 246:3269.
66. Barrell, B. G., Air, G. M., Hutchison, C. A., III. (1976). Nature (Lond.) 264:34.
67. Ling, V. (1972). J. Mol. Biol. 64:87.
68. Fiddes, J. C. (1976). J. Mol. Biol. 1:24.
69. Air, G. M., Blackburn, E. H., Sanger, F., and Coulson, A. R. (1975). J. Mol. Biol. 96:703.
70. Air, G. M., Sanger, F., and Coulson, A. R. (1976). J. Mol. Biol. 108:519.
71. Sinsheimer, R. L. (1959). J. Mol. Biol. 1:43.
72. Seigel, J. E. D., and Hayashi, M. (1967). J. Mol. Biol. 27:443.
73. Min Jou, W., Haegeman, G., Ysebaert, M., and Fiers, W. (1972). Nature (Lond.) 237:82.
74. Fiers, W., Contreras, R., Duerinck, F., Haegeman, G., Merregaert, J., Min Jou, W., Raeymaekers, A., Volckaert, G., Ysebaert, M., Van de Kerckhove, J., Nolf, F., and Van Montagu, M. (1975). Nature (Lond.) 256:273.
75. Fiers, W., Contreras, R., Duerinck, F., Haegeman, G., Iserentant, D., Merregaert, J., Min Jou, W., Molemans, F., Raeymaekers, A., Van den Berghe, A., Volckaert, G., and Ysebaert, M. (1976). Nature (Lond.) 260:500.
76. Denhardt, D. T., and Marvin, D. A. (1969). Nature (Lond.) 221:769.
77. Jazwinski, S. M., Lindeberg, A. A., and Kornberg, A. (1975). Virology 66:282.
78. Jazwinski, S. M., Lindeberg, A. A., and Kornberg, A. (1975). Virology 66:283.
79. Jazwinski, S. M., Marco, R., and Kornberg, A. (1975). Virology 66:294.
80. Weisbeek, P. J., and Van de Pol, J. H. (1970). Biochim. Biophys. Acta 224:328.
81. Hutchison, C. A., III, and Edgell, M. H. (1971). J. Virol. 8:181.
82. Farber, M. B. (1976). J. Virol. 17:1027.
83. Devillers-Thiery, A., Kindt, T., Scheele, G., and Blobel, G. (1975). Proc. Natl. Acad. Sci. USA 72:5016.
84. Burstein, Y., and Schechter, I. (1976). Biochem. J. 157:145.
85. Markert, A., and Zillig, W. (1965). Virology 25:88.
86. Hutchison, C. A., III. (1969). Ph.D. thesis, California Institute of Technology.

178   Barrell

87. Shine, J., and Dalgarno, L. (1975). Eur. J. Biochem. 57:221.
88. Steitz, J. A., and Jakes, K. (1975). Proc. Natl. Acad. Sci. USA 72:4734.
89. Van Mansfeld, A. D. M., Vereijken, J. M., and Jansz, H. S. (1976). Nucleic Acid Res. 3:2827.
90. Ravetch, J. V., Model, P., and Robertson, H. D. (1977). Nature (Lond.) 265:698.
91. Brown, N. L., and Smith, M. (1977). Nature (Lond.) 265:695.
92. Brown, N. L., and Smith, M. (1977). J. Mol. Biol. 116:1.
93. Smith, M., Brown, N. L., Air, G. M., Barrell, B. G., Coulson, A. R., Hutchison, C. A., III, and Sanger, F. (1977). Nature (Lond.) 265:702.
94. Weisbeek, P. J., Vereijken, J. M., Baas, P. D., Jansz, H. S., and van Arkel, G. G. (1976). Virology 72:61.
95. Platt, T., and Yanofsky, C. (1975). Proc. Natl. Acad. Sci. USA 72:2399.
96. Barrell, B. G., and Clark, B. F. C. (1974). Handbook of Nucleic Acid Sequences. Joynson-Bruvvers, Ltd., Oxford.
97. Dickson, R. C., Abelson, J., Barnes, W. M., and Reznikoff, W. S. (1975). Science 187:27.
98. Chen, C. Y., Hutchison, C. A., III, and Edgell, M. H. (1973). Nature (New Biol.) 243:233.
99. Hayashi, Y., and Hayashi, M. (1970). Cold Spring Harbor Symp. Quant. Biol. 35:171.
100. Vanderbilt, A. S., Borras, M. T., Germeraad, S., Tessman, I., and Tessman, E. (1972). Virology 50:171.
101. Smith, L. H., and Sinsheimer, R. L. (1976). J. Mol. Biol. 103:681.
102. Clements, J. B., and Sinsheimer, R. L. (1975). J. Virol. 15:151.
103. Smith, L. H., Grohmann, K., and Sinsheimer, R. L. (1974). Nucleic Acids Res. 1:1521.
104. Grohmann, K., Smith, L. H., and Sinsheimer, R. L. (1975). Biochemistry 14:1951.
105. Smith, L. H., and Sinsheimer, R. L. (1976). J. Mol. Biol. 103:699.
106. Hayashi, M., Fujimura, F. K., and Hayashi, M. (1976). Proc. Natl. Acad. Sci. USA 73:3519.
107. Axelrod, N. (1976). J. Mol. Biol. 108:771.
108. Pribnow, D. (1975). J. Mol. Biol. 92:363.
109. Schaller, H., Gray, C., and Herrman, K. (1975). Proc. Natl. Acad. Sci. USA 72:737.
110. Rosenberg, M., de Crombrugghe, B., and Musso, R. (1976). Proc. Natl. Acad. Sci. USA 73:717.
111. Axelrod, N. (1976). J. Mol. Biol. 108:753.
112. Bertrand, K., Korn, L., Lee, F., Platt, T., Squires, C. L., Squires, C., and Yanofsky, C. (1975). Science 189:22.
113. Sugimoto, K., Sugisaki, H., Okamoto, T., and Takanami, M. (1977). J. Mol. Biol. 111:487.
114. Smith, L. H., and Sinsheimer, R. L. (1976). J. Mol. Biol. 103:711.
115. Burgess, A. B., and Denhardt, D. T. (1969). J. Mol. Biol. 44:377.
116. Godson, G. N. (1971). J. Mol. Biol. 57:541.
117. Gelfand, D. H., and Hayashi, M. (1969). Proc. Natl. Acad. Sci. USA 63:135.
118. Razin, A. (1973). Proc. Natl. Acad. Sci. USA 70:3773.
119. Henry, T. J., and Knippers, R. (1974). Proc. Natl. Acad. Sci. USA 71:1549.
120. Baas, P. D., Jansz, H. S., and Sinsheimer, R. L. (1976). J. Mol. Biol. 102:633.
121. Johnson, P. H., and Sinsheimer, R. L. (1974). J. Mol. Biol. 83:47.
122. Eisenberg, S., Harbers, B., Hours, C., and Denhardt, D. T. (1975). J. Mol. Biol. 99:107.
123. Ikeda, J., Yudelevich, A., and Hurwitz, J. (1976). Proc. Natl. Acad. Sci. USA 73:2669.

124. Friedman, J., and Razin, A. (1976). Nucleic Acids Res. 3:2665.
125. Lee, A. S., and Sinsheimer, R. L. (1974). J. Birol. 14:872.
126. Wiebers, J. L. (1976). Nucleic Acids Res. 3:2959.
127. Baas, P. D., and Jansz, H. S. (1972). J. Mol. Biol. 63:569.
128. Weisbeek, P. J., and Van Arkel, G. A. (1976). Virology 72:72.
129. Ehresmann, C., Skegler, P., Mackie, G. A., Zimmermann, R. A., Ebel, J. P., and Fellner, P. (1975). Nucleic Acids Res. 2:265.
130. Sinsheimer, R. L. (1959). J. Mol. Biol. 1:37.
131. Freifelder, D., Kleinschmidt, A. K., and Sinsheimer, R. L. (1964). Science 146:254.
132. Bartok, K., Harbers, B., and Denhardt, D. T. (1975). J. Mol. Biol. 99:93.
133. Tinoco, I., Borer, P. N., Dengler, B., Levine, M. D., Uhlenbeck, O. C., Crothers, D. M., and Gralla, J. (1973). Nature (New Biol.) 246:40.
134. Maniatis, T., Jeffrey, A., and van de Sande, H. (1975). Biochemistry 14:3787.
135. Peacock, A. C., and Dingman, C. W. (1968). Biochemistry 7:668.

International Review of Biochemistry
Biochemistry of Nucleic Acids II, Volume 17
Edited by B. F. C. Clark
Copyright 1978   University Park Press   Baltimore

6
Chromatin Structure

JEAN O. THOMAS
University of Cambridge, Cambridge, England

The problem of packing long DNA molecules in chromosomes of both eukaryotes and prokaryotes is apparently solved in stages. The first stage is achieved by histones in eukaryotes, and possibly by polyamines in prokaryotes (the small basic protein in the *Escherichia coli* chromosome (1) is not present in sufficient amount to perform the same function as that of histones in eukaryotes)[1]. Despite the difference in structural components between eukaryotic

---

[1]For further discussion, see "Note Added in Proof" at the end of this chapter.

and prokaryotic chromosomes, possible similarities in their primary levels of organization are revealed in the electron microscope, which in both cases shows a beaded fiber 100 Å in diameter. In eukaryotes the "beads" are well defined units of histone and DNA, stable in solution, and the subject of most of this chapter; in prokaryotes the beaded structure (2) has not so far proved amenable to study in solution. Further folding of the primary fiber must occur in both eukaryotic and prokaryotic chromosomes, but evidence on the nature of this folding is scant. Recent studies of the *E. coli* chromosome by electron microscopy, sedimentation, and other physical methods (reviewed elsewhere (3)), reveal large supercoiled loops of DNA whose maintenance may require RNA; the manner of organization of loops is not known, nor is the relationship between a loop as a unit of packing and the gene, or operon, as a unit of function. In the eukaryotic chromosome, the primary fiber may be coiled on itself to give a thicker fiber which may then be further folded, possibly also into loops (see below).

This chapter concentrates on the current status of structural studies of the primary fiber in eukaryotes. It discusses the main lines of evidence in some detail. Considerations of time and space and the patience of reader and author alike have precluded a comprehensive survey; I hope that anyone who feels underrepresented will be tolerant. A wide ranging survey (4) of the field up to 1974 covers the literature up to that time.

THE COMPLEX OF DNA AND HISTONES

Histones

The primary structure and other aspects of histones have recently been reviewed in some detail (4, 5); therefore, only a brief summary is given here. All eukaryotes contain five types of histone: two arginine-rich histones, H3 (MW 15,273) and H4 (MW 11,236); two lysine-rich histones, H2A (MW 13,960) and H2B (MW 13,774); and a very lysine-rich histone, H1 (MW ca. 21,000). Additional cell-specific histones occur in certain cases, for example, in the erythrocytes of nonmammalian vertebrates (see ref. 4) and in some spermatogenous cells (6, 7). Because the weight of the five histones together is equal to the weight of DNA with which they are combined, a structural role for histones in chromatin is strongly favored. The remarkable conservation of amino acid sequence of the arginine-rich histones among higher eukaryotes (two differences between calf thymus and pea seedling H4, and four differences for H3) suggests that the structural role of these two histones has remained constant throughout evolution. Such stringent conservation means that "surface" residues, the most susceptible to alteration during evolution in other proteins, are as important in the histones as "buried" residues that ensure correct folding. One obvious explanation of this feature is that the surface interacts extensively and specifically with other molecules—histones or DNA or both. The lysine-rich histones H2A and H2B are also conserved, although less so than the arginine-rich histones. In contrast, the very lysine-rich histone H1 shows much greater interspecies variation, which suggests a role for H1 that differs fundamentally from that of the other four histones—the four "main" histones. (The rate of evolutionary change in H1 is still relatively low, however, being about the same as that of cytochrome c, a relatively conserved protein (8).)

Until recently, histones were routinely extracted with acid and fractionated by ion exchange chromatography. The availability of chemically pure material enabled complete amino acid sequences to be determined, but the physical properties of native histones were revealed only later, as a consequence of milder extraction and fractionation procedures. In the method of van der Westhuyzen and von Holt (9), histones extracted in 2 M NaCl at pH 5 are partially fractionated by a combination of gel filtration and ammonium sulfate precipitation. The arginine-rich pair of histones so obtained exists as a tetramer (10, 11), $(H3)_2(H4)_2$; the lysine-rich pair as a complex which may be a dimer, H2A-H2B, or a polymer (10, 12); and H1 as a monomer (10). The arginine-rich tetramer was characterized in terms of its subunit structure and molecular weight by sedimentation analysis at pH 5 (10, 11), and by intramolecular cross-linking at pH 8 with the bifunctional amino group reagent dimethylsuberimidate followed by analysis of the products in sodium dodecyl sulfate (SDS)-polyacrylamide gels (10). Although histones extracted from chromatin with strong acid must be denatured, the damage appears to be largely reversible in suitable buffers because H2A and H2B associate to form a dimer (12) and H3 and H4 recombine to give a tetramer (13). Chemical cross-linking revealed no interactions between the H3-H4 tetramer and H2A-H2B complex prepared by salt extraction (10) but weak interactions between acid-extracted lysine-rich and arginine-rich histones were detected by physical methods (13).

A Model for Chromatin Structure

The most significant experimental fact about chromatin structure up to the 1970s was the x-ray evidence for a repeating unit (see also under "Low Angle x-Ray and Neutron Scattering: Evidence for a Structural Repeat"). Both nuclei and chromatin give a low angle x-ray scattering pattern with bands at about 110, 55, 33, and 27 Å, interpreted at that time as arising from a DNA superhelix of pitch 120 Å which was uniformly coated with histones (14). Biochemical evidence for a repeating element of structure emerged with the discovery (15) that an endogenous nuclease in rat liver nuclei made double strand cuts in the DNA at regular intervals. The nature of the repeating unit has been proposed by Kornberg (16) in a general model for chromatin structure which has since found considerable experimental support. The reasoning that led to the proposal was as follows (see ref. 16 for references). Because both the lysine-rich and the arginine-rich histones, but not H1, are required to regenerate the full low angle x-ray pattern in reconstitution experiments with DNA (but see also under "Partial Reconstitution of Chromatin Structure with Selected Histones"), it was assumed that both types of histone, but not H1, are contained in the repeat unit. The discovery of specific lysine-rich and arginine-rich complexes in histones extracted by mild methods (see above) led to the proposal that these oligomers form the basis of the repeat unit, and in particular, that each repeat unit contains one arginine-rich tetramer. Because the four main histones occur in roughly equimolar amounts, for every $(H3)_2(H4)_2$ tetramer there must be two molecules each of H2A and H2B. The stoichiometry of histone and DNA corresponds, for 2 molecules of each histone, to 200 base pairs of DNA. The repeat unit was thought to be roughly spherical, or beadlike, and about 100 Å in diameter on the basis of electron micrographs showing chromatin fibers about 100 Å in diameter (17) and the x-ray evidence for a 100 Å repeat along the fiber axis. An additional feature of the model concerned the relative location of DNA and protein in the bead. Because the histone oligomers are not markedly different from globular proteins, the DNA was thought to be largely on the outside of a histone core.

In summary, the complex of DNA and histones is viewed as a string of closely packed beads, each containing eight histones (two each of H3, H4, H2A, and H2B) and 200 base pairs of DNA; histone H1 is not an integral component of the repeat unit but is associated with it, most probably with DNA on the surface. The folding of DNA in such a structure, which reduces its length by a factor of about 7 (200 base pairs of B-form DNA, 680 Å long when fully extended, are reduced to 100 Å), represents the first level of packing of DNA in chromosomes. The beads are now called nucleosomes (18) and a string of closely packed beads is called a nucleofilament (19).

The beaded character of the chromatin fiber was first visualized in electron micrographs of nuclei lysed in water and centrifuged onto grids (20–22). The chromatin fibers comprised irregularly spaced beads connected by free DNA, with occasional runs of closely spaced beads. While the origin of the extended state, in which beads alternate with free DNA, is still not clear, micrographs of chromatin showing extensive arrays of closely spaced beads have been obtained (18, 23) and are believed to approximate more closely the state of chromatin in vivo (see below).

Formulation of a model for chromatin structure (16) provided a framework for future work. Experiments were devised to test the validity of the model, in particular its quantitative features. The results are discussed in later sections but are summarized briefly here. First, nuclease digestion studies in which cuts are made between nucleosomes confirmed that the length of DNA in a nucleosome is about 200 base pairs (24). Secondly, chemical cross-linking of histones in chromatin with dimethylsuberimidate confirmed that the repeat unit contains a histone octamer and indicated that neighboring octamers are in contact with each other (less than about 11 Å apart) (25). Thirdly, electron microscopy of the minichromosome of SV40 (26), of cellular chromatin (18), and of purified chromatin fragments (27) confirmed a packing ratio of between 5 and 7, strongly supporting a compact structure (as opposed to an extended one with free DNA between beads, which would have a much lower packing ratio (28)). Further evidence for a compact structure has come from determination of the mass per unit length of the fiber by low angle x-ray scattering (29); support for the assignment of the DNA to the outside of the histone core has come from its accessibility to DNase I (30) and from the demonstration that the radius of gyration of the DNA in a single nucleosome is greater than that of protein (31, 32).

Essentially all of the DNA in the nucleus is thought to be packaged in the same manner with histone. The tight binding of histone to DNA may be modulated in vivo, probably by enzymic modification of histones, and such processes may be necessary to allow transcription and replication of DNA. The following account is concerned primarily with evidence on the structure of chromatin, most of it obtained from the enormous output of the last 3 years. The still nebulous area of the relation between structure and function is not discussed in comparable detail.

DIGESTION WITH MICROCOCCAL NUCLEASE

Determination of the Length of DNA in the Repeat Unit

Digestion of DNA in rat liver nuclei with an endogenous nuclease (15) provided the first biochemical evidence of regularity in chromatin structure. Analysis of the extracted DNA in a polyacrylamide gel showed fragments that were roughly multiples of a smallest size, later measured as 180–230 base pairs (33). Another enzyme used in early digestion studies on chromatin (34), the Ca^{2+}-dependent nuclease from *Staphylococcus aureus* (micrococcal nuclease), also cleaves the DNA to multiples of about 200 base pairs (24), and moreover, degrades essentially all of the DNA in nuclei to these multiples or to acid-soluble form. On the Kornberg model, the action of micrococcal and endogenous nuclease is explained by cleavage of the DNA as it passes from the surface of one bead to the next, from one region of tight association with histone to another.

The DNA length in the smallest fragment released by micrococcal nuclease, a nucleosome, should, in principle, be the repeat length of the DNA in the chromatin fiber. In practice, however, this is true only after very brief digestion and at low temperatures, because micrococcal nuclease has exonuclease as well as endonuclease activity and digests the ends of the DNA in liberated nucleosomes or multiples thereof. Thus, Noll and Kornberg showed that the DNA in nucleosomes liberated by brief digestion of rat liver nuclei at 37°C was about 170 base pairs long, whereas digestion at 2°C gave 200-base pair nucleosomes (35). The repeat length in mouse liver (36), for example, estimated from the unit size DNA as 170 base pairs, is an underestimate of the true value of 200 base pairs (37) obtained as described below, probably because of continued exonucleolytic attack on single nucleosomes.

In order to avoid the uncertainties that arise from studies of the DNA content of single nucleosomes, the repeat length has been determined (35) from the differences in size between the unit length DNA and higher multiples. This method eliminates the effect of exonucleolytic digestion because the unit piece of DNA and higher multiples are presumably equally susceptible to such digestion. The method gave a value of 198 ± 6 base pairs for the repeat size in rat liver nuclei, whether digested at 37°C or 2°C, despite the differences in the size of DNA from single nucleosomes and higher multiples produced at the two temperatures. In a convenient application of this method, the sizes of several multiples (up to about 6 or 8) of the unit size DNA from a digest are measured and the repeat size is obtained from the slope of a graph of band size against band number (taking the unit size as band 1) (38, 39). Other ways of overcoming the problem of overdigestion have also been used, but none seems as satisfactory or as convenient as the graphical method described above. For example, Shaw et al. (40) have attempted to compensate for the effect of exonucleolytic digestion by recording a time course of the length of DNA in multiples of nucleosomes up to four as digestion proceeds and extrapolating to zero time. This method is, however, subject to considerable inaccuracies because the slope of the graph (DNA length against time) is greatest, and most difficult to estimate precisely, as zero time is approached. In another method, favored by Chambon and his co-workers (41, 42), the DNA length is measured for a series of multiples of the unit size and is then divided by the band number (the unit size is band 1) to obtain a value for the DNA content of a single nucleosome. A set of values is obtained that increases from band 1 until a roughly constant value is reached beyond about band 4; this is taken to be the true nucleosomal repeat length in chromatin. The rationale here is that the amount of DNA lost by digestion from the two ends of a chromatin fragment represents an increasingly smaller proportion of the total DNA length in the fragment as increasingly larger multiples are considered, until, beyond 4-mer, it falls within the accuracy of estimation of DNA lengths and constant values are obtained.

Estimation of DNA lengths in digested chromatin fragments resolved in a polyacrylamide gel is done by reference to a calibration curve constructed from marker DNA fragments. Frequently used markers are the restriction fragments of the genome of ϕX174 (RF), bacteriophage PM2, or occasionally SV40. In the case of ϕX174, the complete sequence is now known (see Chapter 5), so the sizes of restriction fragments used as molecular weight markers are also precisely known. In the case of PM2 marker fragments, sizes have been assigned by reference to SV40 or other fragments which were themselves not sequenced. The dangers inherent in indirect calibrations of this sort were recently illustrated: correction of the sizes taken for PM2 fragments (ref. 43, note added in proof) brought an early value for the DNA repeat size in yeast chromatin (44) into line with a more recently determined one (38) in which Hae III and Hha I restriction fragments of ϕX174 RF DNA were used as markers. Eco R_{II} restriction fragments of mouse satellite DNA which are multiples of a unit size (45), recently measured as 248 base pairs (35, 46), have also been used as markers for measuring the size of high molecular weight DNA in polyacrylamide gels. A general note of caution should be sounded on determination of DNA sizes by comparison with markers in gels: the electrophoretic mobility of certain restriction fragments of ϕX174 RF DNA is critically dependent on running conditions in the gel, and under certain conditions the relative mobility of two fragments may be reversed (C. Hutchison and M. Smith, personal communication).

Prolonged Digestion with Micrococcal Nuclease: The Core Particle

Nucleosomes released from chromatin by brief digestion with micrococcal nuclease contain DNA of average size 170–200 base pairs (see above) which runs in a polyacrylamide gel as a broad band (width about 80 base pairs). Further digestion shortens the DNA and causes a sharpening of the leading edge of the DNA band at 140 base pairs until eventually a relatively sharp band (width about 10 base pairs) is formed which corresponds to a metastable intermediate in digestion (35). This intermediate is the nucleosome "core particle" (40) from which the most accessible DNA has been removed and in which the remaining 140 base pairs of DNA are protected by tight association with the histone octamer; H1 is no longer present (35, 40, 47, 85) but was removed during digestion of the linker DNA (see under "Location of Histone H1"). The core particle has a sedimentation coefficient of about 11 S, like the less digested nucleosome; its homogeneity with respect to DNA content makes it attractive for study by both physical and chemical methods as discussed later. (The 140 base pair core particle may well be the same as the PS-particle reported previously in extensively digested chromatin (48), although the DNA length reported for the PS-particle at that time was somewhat shorter (110 base pairs).) When digestion of chromatin is allowed to proceed beyond the formation of core particles, internal cuts are made in the bead, and a spectrum of fragments is produced, leading to a rather complex "limit digest" pattern, as discussed below.

Variability in Nucleosomal DNA Content

The DNA repeat length in chromatin may be reliably estimated as described above. A value of about 200 base pairs is obtained for most cell types, for example, rat liver (24, 35, 38, 46); chicken liver (49); rabbit cerebellum and liver and glial cells of the cerebral cortex (39); calf thymus and mouse liver (37); certain plant and animal cells in culture (42, 50); and the minichromosome of SV40 (46). A nucleosomal repeat length of about 200 base pairs as initially proposed (16) thus seems to be the norm.

Genuine differences in DNA repeat length that are not an artifact of the method of estimation are most easily detected by comparison of two micrococcal nuclease digests in a polyacrylamide gel (one digest could conveniently be from rat liver or other nuclei with a known 200 base pair repeat). Corresponding higher multiples of the unit size in the two digests fall increasingly out of register in the gel if the repeat lengths differ (see, for example, gels in refs. 39, 41, 51).

The first indication that a repeat length of 200 base pairs is not universal came in a report (44) of a value of 135 base pairs for *Saccharomyces cerevisiae,* a value now known to be an underestimate attributable to the use of unsequenced PM2 fragments as molecular weight markers (52). The value for the DNA repeat in *S. cerevisiae,* estimated by the use of sequenced ϕX174 fragments, is 165 ± 6 base pairs (38). Values of 170 and 154 base pairs for the repeat lengths in *Neurospora crassa* (43) and *Aspergillus nidulans* (53), respectively, have also been obtained with the use of sequenced ϕX174 fragments. Whether the differences between the three values are real or not could be established by running the three digests in the same gel, but this has not yet been done. A comparable value (about 170 base pairs) has been reported for *Physarum polycephalum* chromatin (54), although a value of 190 base pairs was obtained at shorter digestion times (54).

Values between 178 and 196 base pairs for the DNA repeat length have been reported for several animal cell lines in culture (42), the lowest being for Chinese hamster ovary cells (51). The occurrence of short repeats in fungi and tissue culture cells does not seem to be related to a high rate of cell division, as shown by the results of Compton et al. (42) and by the fact that four cell lines in culture with doubling times of about 14 hr (MOPC21 P3K myeloma cells), 18 hr [rat C6 (Sato) glial cells], and 24 hr (plant crown gall and soya bean cells) all have DNA repeats of about 200 base pairs (50).

Neighboring cell types within the same organism may package the genome in different sized DNA units; the neuronal cells from rabbit cerebral cortex have a repeat of 162 ± 6 base pairs, whereas nonastrocytic glial cells from the same tissue have a repeat of 200 ± 6 base pairs (39). A short repeat is not a general property of neuronal cells because neurons of the cerebellum have a repeat of about 200 base pairs (39); there is, however, a difference in function between neurons of the cerebral cortex (principal neurons with long axons) and neurons of the cerebellum (intrinsic granule cell neurons that act over much shorter distances).

Repeat lengths longer than 200 base pairs are also found; for example, about 210 base pairs in chicken erythrocytes (42, 49); 218 base pairs in sea urchin gastrula, 241 base pairs in sea urchin sperm (41); and 220 base pairs in the macronucleus of *Stylonychia* (55). In some cases, a correlation seems to exist between extreme transcriptional inactivity (as in chicken erythrocytes) and a DNA repeat length greater than 200 base pairs on the one hand, and a high level of transcriptional activity (as in cerebral cortex neurons) and a repeat length shorter than 200 base pairs on the other hand. Sea urchin sperm, which are transcriptionally inert, also have a substantially longer repeat length than sea urchin gastrula. In contrast, however, the transcriptionally inactive micronucleus of *Stylonychia* has a shorter repeat length (202 base pairs) than the active macronucleus (220 base pairs) (55). Differences in transcriptional activity might arise in some way from differences in higher levels of folding of chromatin, determined by repeat length.

It is possible that cell-specific repeats are a general phenomenon but that the methods of measurement are not sufficiently sensitive to detect differences in lengths of, for example, five base pairs or less. Whatever the case for the nucleosome repeat length, it is a remarkable fact that the size of the DNA in the core particle produced by prolonged digestion (see above) is always 140 base pairs (35, 38–43, 49, 51–53); variation therefore resides in the linker DNA joining 140 base pair nucleosome cores. This variation may reflect differences in the histones. A long repeat length might be associated with a more basic histone octamer, or more basic H1, or an increased amount of H1; the converse would be expected for a short repeat (43, 49). In chicken erythrocytes, for example (repeat length 210 base pairs (49)), some of the H1 is replaced by the more basic H5.[1] Sea urchin sperm, with the longest known repeat length (41), also contains a more basic H1 type of histone than somatic tissues, as well as a more basic H2B (56). (Nonallelic variants of H2A and H3, containing neutral amino acid substitutions, have also been detected in the course of development in the sea urchin (see under "Histone Variants").) Increased acetylation in neuronal compared with glial nuclei (57) would have the effect of making the histones less basic and· might perhaps be related to the shorter neuronal DNA repeat length.

In summary, at present neither the structural reason for, nor the functional significance of, variation in nucleosomal DNA repeat length is clear. Reconstitution experiments might be expected to shed some light on the former (see under "Reconstitution of Chromatin").

Heterogeneity in Repeat Length within a Cell Type

The DNA fragments from chromatin digested with micrococcal nuclease form quite broad bands in polyacrylamide gels (a breadth of about 80 base pairs for the monomer fragment), possibly reflecting the breadth of the site for potential cleavage between more protected regions of adjacent nucleosomes. Alternatively, the dispersion in the size of DNA fragments in a band could arise from a heterogeneous population of repeat units, the range of sizes being small enough to preserve a discrete band pattern up to multiples of 7 or 8 times the unit size.

It has been claimed that support for heterogeneity in repeat length comes from the dependence of the value determined for the repeat on digestion time (52). This value was longer at early times of digestion than at later times, a fact attributed to increased accessibility of the longer linker regions to micrococcal nuclease, shorter regions being attacked later. Other studies, however, have not shown such a dependence of repeat size on extent of digestion (35, 38, 46, 51) (but see ref. 54). It has further been argued that the widths of the bands given by DNA fragments in a polyacrylamide gel increase with increasing band number (52), as expected for heterogeneity in repeat length, but this evidence has been questioned (58) on the grounds that band widths are difficult to measure with the required accuracy and that the relative contributions of heterogeneity and, for example, variable cleavage in a large potential cutting site cannot be distinguished.

Although the results from micrococcal nuclease digestion studies do not settle the question of heterogeneity of the repeat length within a cell type, other evidence (59) provides a clear indication of such heterogeneity. On treatment of nucleosome monomer from rat liver (generated by the action of micrococcal nuclease) with a mixture of *E. coli* exonuclease III and single strand-specific nuclease S1, the DNA band in a polyacrylamide gel is shifted from a mean size of about 170 base pairs to 140 base pairs and is narrowed from a breadth of about 80 base pairs to near homogeneity. When a nucleosome dimer is treated in the same way, the DNA band is shifted to a similar extent, but its breadth, initially about 150 base pairs, is unchanged, suggesting that the linker DNA in a nucleosome dimer may vary in length by as much as 150 base pairs, which represents a marked heterogeneity of nucleosome DNA content (59).

Preparation of Native Chromatin

The structure of chromatin, with DNA wound around an array of histone cores, is susceptible to damage by shear forces such as those generated by homogenization or other mechanical stress. This damage is revealed as a loss of the regular pattern of DNA bands in a polyacrylamide gel after brief digestion with micrococcal nuclease (60). Although the regularly repeating structure is damaged, some vestige of the structure remains because the "limit" micrococcal digest pattern and DNase I digestion pattern (see below) are still observed.

Digestion with micrococcal nuclease provides a method of preparing large pieces of chromatin in which the native structure, defined as a regular linear close-packed array of nucleosomes (a nucleofilament), is preserved (60). Brief treatment of nuclei with nuclease is followed by termination of digestion with EDTA and lysis of the pelleted nuclei in hypotonic medium; clarification of the lysate by centrifugation results in a supernatant population of chromatin fragments containing between 15 and 150 nucleosomes (3,000–30,000 base pairs DNA). When this material is digested further with micrococcal nuclease, all the DNA appears in bands which are multiples of a unit size. This "native" chromatin has been used in studies of the association of histones in chromatin by cross-linking (25, 61) (see under "Associations of Histones in Chromatin"), in electron microscope studies of higher orders of folding (19), (see under "Higher Order Structure in Chromatin"), and in low angle x-ray scattering studies of chromatin in solution (29).

Other procedures have also been used to avoid shearing of chromatin. For example, autolysis of rabbit thymus nuclei was used to obtain a soluble product of molecular weight 10^7–2×10^8 (i.e., containing about 15,000–300,000 base pairs of DNA) (63); transcriptionally active chromatin was prepared without shearing by digestion with DNase II (64); and the use of 0.1 M NaCl is reported to confer some protection against damage caused by shear (65).

ASSOCIATIONS OF HISTONES IN CHROMATIN

Chemical cross-linking has revealed histone-histone associations at two levels. First, it has demonstrated the presence of a repeating array of histones in chromatin; and secondly, it has defined some, if not all, of the histone-histone contacts within this array. A number of reagents have been used, some better understood chemically than others. The reactions are carried out in dilute solution so that intramolecular (intrafiber) reaction will be favored over intermolecular reaction.

Dimethylsuberimidate, a bifunctional imidoester of span 11 Å, reacts with the amino groups of proteins at slightly alkaline pH (8–9) to form amidine derivatives (pK_a ca. 13). Intramolecular cross-linking of oligomeric proteins with this reagent, followed by analysis of the cross-linked products in SDS-polyacrylamide gels, reveals the subunit structure of the proteins (66). By the use of this method, Kornberg and Thomas (10) showed that the arginine-rich histones extracted from chromatin by mild methods existed as a tetramer $(H3)_2(H4)_2$ (see above). Procedures for cross-linking of histones free in solution and of chromatin-bound histones with dimethylsuberimidate have recently been reviewed (67).

When chromatin was cross-linked with dimethylsuberimidate at pH 9 and the products analyzed in an SDS-gel, bands were obtained corresponding to histone dimers, trimers, etc., up to octamer (consistent with incomplete cross-linking of a histone octamer), as well as weak bands corresponding to 16-mer and 24-mer. This result (25) provided the first evidence that the histones in chromatin were organized in sets of eight. Cross-linking of isolated nucleosomes also gave bands up to 8-mer, indicating that a set of eight histones is associated with 200 base pairs of DNA. The pattern of bands found after reaction with dimethylsuberimidate at pH 8 differed from that obtained at pH 9 by revealing contact between neighboring sets of eight histones (25). (This would seem to argue against models for chromatin in which nucleoprotein beads are separated along the chromatin fiber by bridges of free DNA (28).)

Cross-linked histone dimers may be separated into five bands in a high resolution SDS-gel system (25). The composition of these bands has been determined by forming disulfide-containing cross-links which are cleaved by soaking the gel in 2-mercaptoethanol, thereby allowing separation of the component histones by electrophoresis in a second dimension at right angles to the first (61, 67). The dimers identified in this way were H3-H3, H3-H4, H2A-H2B, H2A-H4, and H2B-H4, consistent with the presence in chromatin of an $(H3)_2(H4)_2$ tetramer and H2A-H2B dimer, previously identified for histones in free solution (see above) and indicative of contacts between the tetramer and dimer. In a similar study involving slightly longer cross-links and a less direct method than two-dimensional gel electrophoresis for determining dimer compositions, the additional dimers H3-H2B and H3-H2A were detected (68). Still further histone-histone contacts may occur in chromatin but do not lead to cross-linking because cross-links can be formed only if the orientation of two lysine residues, as well as the juxtaposition of two histones, is favorable.

One of the advantages of imidoesters for cross-linking is that the chemistry of their reaction with proteins is well studied (69, 70) (see ref. 67); another is that the modified protein has unaltered charge at neutral pH; and a third is that imidoesters may be used to generate cleavable cross-links, as described above. In contrast, the reactions of reagents such as formaldehyde and glutaraldehyde, which have been used in several studies of chromatin (71–76), are poorly understood (but probably involve Schiff base formation at amino groups), complicated by the tendency of the reagents to form self-condensation products, and generally irreversible. Moreover, there is a danger that histone may be lost by cross-linking to DNA. Chalkley and his co-workers (73, 74) have shown that treatment of chromatin with 2% formaldehyde causes 80% of the histone to become resistant to acid extraction after 30 min, and all of it after 120 min (see also ref. 71). According to Hyde and Walker (75) histones are completely released from formaldehyde-fixed chromatin by treatment with DNase I. Stable histone-histone cross-links can be formed with formaldehyde (but not glutaraldehyde) under mild conditions such that histone-DNA cross-links simultaneously formed are apparently labile on dialysis against water for 48 hr at 37°C (73).

Other reagents have also been used to cross-link histones in chromatin. A water-soluble carbodiimide, 1-ethyl-3-(3-dimethylaminopropyl)carbodiimide, generates an H3-H4 dimer (77), presumably by formation of an amide bond between one histone and another. Tetranitromethane can also be used to generate "zero length" cross-links in proteins, probably between suitably disposed tyrosyl side chains (the reaction being a frequently encountered side reaction in nitration of the tyrosine residues of proteins). The cross-linked dimer H4-H2B results from tetranitromethane treatment of chromatin, whole cells in culture, and reconstituted chromatin (78). Cross-linking after reconstitution with different combinations of histones showed that H2A (but not H3 or H1) was essential of H4 and H2B, and successive treatments with tetranitromethane and ultraviolet light generate the trimer H2A-H2B-H4 (80). The same H2B molecule, therefore, interacts with both H2A and H4. Elucidation of the positions of the cross-links in the histone primary structure is in progress; the indications from analysis of the cyanogen bromide-generated peptides are that H2A interacts with the NH_2-terminal half of H2B, whereas H4 interacts with the COOH-terminal portion (80).

Interactions involving histone H1 have also been detected in cross-linked chromatin. In general, H1 seems to be preferentially cross-linked to itself rather than to the other histones, for example, by glutaraldehyde (72, 74), carbodiimide (77), dimethylsuberimidate (25), and disulfide-containing cross-links (61, 68). The results suggest that chains of H1 molecules may exist, either by juxtaposition of the molecules along a chromatin fiber or, perhaps, by juxtaposition arising from a higher level of coiling of the fiber.

LOCATION OF HISTONE H1

Because H1 does not contribute to the x-ray diffraction pattern of concentrated gels of chromatin (10, 81) and because its amino acid sequence is variable between species and tissues (see ref. 4), it was not assigned a role in the repeating structure of histones and DNA proposed by Kornberg (16). H1 is therefore assigned to the outside of the repeat unit. The number of H1 molecules is in some doubt and may be one (16, 82) or two (83) per pair of the other histones, i.e., per bead; symmetry considerations might, perhaps, argue for two per bead. The H1 molecule appears to be elongated (12, 84) and to interact with other H1 molecules in chromatin, as shown by several cross-linking studies (see above).

It appears likely that H1 is associated with the linker DNA in chromatin (35, 40, 47, 85), in other words, the DNA that passes from the surface of one bead to the next (this does not mean DNA extended between 140-base pair core particles; see above). H1 is not present in the 140-base pair core particle, which is formed as a metastable intermediate on prolonged micrococcal nuclease digestion of chromatin (35, 40) (see above), whereas ''untrimmed'' nucleosomes containing a full complement of DNA do contain H1. Exonucleolytic digestion of an excised nucleosome containing about 200 base pairs of DNA and H1 by micrococcal nuclease proceeds first to a DNA length of 160–170 base pairs, and then after a brief ''pause,'' to a DNA length of 140 base pairs, with loss of H1 in this final step (35). Conversely, if H1 is removed from chromatin by binding to tRNA (86), nucleosomes are rapidly trimmed from about 200 to 140 base pairs with no pause at 170 base pairs (35). Both of these observations suggest that H1 is associated with the DNA at the ends of nucleosomes, in other words, the DNA that, in chromatin, connects one bead to the next. A separate line of evidence leads to the same conclusion. The sedimentation coefficient of a run of nucleosomes (two up to six) falls when H1 is removed (removal of H1 has no effect on the sedimentation coefficient of a single nucleosome), suggesting that H1 is involved in holding nucleosomes together in chromatin (35). (Removal of H1 may only cause a tendency for nucleosomes to come apart, for example, in the ultracentrifuge, because chromatin depleted of H1 with tRNA showed beads in contact by electron microscopy (87).) This again suggests association of H1 with linker DNA between one bead and the next.

Because all nuclei so far examined, whatever the DNA repeat length, form a 140-base pair core particle as a metastable intermediate in digestion (see above), variation in repeat length must be a consequence of variation in length of the linker DNA. Part, at least, of this DNA is associated with H1, and because H1 displays sequence variability between species and even between tissues, it has been suggested that the nature of H1, in particular its content of basic amino acid residues, determines the length of the intercore linker DNA and hence the length of the DNA repeat in a given cell type (49).[1] It has been speculated that this plays some role in regulating gene expression (49).

VIRAL CHROMATIN: ITS CONTRIBUTION
TO UNDERSTANDING CELLULAR CHROMATIN

Simian virus 40 (SV40) contains a complex of 5,100 base pairs of circular double-stranded DNA and cellular histones enclosed in a capsid. The complex, a "minichromosome," may be obtained either by disruption of virions at alkaline pH or from the nuclei of infected cells. Histones H3, H4, H2A, and H2B are present in roughly equimolar amounts, and recent reports suggest that H1 is also present in the minichromosome isolated from nuclei (88, 89), although not in virions disrupted under comparatively mild conditions at pH 9.8 (89). Studies of the SV40 minichromosome have played a substantial role in consolidating the string-of-beads model for chromatin (16) and have also revealed an important property of nucleosomes, namely the torsional constraint they impose on the DNA (see below). Whether or not the SV40 minichromosome is a perfect model for chromatin structure is not, however, altogether clear, as discussed below.

Griffith (26) was the first to show that SV40 chromosomes from infected cell nuclei contained circular arrays of nucleosomes. At low ionic strength (I = 0.015), a structure containing 21 beads separated by bridges of free DNA was seen. (The values given were 170 base pairs of DNA in the bead and 40 in the bridge, but these are probably underestimates because they do not account for the DNA in the whole genome.) At higher ionic strength (I = 0.15), a more compact structure was observed (doughnut-like) in which no bridges were visible. The packing ratio for the DNA in this condensed form (circumference of free DNA to circumference of chromosome, measured by electron microscopy) was 7 ± 0.5:1, in excellent agreement with that proposed (16) in the model. Other workers (46, 90) subsequently failed to observe this condensed form of the minichromosome and saw discrete beads in the electron microscope, irrespective of ionic strength, although addition of exogenous (calf thymus) H1 to the beaded minichromosome caused a dramatic contraction of the structure. A likely explanation, in view of recent findings (88, 89), is that the highly condensed material (26) contained H1 and that this was not present in the other experiments. It is not clear however whether the H1 found in minichromosomes is a genuine constituent or a contaminant from cellular chromatin.[1]

The beaded form of the minichromosome contains an average of 20 ± 2 nucleosomes each containing 190–200 base pairs of DNA, roughly half of which are close together (separation less than 15 base pairs), but the remainder of which are connected by variable short lengths (15–74 base pairs) of apparently free DNA, accounting in total for about 15% of the viral genome (46). The appearance of the 20-beaded structures seen in the electron microscope correlates well with the results of nuclease,digestion studies; in particular, the absence of more than about four closely juxtaposed beads in electron micrographs accounts for the absence of DNA multiples greater than 4 times the unit size in micrococcal nuclease digests (46). Because trinucleosome and tetranucleosome DNA fragments are roughly multiples of 200 base pairs, about 200 base pairs must be contained in the biochemical repeat unit. The average value for the nucleosomal DNA content determined by electron microscopy (obtained from the difference between total bridge length and contour length of free DNA molecule, divided by the number of nucleosomes) is 194 ± 14 base pairs (or 200 ± 15 base pairs) in the absence (or presence) of salt, indicating that the nucleosomal DNA content estimated by electron microscopy is identical with that determined biochemically from nuclease digestion experiments. This corroborates the earlier finding of Finch et al. (27) for purified fragments of rat liver chromatin that the biochemical repeat unit and the bead seen in the electron microscope were synonymous.

The agreement between the values estimated for nucleosomal DNA content from electron microscopy and nuclease digestion studies suggests that bridges of free DNA between about half the total nucleosomes are unlikely to have arisen by unraveling of the nucleosome DNA (46). The packing ratio for the intranucleosomal DNA (estimated by electron microscopy) is 5.2–5.5:1, whereas the presence of bridges accounts for an overall packing ratio in an SV40 minichromosome containing 20 nucleosomes of about 3.1–3.5:1. Beaded structures for SV40 and polyoma minichromosomes similar to those described above (46) for SV40 have been observed independently (90); in this study, SV40 had 21 ± 1.5 nucleosomes, each containing 175 base pairs separated by 55 base pairs, and polyoma had 20 ± 1.5 nucleosomes containing 205 base pairs, separated by 55-base pair bridges.

It is possible that a 19- or 20-nucleosome SV40 minichromosome might have arisen from a native form comprising, for example, 23 or 24 nucleosomes and no bridges (46). Alternatively, however, it might be a direct consequence of random assembly of histones onto the DNA (recent evidence from the use of a cell-free system for assembly of SV40 chromatin argues against a cooperative process in that instance (91)), with the spacing of two already assembled nucleosomes determining whether a third fits in between. The close packing of nucleosomes in cellular chromatin argues for its assembly by a different process, and it might then be unwise to take the analogy between cellular chromatin and the SV40 minichromosome too far. However, it is possible that the native minichromosome contains H1 and closely juxtaposed beads and that it is indeed a good model for cellular chromatin. There is also some evidence (92) for the formation of 25-nucleosome structures, with virtually no internucleosomal DNA, after in vitro reconstitution of SV40 chromatin in the absence of H1.

One important result demonstrated for the SV40 minichromosome, which appears to be of general relevance to chromatin structure, is that nucleosome formation appears to unwind the DNA double helix by about one turn. In in vitro reconstitution experiments with SV40 DNA and histones H3, H4, H2A, and H2B from calf thymus, Germond et al. (92) showed that each nucleosome generated on relaxed, closed, circular SV40 DNA inserted approximately one apparent negative superhelical turn into the DNA. With relaxed DNA and a histone to DNA ratio of 2:1, a highly twisted beaded structure containing about 25 nucleosomes was formed; nicking-closing (relaxing, untwisting) enzyme relaxed the beaded structures by removal of constraints on the internucleosomal DNA, but the DNA extracted from such relaxed complexes was highly supercoiled. In the relaxed complexes, the torsional constraint on the DNA is, therefore, confined to the nucleosomes and after deproteinization spreads over the whole DNA molecule to induce superhelical turns. The apparent number of superhelical turns in the DNA, estimated by the elegant Agarose gel method of Keller (93), corresponds to the number of nucleosomes present before deproteinization, and suggests that formation of one nucleosome generates the equivalent of 1–1¼ superhelical turns in the DNA.

Nucleosomes also form in vitro on the negatively supertwisted, closed, circular DNA extracted from SV40 virions or infected cells (92). In this case, the beaded complex is relaxed (like the minichromosome extracted from virions or infected cells) and contains up to 25 nucleosomes. DNA extracted from such a complex is as highly supercoiled as the initial DNA, and this supercoiling is retained if the complex is first treated with nicking-closing enzyme, showing that no internucleosomal constraints remain in the 25-nucleosome structure. (In fact, the final product sometimes contained a few more superhelical turns than the initial DNA.) The findings that nucleosomes form preferentially on negatively supercoiled SV40 DNA in vitro and that increased nucleosome formation on supercoiled SV40 DNA reduces the number of superhelical turns in the internucleosomal DNA, indicate that the free energy contained in negative superhelical turns favors nucleosome formation (92). Insertion of one negative superhelical turn into DNA is equivalent to unwinding the double helix by one turn (94). The relative contributions of supercoiling, unwinding (melting), bending, and kinking (see under "DNase I-Susceptible Sites") to constraint on nucleosomal DNA are unknown at present. H1 binds preferentially to supercoiled DNA (95, 96), a property which might be responsible for its binding to nucleosomes. It is not yet known whether H1 has any effect on the superhelix density because it was not included in the SV40 experiments described above (92), but it could well have none in view of the apparent close similarity between viral nucleosomes and those of cellular chromatin.[1]

A cell-free system from eggs of *Xenopus laevis* is capable of reassembling histone and exogenous relaxed, circular SV40 DNA into SV40 minichromosomes whose DNA has the same number of superhelical turns as natural SV40 chromatin (91). This system should be of great value in the study of chromatin assembly.

HIGHER ORDER STRUCTURE IN CHROMATIN

The contraction in the length of the DNA (about 7-fold) achieved in the string-of-beads (nucleofilament) structure for chromatin is clearly insufficient to account for the packing of DNA even in interphase chromatin, and higher levels of coiling or folding must exist. It has long been known (17) that electron micrographs of chromosomes (prepared by critical point drying) show 100 Å diameter fibers in the presence of chelating agents and 200 Å diameter fibers in their absence. On the basis of electron micrographs of negatively stained and freeze-etched/shadowed native chromatin (prepared as described above), Finch and Klug (19) have proposed that 100 Å diameter nucleofilaments are coiled in the presence of Mg^{2+} ions (0.2–1.0 mM) to give supercoils or "solenoidal" structures, 300–500 Å thick, with cross-striations spaced at 120–150 Å, which could correspond to the thick fibers seen in chromosomes. They make a tentative estimate of about six or seven nucleosomes per turn of the solenoid. It seems likely that H1 is needed for solenoid formation or, at least, for its stabilization on the electron microscope grid.

Solenoidal structures might correspond (19) to the "superunit threads" seen by Davies and his colleagues (97) (summarized by Davies and Haynes (98)) in condensed interphase chromosomes (chromatin bodies) from several tissues. Electron micrographs of thin sections of nuclei preferentially stained for protein or DNA showed these threads, which were lined up in layers at the nuclear envelope, to have a diameter of 280 Å and an inner DNA-rich core about 170 Å in diameter. By equating the superunit thread and the solenoid, Finch and Klug adopted a value of about 300 Å for the outer diameter of the solenoid, which would then have about six nucleosomes per turn and a central hole about 100 Å in diameter. The central hole, which seems from Davies' work to be DNA-rich, could perhaps (99) accommodate the small proportion of "naked" DNA that seems to exist in chromosomes (100), or possibly H1 or nonhistone proteins; alternatively, it might just be a stain-absorbing hole. The relationship between the electron microscopic evidence for solenoids and the x-ray and neutron scattering patterns of chromatin is discussed in the next section.

The packing ratio for the DNA in a solenoid having six nucleosomes per turn would be 40:1. This differs by only a factor of two from the packing ratio of DNA in interphase chromatin (estimated as about 80:1 in the salivary X chromosome of *Drosophila melanogaster* (101)), but is still a long way short of explaining the roughly 10^4-fold compaction of DNA in metaphase chromosomes which must require additional levels of folding. Solenoids folded into loops may be one possibility (99), and there is some evidence to support the existence of loops in chromosomes (see below).[1]

Technical considerations make the study of higher orders of folding of chromatin difficult. Solenoid formation (19) was achieved with lengths of native chromatin up to 40 nucleosomes long (see under "Preparation of Native Chromatin"); however, generation of higher levels of folding will almost certainly require much longer lengths, whereas 40-nucleosome chromatin is the longest that can be obtained in homogeneous solution by the nuclease method (60). Clearly, a different approach, such as the isolation of intact loops, will be needed, and in this context the use of restriction enzymes might prove useful.

Folding domains consisting of supercoiled DNA loops, first described in the *E. coli* chromosome (102), have now been described in the DNAs of cultured human cells (103), mouse cells (104), and *D. melanogaster* cells (105). In the latter case, the intact interphase genome was isolated on sucrose gradients after gentle lysis of the cells in 0.9 M NaCl-0.4% Nonidet P40 (neutral detergent) and was shown to contain histones H2A, H2B, H3, and H4 (despite the high salt concentration) as well as RNA. The DNA contains the same superhelix density as *E. coli* and SV40 closed circular DNAs, namely about one negative turn per 200 base pairs in 0.15 M-0.2 M NaCl at 26–37°C. Nicking of the folded genome by DNase I removes the superhelical turns. Intercalation by ethidium bromide also destroys the superhelicity and displaces all four histones simultaneously. A precise correlation between the two effects suggests that histone loss is a direct result of the untwisting of the double helix that accompanies intercalation and can be understood in the light of the relationship between nucleosomes and superhelical turns (see above).

This partial characterization of supercoiled nucleosome-containing DNA loops in the *Drosophila* chromosome (105) is a step toward understanding DNA packing in chromosomes. The loops in *Drosophila* contain, on average, 400 nucleosomes but may be heterogeneous in size; the nature of the loop stabilizers, whether RNA or protein, or both, is as yet unknown. Similarly unknown at present is the relationship of loops to units of genetic activity, because the gene for an average protein could be accommodated in about five nucleosomes. Crick (99), however, has raised the possibility that there may be one loop per band of a *Drosophila* polytene chromosome, and one solenoid per loop.

LOW ANGLE X-RAY AND NEUTRON SCATTERING: EVIDENCE FOR A STRUCTURAL REPEAT

X-ray diffraction patterns of concentrated gels of chromatin provided the first indication of a repeating structural element. The pattern comprised a series of rings at spacings of about 110, 55, 37, 27, and 22 Å (81, 106, 107) and the 110 Å reflection was oriented in the fiber direction (81).

In proposing the bead model for chromatin structure, Kornberg (16) suggested that the 110 Å repeat arises from the repeat of beads along the fiber; i.e., the length of the repeating unit in the fiber direction is about 110 Å. The 110 Å band in the x-ray pattern is enhanced relative to background scatter as the concentration of chromatin is increased, and it was reasonable to suppose that this was compatible with a flexibly jointed structure in which closer packing at higher concentrations resulted in straightening of the beaded chain (108).

An alternative explanation for the origin of the 110 Å reflection has recently been proposed by Finch and Klug (19), who suggest that the 110   reflection arises from a higher level of folding of the string of beads, in particular from the spacing between the turns of a solenoidal type of structure of low pitch angle, such as that deduced from electron micrographs (see above). In support of this interpretation, chromatin in the presence of Mg^{2+}, which gives solenoids in the electron microscope, gave a good diffraction pattern at a concentration of 25% wt/wt or less, with more order than normally observed for solutions almost twice as concentrated (62). Finch and Klug (19) suggest that the well established concentration dependence of the x-ray pattern of chromatin is attributable to an increased tendency to form solenoids as the string-of-beads structure loses water. Because this could also diminish the spacing between turns of the solenoid, it would account for the decrease in x-ray spacings by about 10% as the chromatin concentration is increased from 30 to 50% (wt/wt) (109).

A recent neutron diffraction study of chromatin depleted of H1 (110) in which the 110 Å reflection is off-meridional by a very small amount (8–9°) has also been interpreted in terms of a helical rather than a linear array of nucleosomes, in agreement with the observations of Finch and Klug (19) on chromatin. (However, H1-depleted material viewed in the electron microscope does not give solenoidal structures; see previous section). A theoretical discussion of the origin of the 110-Å x-ray spacing in terms of the packing of 110 Å diameter spheres has also been given (111).

What is the origin of the reflections at 55, 37, and 27 Å? Do they arise as higher orders of the 110 Å reflection (14) or do they arise from the internal structure of the bead? Harrison and Kornberg (112) have recently reported results in support of this second interpretation. They show that, except at very low angles, the same diffraction pattern is obtained from a solution of isolated beads as from intact chromatin, and they rule out aggregates as a possible explanation. Thus, the beads in chromatin diffract essentially independently at low concentration. It is suggested that the maxima at 55, 37, and 27 Å in the diffraction pattern of isolated beads could arise from the shape of a spherical bead (for example, the diffraction pattern expected from a uniform sphere of protein, 35 Å in diameter, surrounded by a uniform shell of DNA 20 Å thick, would be similar to the pattern observed), although the maxima are not as distinct as would be expected for a uniform sphere, possibly because of radial density variation in the sphere. However, other models are of course possible.

Neutron diffraction studies of chromatin (113) have also attempted to identify the origin of the reflections at 55, 37, and 27 Å.

INTERNAL STRUCTURE OF THE NUCLEOSOME

Information about the internal structure of the nucleosome has come both from studies on whole chromatin and from work on isolated nucleosomes and core particles. The advantage of the latter for some purposes is that, unlike larger, less digested nucleosomes which contain between 160 and 200 base pairs of DNA, the core particles have a homogeneous DNA length of 140 base pairs.

Preparation and Characterization of 140-Base Pair Core Particles

Preparation   Core particles containing a histone octamer $(H_3)_2$ $(H_4)_2$ $(H2A)_2$ $(H2B)_2$, 140 base pairs of DNA, and no H1 may be prepared by prolonged digestion of nuclei or chromatin and are purified either on sucrose gradients (35, 47), or on a larger scale in a zonal rotor (114, 115) or by gel chromatography (116, 117). Digestion of nuclease-generated chromatin (60) in the presence of 0.1 M NaCl is reported to reduce the probability of internal nicks in the final core particle (47).

The protein content of the product may be checked in 18% SDS gels (25), which demonstrate the presence of roughly equimolar amounts of the four histones, H3, H4, H2A, and H2B, and no H1. The DNA content may be measured in 6% acrylamide gels (118), which show a single, relatively sharp, double strand fragment (band width about 10 base pairs); any nicks made in the DNA during preparation of the core particle may be detected by analysis in single strand form in polyacrylamide gels containing formamide (119) or 7 M urea (120) as denaturant. In general, a preparation of core particles may be produced which is much more homogeneous (notably with respect to DNA length) than a preparation of nucleosomes isolated from nuclei after brief digestion (band width about 80 base pairs).

Hydrodynamic Properties   The hydrodynamic properties of core particles (114, 117) show them to be compact structures. The core particle sediments at about 11 S as does the 200-base pair nucleosome (24), despite having about 60 base pairs less DNA and no H1, and must, therefore, have a lower frictional coefficient than the complete nucleosome. The weight average molecular weight of a core particle measured by sedimentation equilibrium, for an assumed (117) or measured (114) partial specific volume of $0.66\,ml.gm^{-1}$, is 196,000–210,000; from the frictional coefficient $(f^0_{20,w})$, the diameter of the hydrated core particle is calculated as 107–110 Å, compared with about 75 Å calculated for the unhydrated particle diameter from the molecular weight and partial specific volume. The dimensions of the hydrated particle agree well with values for the radius of gyration determined by low angle neutron scattering, as discussed below under "External Location of DNA."

Homogeneity   Strong support for the homogeneity of core particles comes from the demonstration that they can be crystallized (115, 121).[1] The simplest working hypothesis is that a preparation of core particles is homogeneous with respect to histone composition, containing two molecules each of H3, H4, H2A, and H2B. Measurement of histone stoichiometry in core particles (83), like the determination of histone composition for the octamer in 2 M NaCl (25) (see under "Protein Core of the Nucleosome"), indicates roughly equimolar amounts of the four histones and is entirely consistent with an unique octamer; however, it does not rule out a mixture. An immunochemical study (122) of the sedimentation of nucleosome-antihistone complexes showed that anti-H2B (7 S IgG antibody) bound to all molecules in a population of trimmed nucleosomes (containing 140–180 base pairs of DNA), indicating that each nucleosome contained H2B. Binding of between one and three antibody molecules per nucleosome was attributed to a normal distribution during histone-antibody titration, arising as a consequence of expected heterogeneity in the antibody population, rather than to heterogeneity in histone composition of nucleosomes, because complexes containing different amounts of anti-H2B had identical values ($\pm$ 8%) for the ratio of H2A to H4 and of (H2B + H3) to H4, estimated by densitometry of stained gels. The results are entirely consistent with the proposition (16) that each nucleosome has an identical histone complement of two molecules each of H3, H4, H2A, and H2B.

Reported heterogeneity of nucleosomes with regard to H1 content (85) is attributable to variable exonucleolytic digestion of nucleosomes (see under "Location of Histone H1"). True microheterogeneity of nucleosomes could, however, occur as a result of amino acid sequence variants of some of the core histones or as a result of histone modification (see under "Histone Variants" and "Histone Modifications"), and could well represent a modulation of structure for some functional purpose; but this does not represent heterogeneity of composition with respect to histone types.

Arrangement of Histones in the Nucleosome

Cross-linking with dimethylsuberimidate (see above) revealed the same histone-histone associations in isolated nucleosomes (chromatin monomer) as in whole chromatin (25). The cross-linked dimers formed were consistent with (although, of course, they do not prove) the presence within the nucleosome core of a tetramer, $(H3)_2(H4)_2$, and two H2A-H2B dimers. The composition $(H3)_2(H4)_2(H2A)_2(H2B)_2$ assumed for the octameric core suggests a 2-fold axis of symmetry in the structure, which might be achieved by having a core of the arginine-rich tetramer $(H3)_2(H4)_2$, with two dimers H2A-H2B symmetrically disposed about it (25).

Arrangement of DNA in the Nucleosome

External Location of DNA   Intranucleosomal DNA in intact chromatin or in 140-base pair core particles is nicked extensively by the enzyme DNase I (pancreatic DNase), and the high accessibility of the DNA (30) provided the first evidence in support of the surface location of the DNA in the nucleosome (16). The enzyme makes single strand cuts, and electrophoretic analysis of the DNA fragments in a denaturing gel gives a pattern of bands corresponding to fragments differing in size by 10 bases, indicating cuts at intervals or multiples of intervals of 10 bases on each strand (30). This cutting pattern is discussed further below.

Direct evidence for the location of the DNA on the outside of the particle has now come from low angle neutron diffraction studies on core particles (31, 32) with the use of the technique of contrast variation in H_2O-D_2O mixtures. (In this method, the scattering density of the D_2O-H_2O solvent is matched to the scattering density of protein or DNA in turn so that diffraction from the DNA and protein, respectively, can be studied separately.) The radius of gyration is 50 Å under conditions in which scattering from the DNA dominates and 30 Å when scattering from the protein dominates, showing that the particle has a protein-rich core surrounded by a DNA-rich shell. It should be noted that so far the data are at a relatively gross level of resolution and preclude any firm conclusion about the uniformity of distribution of protein and DNA or, for instance, about internal cavities, surface depressions, or protuberances. Pardon et al. (31) suggest two arrangements for protein and DNA consistent with the available neutron scattering data, but point out that data at lower values for the contrast and at higher scattering angles are now necessary in order to distinguish between possible models. It should also be noted that the use of D_2O to vary the solvent scattering density for neutron work assumes that the densities of protein and DNA as seen in neutron diffraction are uniform, which may not be true (123). However, it is a better approximation for x-ray scattering, so this method may prove more reliable in the end.

Folding of the DNA   About 200 base pairs of DNA are contained in a nucleosome about 100 Å in diameter, but the path of the DNA and the nature of its folding are as yet unknown. It is known to have the equivalent of one left-handed superhelical turn (see above), but it is not known, for instance, whether there is actually one physical superhelical turn of DNA around the protein core (105, 124), although this seems unlikely from the dimensions of the nucleosome, or whether the DNA follows another path, perhaps making two turns around the protein core (99, and ref. 19, Figure 7)[1] and perhaps being "kinked" (125, 126). It is plausible that kinks in the DNA could be the cutting sites for DNase I (125), but this is as yet unproven.

In a kinked structure, stretches of undistorted double helix alternate with sharp bends (kinks). (The alternative to kinks would be continuous bending of the DNA double helix about 20 Å in diameter to a radius of curvature about 30–50 Å.) Two types of kink have been proposed. Crick and Klug (125) have demonstrated by model building studies the stereochemical feasibility of a kink of 98° in which the double helix is bent toward the side of the minor groove. There is no unpairing of bases, but at the kink one base pair is completely unstacked from the neighboring one so that some stacking energy is lost. Kinks of this sort every 20 bases would generate a shallow, kinked, left-handed DNA superhelix having 10 undistorted (interkink) stretches of DNA per 200-base pair nucleosome, accommodated in about two turns of the superhelix. Sobell and his co-workers (126) have proposed an alternative stereochemically acceptable kink through an angle of 40° toward the major groove of the double helix. They suggest that intercalating drugs may bind at such kinks, giving rise to structures established by x-ray crystallographic analysis for drug intercalation into three dinucleoside monophosphates. B-form DNA kinked in this way every 10 base pairs is coiled into a left-handed, kinked, toroidal superhelix with a diameter of about 100 Å, the diameter of the nucleosome. There would be roughly two superhelical turns per 200 base pairs. Whether nucleosomal DNA will prove to have one or other of these kinked structures remains to be seen, perhaps not until the crystal structure of the nucleosome is solved.

Benyajati and Worcel (105) and Weintraub et al. (124) propose one superhelical turn per se (a 360° left-handed toroidal coil) in each nucleosome, rather than the equivalent in kinking, bending, and/or melting of the double-stranded DNA. However, a single toroidal coil containing 200 base pairs of DNA would be too large to account for the diameter of the nucleosome observed in electron micrographs; it is suggested (105), therefore, that less than 200 base pairs of DNA are coiled around a histone core and that relaxed "internuc-leosomal DNA" (a misnomer because the nucleosome is the whole biochemical repeat unit (18)) accounts for the remainder of the DNA in the repeat unit of the structure.

DNase I-Susceptible Sites   Generation by DNase I of a set of fragments differing in size by 10 bases does not necessarily mean that the nuclease attacks every 10 bases in each strand; the fragments could arise in a number of ways. For instance, an 80-base fragment could be generated from the 140-base pair DNA of a core particle by cleavage of a strand 60 bases from one end or 30 bases from each end. The probability of cutting at potential cutting sites can be expected to reflect the arrangement of the DNA in the nucleosome and/or certain features of DNA-histone interactions. Potential cutting sites every 10 bases on each strand might be a simple consequence of having the B-form of the DNA double helix (pitch 10 base pairs) wrapped around a protein core so that one side is protected and each strand is accessible to the nuclease as it rises from the surface. Alternatively, the regular pattern of sites may reflect periodic distortions or kinks in the DNA where the molecule is either especially sensitive or especially insensitive to nuclease action (30).

Several investigations have recently been concerned with mapping the DNase I-susceptible sites in isolated core particles (127–129). In each case, the approach has been to label the 5'-end of the DNA in 140-base pair core particles with [γ-^{32}P]nucleoside triphosphate, to digest with DNase I, and then to analyze the distribution of end label among the single strand fragments produced. One study (127) showed that potential cleavage sites exist every 10 bases, with the exception of the site 80 nucleotides from the 5'-end (no label in the 80-base fragment); sites 20, 40, 50, 100, 120, and 130 nucleotides from the 5'-end are readily cleaved; those 30 and 110 bases from that end are relatively resistant; sites 60, 70, and 90 bases from the 5'-end show intermediate reactivity. A central region 50–110 bases from the 5'-end is thus relatively resistant to digestion. All three studies (127–129) are in general agreement, although there are differences in the extent of cleavage found at certain sites. It is not clear at the moment whether such differences arise from different procedures for preparing core particles or from different methods of analysis. Although the pattern of cleavage in a single strand of the 140-base pair DNA is not symmetrically disposed about the midpoint of the DNA, the distribution of cleavage sites within the nucleosome could still be symmetrical (in keeping with the presumed symmetry of the particle) provided both strands contain the same cleavage sites. Whether or not this is so should emerge from an analysis of the kinetics of digestion and a 3'- end labeling experiment now in progress (L. Lutter, personal communication).[1]

It may be possible to distinguish various alternatives for the origin of the DNase I susceptibility (kinks of one kind or another or merely protection of one side of a regular B-form double helix) by the nature of the cut ends in the products of DNase I digestion isolated in double strand form. For example, cutting at the sites most susceptible to nuclease in a B-form double helix wound regularly around a protein core should give double strand fragments with ends staggered by six bases. One way of investigating the stagger in double strand fragments is by gel analysis of these fragments in single strand form, either after trimming down the single strand tail or after filling in the single strand gap; the results so far suggest a stagger of two bases (129, 130). However, such experiments should not necessarily be expected to provide a definitive answer to the DNA folding pattern nor to distinguish one sort of kink from another because, for example, the cutting site for DNase I may not be at the kink point itself but some distance away.

A DNase II-Susceptible Site in the Nucleosome   Digestion of nuclei or chromatin at 100-base pair intervals with the endonuclease DNase II (131) (which has no metal ion requirement) suggests (although, of course, it does not prove) that there may be an axis of 2-fold rotational symmetry in nucleosomal DNA. In the case of chromatin in 10 mM Tris-Cl, pH 7, the enzyme cuts between nucleosomes, presumably at the same sites as micrococcal nuclease, but in the presence of divalent (1 mM $CaCl_2$) or monovalent (150 mM NaCl) cations it makes additional double strand cuts in the middle of the nucleosome, giving DNA fragments that are multiples of 100 base pairs. The accessibility of the intranucleosomal site to cleavage depends on temperature (132) (accessible at 37°C but not at 0°C), ionic conditions (accessible in the presence of divalent cations), and the presence of H1 (no internal cuts when H1 is removed). Although the observed effects might arise from conformational changes within the nucleosome, they could also be a consequence of interactions between nucleosomes and the formation of higher order structures (131). DNase II could thus be a useful and sensitive tool for probing further folding of the nucleofilament. (The discovery of sites cleaved by DNase II at 100-base pair intervals could well explain the results of a previous study (133) in which digested DNA fragments were estimated by electron microscopy to be multiples of about 400 Å (i.e., 120-base pairs DNA), but in which the possibility of intranucleosomal cleavage was not considered.)

Cleavage by DNase II at the intranucleosomal site does not lead to a release of half-nucleosomes. The particles produced by DNase II have the same sedimentation properties in sucrose gradients as those generated by micrococcal nuclease, showing that the internal cut does not disrupt the nucleosome (131).

There is, however, some indication that nucleosomes might have a weak point at which they can fall apart into two halves under some conditions in vitro. Chambon and his colleagues (134) have made the striking observation that the SV40 circular minichromosome, normally a 20-beaded structure (see above), appears in electron micrographs of samples at very low ionic strength (0.2 mM EDTA) as a 40-beaded structure, with beads about half the normal size. The beads, which have so far been observed only on the microscope grid, could well be "seminucleosomes" arising from symmetrical splitting of the nucleosome, possibly along a dyad axis, and containing one each of the four histones H2A, H2B, H3, and H4. They have also been observed in cellular chromatin (134).

   Micrococcal Nuclease-Susceptible Sites in the Nucleosome: The Limit Digest   Micrococcal nuclease makes double strand cuts in the DNA within the nucleosome after cleavage at its preferred sites of action, i.e., between nucleosomes. On prolonged digestion, a precipitated limit digest is produced (135) in which the DNA consists of eight fragments between 45 and 145-base pairs long. The existence of so many fragments after exhaustive digestion suggests that a given nuclease-susceptible site within the nucleosome is not cleaved with the same efficiency in all nucleosomes. The reason for this is unclear; no hard evidence exists to support the notion of heterogeneity among nucleosomes (136), although, of course, there are almost certainly different degrees of histone modification (acetylation, methylation, etc.). The explanation could instead simply be that initial cleavages in the nucleosomal DNA influence subsequent ones, so the heterogeneous pattern might simply reflect different pathways of degradation of a single nucleosome type. An alternative explanation (136a), which would account for the constancy of the digestion pattern with time, suggests that cleavage occurs only once per nucleosome. To account for this, specific structural heterogeneity of a novel type is invoked, arising fom entry and exit of DNA at any one of a number of distinct sites in the nucleosome.

DNA-Histone Interactions

In an approach to the study of DNA-histone interactions within the nucleosome, nucleoprotein fragments obtained after successive nuclease and protease treatment have been examined (137). Proteolysis of the nuclease-generated limit digest with trypsin releases three of the characteristic DNA fragments already mentioned as free DNA, and the other five as soluble nucleoprotein complexes sedimenting at 5 S which have previously been described (48). It is concluded (137) that the three free DNA fragments were probably initially associated with the 20–30 NH_2-terminal residues of histones removed by trypsin (136), whereas four of the other DNA fragments were associated with COOH-terminal fragments from all four main histones (H3, H4, H2A, H2B); the remaining DNA fragment was associated with peptides from only H3 and H4. Although such studies could indeed lead to an understanding of DNA-protein interactions in chromatin, the results of trypsin digestion cannot be interpreted with any confidence in view of the possibility that generation of the insoluble limit digest might involve a greater or lesser degree of structural rearrangement. Similar caution must be exercised in any interpretation of the relative rates of histone digestion in chromatin (over which there is some disagreement (60, 136)) in view of the possibility of secondary structural changes.

Susceptibility of histone NH_2-terminal regions in chromatin to trypsin has been interpreted (136) in terms of NH_2-terminal "arms" extending from a globular histone core and enveloping the DNA. Nuclear magnetic resonance evidence for conformationally free NH_2-terminal regions in histones is discussed later under "Histone Interactions and Conformations." Alternative explanations of NH_2-terminal digestion should, however, be considered—one being protection of the NH_2-terminal basic stretch (by DNA in the case of chromatin and by normal protein-protein interactions in histone oligomers free of DNA) except at a small susceptible region (cf. RNase A and micrococcal nuclease (see under "Histone Interactions and Conformations")). The latter situation in histones could result in generation of large NH_2-terminal peptides (20–30 residues), but these would be lost in an analysis of fragments in polyacrylamide gels (137).

In a well conceived approach to the study of DNA-protein interactions within the 140-base pair core particle, the ends of the DNA have been covalently attached to neighboring histones with the aim of identifying the relative location of the various histones and DNA (138). The 5'-ends of the DNA were labeled with ^{32}P (with the use of $[\gamma-^{32}P]ATP$ and polynucleotide kinase), and the labeled end was cross-linked to a histone amino group. (Cross-linking was achieved by sequential methylation of the DNA, depurination to form an aldehyde, imine formation between aldehyde and lysyl amino group, and reduction of the imine (139).) After the DNA was exhaustively digested by nucleases, the label was found bound to histones H3 and H4 equally. Certain arrangements of the 8 histone molecules within the core particle compatible with these results have been suggested (138). An additional possibility, however, is that the arginine-rich tetramer organizes the whole length of the 140-base pair DNA in the core particle (see also under "Partial Reconstitution of Chromatin Structure with Selected Histones") and that the lysine-rich histones interact with this "primary particle" in some way.

THE PROTEIN CORE OF THE NUCLEOSOME

It is now generally agreed that the protein core of the nucleosome is an octamer, probably containing two each of histones H3, H4, H2A, and H2B, as proposed by Kornberg (16). The first evidence for this came from chemical cross-linking studies on chromatin (25) in which the molecular weights of the cross-linked products were consistent with a repeating array of such octamers. The molecular weight of the protein core of the nucleosome, calculated from the difference in molecular weights of the 140-base pair core particle and its DNA component, was also consistent with the proposed octameric structure (117).

Is the octameric core protein of the nucleosome capable of an independent existence in free solution when the DNA is removed in 2 M NaCl? There is good evidence that it is (25), but it has also been suggested that it exists instead as the half-molecule (a heterotypic tetramer) (140). There are three lines of support for an octameric rather than a tetrameric species in solution in 2 M NaCl. First, complete cross-linking of the core protein in 2 M NaCl at pH 9 gives a cross-linked octamer (25). A time course of cross-linking, analyzed in SDS-gels, shows a smooth conversion of histones into products of increasingly higher molecular weight, i.e., from dimer, trimer, etc., at short times to octamer after about 45 min, without detectable build up of the tetrameric intermediate which might be expected if formation of a cross-linked octamer were an artifact of the association of two tetramers (82). Second, dilution of the core protein before cross-linking gives cross-linked dimers and hexamers. The compositions of the cross-linked octamers, hexamers, and dimers, determined by the use of cleavable (disulfide-containing) cross-links and two-dimensional gel electrophoresis, were consistent with dissociation of an octamer $(H3)_2(H4)_2(H2A)_2(H2B)_2$ by loss of a dimer H2A-H2B (25, 61). Further dilution generated a tetramer of hitherto undetermined composition. Because H3 and H4 exist in solution at low ionic strength as a stable tetramer $(H3)_2(H4)_2$ (10) (association constant (13) 0.7×10^{21} M^{-3}), it seems not unreasonable to suppose that this is also the tetramer generated by dilution of the octamer at high ionic strength and that the tetramer is an integral component of the octamer (25). Third, direct measurement (141) of the molecular weight of the core protein from rat liver by sedimentation equilibrium methods in 2 M NaCl gives a value of 107,500 (calculated molecular weight of $(H3)_2$ $(H4)_2$ $(H2A)_2(H2B)_2$ is 109,000) by using a directly measured value for the partial specific volume of 0.767 ml.g^{-1}. The molecular weight is independent of pH in 2 M NaCl over the range pH 7–9, and is the same whether or not the core protein is cross-linked. The sedimentation coefficient $(s_{20,w}^0)$ is 4.77 S for uncross-linked material at pH 7 and 9, and 5.24 S for cross-linked material at pH 9, suggesting that cross-linking generates a structure which is in some way more compact. The above evidence taken as a whole suggests strongly that the core protein released from DNA in 2 M sodium chloride is an octamer of histones.

The alternative view (140), that the histones released from DNA in 2 M NaCl exist in solution as half-octamers or heterotypic tetramers of composition H3·H4·H2A·H2B, is based on two types of molecular weight measurement: sedimentation equilibrium analysis of core protein from chicken erythrocytes giving a molecular weight of 51,000, and gel filtration of the same material, calibrated with globular proteins, giving an apparent molecular weight of 100,000, which is corrected to about 60,000 on the basis of a sedimentation coefficient $(s_{20,w}^0)$ of 3.8 S (measured in sucrose gradients with reference to standard proteins). Similar molecular weights have been obtained by laser light scattering studies of core protein from chicken erythrocytes and calf thymus (142). The reason for the discrepancy between these values indicative of a histone tetramer and that indicative of an octamer is unclear; possible reasons (141) include dissociation of the material derived from chicken erythrocytes in the course of its preparation and, perhaps less likely, species differences (octamer from rat liver, but tetramer from chicken erythrocyte and calf thymus).

Dissociation of core protein in 2 M NaCl after lowering of the pH from 7.1 to 5.5, which is known to give complexes containing lysine-rich histones only and complexes containing arginine-rich histones only (140), would be easily understood in terms of the dissociation of an octamer:

$$(H2A-H2B)\ (H3)_2(H4)_2(H2A-H2B) \quad \overset{pH\ 5.5}{\underset{pH\ 7.1}{\rightleftharpoons}} \quad (H3)_2(H4)_2 + 2\ (H2A-H2B)$$

In this scheme, overall symmetry is preserved and only one class of contacts broken, those between H2A-H2B and the arginine-rich tetramer. Several criteria for comparing the nucleosome core protein extracted in 2 M NaCl with histones in situ in chromatin (sensitivity to trypsin, pattern of iodination, etc. (140)) do not distinguish between heterotypic tetramers and octamers.

In summary, there is general agreement on the occurrence of octamers in chromatin, but some dispute over whether the octamer can lead an independent existence in 2 M NaCl in the absence of DNA. Does this issue have any functional relevance? It depends partly on whether histones are ever released from DNA at physiological ionic strength, for example, during replication. If they are, then it seems likely that they will exist as arginine-rich tetramers and lysine-rich dimers, and the state of the core protein in 2 M NaCl assumes no more than academic physicochemical interest; octamers are presumably reformed when tetramers and dimers recombine on the DNA after the completion of DNA replication. On the other hand, it has been suggested that histones may never be released from the DNA during replication, but that instead the octamer in the nucleosome splits symmetrically as DNA strands come apart, and one heterotypic tetramer remains bound to each strand (124, 143). This model, therefore, predicts that histones are distributed semiconservatively during rep- lication. There is some evidence for this (144), although there is also evidence suggestive of a random redistribution of histones after replication (145) (see under ''Chromatin Replication''). A distinct functional role is thus proposed for the heterotypic tetramer (124, 143). The issue of its existence in free solution (albeit in 2 M NaCl) therefore assumes added interest.[1]

It must be emphasized, however, that whether the octamer splits in half in certain states of chromatin, for example, during the course of DNA replication, and whether the octamer dissociates into heterotypic tetramers in solution are separate questions. In this context, the seminucleosomes recently observed in electron micrographs by Chambon and his colleagues (134) may contain one each of H2A, H2B, H3, and H4, but this does not mean that a heterotypic tetramer must be found free in solution.

RECONSTITUTION OF CHROMATIN

It is sometimes assumed that reconstitution of chromatin is easily achieved, largely because of some success in studies of transcription in the late 1960s (146, 147). Histones and nonhistones were dissociated from DNA in 2 M NaCl and 5 M urea (147) and subsequently reassociated by gradient dialysis from high to low salt concentration, usually in the presence of urea, and then to zero urea, to give reconstituted chromatin whose transcript, formed by *E. coli* RNA polymerase, was similar to nuclear RNA. (These transcripts were subsequently found to be largely from repetitive sequences.) It was further shown that the specificity of transcription of reconstituted chromatin from different tissues could be altered by interchange of nonhistone proteins (148, 149). More recently, cDNA probes have been used to study transcription of unique sequences and it has been shown, for example, that the nonhistone proteins of erythroid cells induce transcription from globin genes when reconstituted with histones and DNA of nonerythroid cells (150, 151). Low reproducibility has been a major problem in such transcription experiments, and in experiments on unique sequences contamination with endogenous mRNA, which gives a false appearance of specificity, has been a worry. An additional problem has been non-specific transcription by the bacterial polymerase. Nonetheless, all the results suggest that it is possible to reconstruct the structure sufficiently well to reproduce some, at least, of the original template properties. This work has recently been summarized (152).

It is a reasonable assumption that precise reconstitution of the function of chromatin as template in transcription depends on accurate reconstitution of its structure, which includes both histones (probably combining without any sequence preference) and nonhistones (which may be sequence-specific). Present knowledge that chromatin structure is a repeating array of histone octamers surrounded by about 200-base pair lengths of DNA provides criteria for assaying the fidelity of reconstitution of histones and DNA. At present there is no structural test for correct reassembly of nonhistones. The criteria for correct assembly of histones and DNA (already discussed in relation to the native structure) are: an x-ray diffraction pattern characteristic of nuclei; a regularly repeating DNA structure as judged by the products of micrococcal nuclease digestion analyzed in polyacrylamide gels; a chain of closely juxtaposed particles, 100–120 Å diameter, in the electron microscope; a DNase I digestion pattern with a set of single strand fragments differing in size by 10 bases, showing regeneration of DNA-histone interactions within the nucleosome, and, serving a similar purpose, a typical limit digest pattern on exhaustive digestion with micrococcal nuclease; and, finally, indicating formation of correct histone-histone contacts, a histone cross-linking pattern with dimethylsuberimidate comparable with that obtained for native chromatin. The results so far indicate that individual beads may reform relatively readily but that a long, regularly repeating array of nucleosomes is more difficult to achieve (see below).

Reconstitution Procedures

In early reconstitution experiments, acid-extracted histones and DNA were mixed in high concentrations of sodium chloride (2 M) and urea (5 M) and then dialyzed first to low salt and then to low urea concentration (147). Histone H1 is the first to bind (at 0.4 M NaCl) in 5 M urea, and all five histones are bound at 0.1 M NaCl in 5 M urea (153). Urea could conceivably be useful in promoting correct refolding of acid-extracted histones but is unnecessary, and possibly harmful, if salt-extracted histones are used instead. The x-ray pattern of native chromatin is not in fact obtained for chromatin reconstituted in the presence of urea (154), but is obtained in its absence (10, 154). Regeneration of a micrococcal nuclease digestion pattern is also adversely affected by urea (155). A mixture of all five histones for use in reconstitution is readily prepared by extraction of chromatin with 2 M NaCl; a mixture of the four core histones H3, H4, H2A, and H2B can be obtained by extraction of H1-depleted chromatin.

Regeneration of a Beaded Structure

Reconstitution of the four core histones by salt gradient dialysis regenerates nucleosomes of normal size and containing 200 base pairs of DNA, as judged by electron microscopy (18). There seems to be no sequence specificity in nucleosome reconstitution because a beaded structure can be regenerated equally well from calf thymus histones and adenovirus DNA (18) (adenovirus does not contain histones) or SV40 DNA (46) (normally associated with histones). An excess of histone is required to obtain reconstituted material showing closely spaced nucleosomes; for example, a fully beaded SV40 minichromosome with 25 nucleosomes could be obtained only with a 2:1 (by weight) ratio of histone to DNA (compared with the 1:1 ratio found in chromatin) (46, 92). There could be several trivial explanations for this requirement for an excess of histone (e.g., less than optimum conditions of pH, time, ionic strength, etc. for reconstitution, or partially damaged histones), but it is also possible that a high efficiency of assembly of histones and DNA into nucleosomes is achieved in vivo through the action of an assembly factor, possibly protein(s). A strong indication that this may be so comes from the recent report (91) of a cell-free system from eggs of X. laevis that will assemble SV40 DNA and histones into chromatin whose DNA is supercoiled in the same way as that of natural SV40 chromatin (see under "Viral Chromatin: Its Contribution to Understanding Cellular Chromatin"). In addition to a requirement for nicking-closing enzyme, a fully active assembly system has a requirement for a thermolabile supernatant factor whose future characterization will be of considerable interest.

Regeneration of a Regular DNA Repeat

A regularly repeating DNA structure, as judged by micrococcal nuclease digestion, has been regenerated both by recombining pure DNA with pure histones and by recombining the unfractionated products of chromatin dissociation at high ionic strength. In experiments of the first type, use of the four histones H2A, H2B, H3, and H4 regenerated a DNA repeat of about 140 base pairs, suggesting that the repeating structure is probably an array of closely packed core particles (156, 157). (In contrast, nucleosomes spaced on the DNA rather than close-packed contained 200 base pairs of DNA as judged by electron microscopy; see previous section.) When H1 is present in reconstitutions carried out by this procedure, a regularly repeating structure is apparently not obtained, although the formation of individual nucleosomes containing H1 some distance apart on the DNA may be inferred from the protection of 185-base pair lengths during nuclease digestion (156). In an experiment of the second type (155), all five histones were apparently present, and a repeating structure was generated, although the repeat length was again considerably smaller than in the starting material. Improvement of the conditions for reconstitution in the presence of H1 is of considerable interest in view of current ideas (see below).

It is likely that where the repeat is about 140 base pairs (156, 157), octamers are in contact along the fiber, as they also seem to be in native chromatin when H1 is present, and the repeat is about 200 base pairs (25). One possibility, therefore, is that in a 200-base pair nucleosome a 60-base pair length of DNA is looped out, around H1, probably in the contact region between one octamer and the next. This DNA is the linker DNA which is assumed to accommodate the variability found in chromatins with different DNA repeats (see under "Location of Histone H1"). It has been suggested (53) that its length may be determined by the amino acid composition of H1, in particular by its content of basic residues.

Future Experiments

When conditions for reconstituting in the presence of H1 can be defined such that linker and core particle DNA are both represented in the repeat, a priority will be the reconstitution of chromatins whose natural DNA repeat lengths differ, for example, the neuronal and glial chromatins of the cerebral cortex (39) or chromatin from chicken liver and chicken erythrocytes (53) (see under "Variability in Nucleosomal DNA Content"). For instance, interchange of H1 molecules in reconstitution experiments with DNA and all five histones might resolve the question of whether, in vitro at least, H1 determines the repeat length. A further possibility is that attainment of a higher level of folding in some way influences repeat length. The number of possible variables (e.g., ionic strength, buffer, time, temperature, concentration, etc.) in a reconstitution experiment is enormous and perhaps some lessons from ribosome reconstitution would be well taken (158).

Partial Reconstitution of Chromatin Structure with Selected Histones

A low angle x-ray diffraction pattern with rings at spacings of 55, 33, and 27 Å, similar to that of chromatin, has been obtained by mixing DNA in 2 M NaCl, pH 7, with equal weights of the arginine-rich tetramer $(H3)_2(H4)_2$ and the lysine-rich complex H2A-H2B, prepared according to the method of van der Westhuyzen and von Holt (9), in a ratio of 5 parts of DNA to 4 parts by weight of total histone, and then dialyzing directly to 0.15 M NaCl (10). Both pairs of histones were required to generate the complete diffraction pattern. Acid-extracted H3 and H4, combined with DNA by a salt gradient dialysis procedure, produce diffraction maxima at 59, 35, 28, and 22 Å, but of lower intensity than those from chromatin (154). No other combination of two histones would suffice, and the presence of a third histone had a deleterious effect. Again, all four histones were required to regenerate the full diffraction pattern.

Felsenfeld and his co-workers (159) have assessed the extent of reconstitution of nucleosome structure by monitoring the regeneration of a nuclease digestion pattern in gels. The pattern results from cleavage at 10-base pair intervals on exhaustive treatment with micrococcal nuclease and extends from 50 up to 140 base pairs. In reconstitution experiments with arginine-rich and lysine-rich complexes fractionated according to van der Westhuyzen and von Holt (9), only the H3-H4 complex produced bands; these extended up to 70 base pairs, although fragments up to 160 base pairs were detected at very early times of digestion, suggesting that the complete DNA length of a core particle can be stabilized by H3 and H4 alone. All four histones were required for regeneration of the normal limit digest pattern. Similar conclusions concerning the role of the tetramer were drawn from a study using proteases and other nucleases to monitor nucleosome reconstitution (160).

Reconstitution experiments with different combinations of histones and DNA, monitored by electron microscopy, have shown that all four histones are required to form nucleosomes (125 Å diameter particles) and that the histones must be present in equimolar amounts. However, H3 and H4 alone, with DNA, give rise to particles about 80 Å in diameter, containing 129 ± 8 base pairs of DNA (134). Furthermore, the H3-H4 complex introduces negative superhelical turns into relaxed SV40 DNA (134).

The electron microscopy (134) and digestion studies (159, 160) thus suggest that the tetramer $(H3)_2(H4)_2$ is capable of re-establishing some part of the structure of a nucleosome, although all four histones are required for complete reconstitution, in accord with the evidence from x-ray diffraction (10, 154). These results strengthen the case in favor of a central structural role for the arginine-rich tetramer as the core of the nucleosome (16, 25).[1]

ARE ALL DNA SEQUENCES PACKAGED IN NUCLEOSOMES?

Nuclease digestion results indicating that most nuclear DNA can be obtained in fragments that are multiples of a unit size provide evidence that the bulk of the DNA is packaged in nucleosomes. However, a small proportion of the genome might not be so arranged, for instance, actively transcribed regions (or even coding sequences in general), simple sequence DNA such as centromeric heterochromatin, or certain classes of repeated genes such as those coding for ribosomal RNA. Good evidence from nuclease digestion studies now exists which suggests that nucleosomal organization is ubiquitous in DNA, with the possible exception of genes in the act of transcription, for which there are conflicting results from electron microscopy.

Transcribed Sequences: Nuclease Digestion Studies

There can be little doubt that coding sequences are organized in nucleosomes. Preparations of DNA complementary to various messenger RNAs (cDNA) have been shown to hybridize to the DNA fragments generated by internucleosomal

cleavage with micrococcal nuclease. This was first evident in experiments showing that cDNA to duck globin mRNA hybridized to the protected DNA of the
limit digest produced by exhaustive digestion of chromatin from duck red blood
cells with micrococcal nuclease (161). The existence of coding DNA in nucleosomes was further indicated by the demonstration that virtually all liver sequences transcribed into mRNA are present in micrococcal nuclease-resistant
DNA (162); other workers have reached the same conclusion (163). The most
direct evidence that nucleosomes contain coding sequences has come from the
hybridization of ovalbumin message cDNA to monomer, dimer, and trimer DNA
fragments, etc. produced by micrococcal nuclease digestion of oviduct nuclei
and red blood cell nuclei (P. Chambon, personal communication).[1] There is more
hybridization to monomer DNA than to higher multiples for nuclei from oviduct,
whereas the reverse is true for red blood cells. This may reflect increased accessibility to nuclease in actively transcribed regions (P. Chambon, personal communication). Finally, it has been shown that the transcriptionally active and
inactive fractions obtained from rat liver chromatin by a procedure (164) involving brief digestion with DNase II both give nucleoprotein particles with a DNA
content of about 200 base pairs on redigestion with the same enzyme (165).
However, the particles from the template-active material sediment at 14 S as
opposed to the value of 11 S usually found for nucleosomes; they contain nonhistones and RNA as well as histones; and they differ from normal nucleosomes in
sensitivity to DNase I and single strand-specific nuclease S1, and in optical
melting behavior (165). Other evidence suggesting that chromatin subunits are
altered during the process of transcription is discussed below.

Ribosomal genes in *Tetrahymena pyriformis* have been shown in two separate studies to have the same nucleosomal organization as the bulk chromatin
(166, 167). Bulk DNA of macronuclei and ribosomal DNA (rDNA), which is
extrachromosomal, were differentially radiolabeled (by making use of the fact
that rDNA replicates ahead of the bulk DNA) and shown to give the same pattern
of fragments on digestion with micrococcal nuclease. Because *Tetrahymena*
rDNA has very little nontranscribed spacer region, the evidence suggests that
transcribed DNA is organized in nucleosomes. One interpretation of a study
(168) in which ribosomal RNA was hybridized to chromatin monomer DNA
from both transcriptionally active and transcriptionally inert cells of *Xenopus* is
that actively transcribed genes might not be packaged in nucleosomes (see next
section). Precise interpretation, however, is difficult (168) because the fraction
of the reiterated ribosomal genes being actively transcribed is not known.

Packaging a DNA template into a nucleosome by tight association with
histones presents a mechanical problem for the transcriptional apparatus and it is
likely that an altered conformation of the nucleosome is necessary for transcription to proceed. Several lines of evidence, direct and indirect, lead to this conclusion. Axel et al. (161) have used duck globin cDNA as a probe of the accessibility of the globin gene in duck reticulocytes (active in synthesis of globin mRNA)
and erythrocytes (inactive). For erythrocytes, the probe hybridized equally to
"covered" sequences (micrococcal nuclease-resistant in a limit digest) and

"open" sequences (accessible to polylysine binding); i.e., all globin gene sequences were represented in both covered and open sequences. However, in reticulocytes, all gene sequences were represented in covered DNA, but a specific portion of the gene was missing from open DNA, possibly a reflection of an altered arrangement of proteins near the actively transcribed globin gene. Evidence for an altered conformation of nucleosomes in active gene regions has been obtained by Weintraub and Groudine (169). Hybridization to cDNA probes showed that brief digestion of nuclei with DNase I (to generate 10% acid soluble DNA) results in preferential degradation of globin sequences in erythroid, but not nonerythroid, cells. Moreover, embryonic but not adult globin sequences are destroyed in embryonic cells, and adult but not embryonic sequences in adult cells. Similarly, the ovalbumin gene is preferentially degraded by DNase I in ovalbumin-producing tissues (e.g., oviduct) but not in those that do not produce ovalbumin (170). In contrast, treatment of red cell nuclei with micrococcal nuclease does not result in preferential digestion of active globin genes, and 11 S monomer nucleosomes contain all the globin sequences. However, digestion of these monomers with DNase I again results in preferential loss of active globin genes (169). Thus, nucleosomes containing active genes seem to adopt a conformation in which the DNA is particularly sensitive to DNase I. [However, the ovalbumin gene in oviduct monomers was not found to retain its sensitivity to DNase I (170).] The work of Gottesfeld also suggests that nucleosomes in template-active regions may have an altered conformation (see above), and the results of Chambon suggest that actively transcribed sequences show increased accessibility to micrococcal nuclease (see above). The way in which a whole gene may acquire an altered association with histones has been a subject of speculation (143).

Transcribed Sequences: Electron Microscopy

Electron microscopy of transcriptional complexes spread by the technique of Miller and Beatty (171) seems to contradict some of the conclusions described above concerning packaging of active genes in nucleosomes. One study (23) suggests that no nucleosomes are present in transcriptionally active lampbrush chromosome loops or in nucleoli from *Triturus,* and that this is true not only for the transcribed sequences themselves but also for the intervening spacers. Particles seen in the spacers are thought not to be nucleosomes because they cause no contraction in the length of the DNA. Another study (172), however, is interpreted as showing nucleosomes in the nontranscribed spacer region between actively transcribed nonribosomal genes, and perhaps even between growing fibers in the transcribing regions. It is further suggested that particles present in the spacer regions of ribosomal cistrons are nucleosomes (rather than polymerase molecules). Both studies (23, 172), however, agree on the absence of nucleosomes on actively transcribed ribosomal genes.[1]

The apparent conflict between some of the results of nuclease digestion (166, 167) and electron microscopy (23, 172) concerning nucleosomal organization of ribosomal genes is as yet unresolved. One possible explanation might be that gene sequences studied in nuclease digestion experiments are those that are not being actively transcribed, whereas electron microscopy focuses on actual transcription complexes. Another might be that a 200-base pair periodicity in the DNA arises somehow other than from packing in nucleosomes.[1]

Simple Sequence DNA

A direct demonstration that mouse satellite DNA is organized in nucleosomes has come from hybridization of ^{3}H-labeled DNA complementary to purified mouse satellite light strand with the DNA fragments from a micrococcal nuclease digest of whole nuclei (36). Satellite DNA in the kangaroo rat (52% of the total DNA) also exists in nucleosomes (173).

When mouse nuclei (10% satellite DNA) were digested with the endonuclease Eco R_{II}, known to digest naked mouse satellite DNA to fragments that are multiples of about 245–248 base pairs (35, 45, 46), the relative amounts of the fragments led to the conclusion that "cleavage of satellite DNA in chromatin is subject to two periodicities, the spacing of the cleavage sites on the DNA and the spacing of the nucleosomes" (174), and that, broadly speaking, cleavage is possible only where the two periodicities coincide (in fact, quantitative explanation of the intensities of various fragments requires a certain degree of cleavage within the nucleosome). The frequency with which Eco R_I cleaves bovine satellite I (7% of the genome) has also been measured (175); this DNA contains Eco R_I sites at 1,400-base pair intervals. A third of the satellite I DNA in intact calf thymus nuclei is liberated as a 1,400-base pair fragment. Assuming a random distribution of nucleosomes around potential cleavage sites, it is estimated that about 57% of the total Eco R_I sites in nuclei are accessible to cleavage. The frequency of generation of particular fragments indicates that if all sites in the linker DNA are accessible, 45% of the sites in the 140-base pair cores are also cleaved (175).

THE PHASE PROBLEM

The question has been raised whether or not there is a specific phase relationship between nucleosomes and base sequences in DNA (16, 108). For example, is a certain base sequence always found in the linker DNA between 140-base pair core regions, or is it always found in a core, in cells of a given type?

In the case of SV40 chromatin, it seems clear that the relationship of nucleosomes to base sequences in the DNA is random. Cleavage of SV40 chromatin by Hind III results in all five of the DNA fragments produced by digestion of naked DNA, but in reduced amounts (176). Similarly, the single Eco R_I site in SV40 DNA is neither fully blocked by histones nor fully free, but available for cleavage in 20% of the molecules (90, 176, 177).[1]

Although studies of viral chromatin may have no direct bearing on the phase problem in cellular chromatin, recent work has shown that the relationship of nucleosomes to base sequences is random for the single copy sequences in rat liver cells (59). In this work, 140-base pair core particle lengths of DNA were digested to a limit with *E. coli* exonuclease III, which degrades DNA from the 3' terminus and is specific for double strand DNA. The resulting single strands, about 70 residues long (half from each of the original strands), are expected to reassociate with each other if the phase relationship is random, and to remain single-stranded in the case of a specific phase. Extensive reassociation was found, suggesting a random phase relationship (59).

ASPECTS OF HISTONE STRUCTURE

Histone Interactions and Conformations

Several studies have been made of the association behavior of histones in solution (10–13, 178–184) (see also under "Histones"). Most of the pairwise associations found also seem to exist in chromatin, as indicated by the composition of cross-linked dimers (see under "Associations of Histones in Chromatin"). The native state of the arginine-rich histones in solution at physiological ionic strength is probably the tetramer $(H3)_2(H4)_2$ (10, 11); the lysine-rich histones also exist as a specific oligomer (10, 12) (see under "Histones"). The octamer of histones isolated from chromatin in 2M NaCl (25, 141) (see under "The Protein Core of the Nucleosome") may comprise a tetramer and two lysine-rich dimers (25).

The precise molecular details of the interactions of histones with each other, and with DNA, in the nucleosome will come ultimately from (and probably only from) x-ray crystallographic analysis of nucleosome core particles. In the meantime, attempts are being made to approach the problem through study of histone conformations, mainly by nuclear magnetic resonance (NMR) methods. Earlier studies of individual acid-extracted histones (185) are now of less interest in relation to chromatin than are more recent studies of gently extracted histone complexes, in particular the arginine-rich tetramer $(H3)_2(H4)_2$. Work on the tetramer (see below) has been largely directed to the question of whether the NH_2-terminal region of the histone has a conformation different from that of the rest of the molecule. The question arises from the fact that the basic amino acids are located primarily in the NH_2-terminal regions of the histones and from the demonstration that trypsin cleaves 20–30 amino acid residues from the NH_2-terminal regions of the histones in chromatin (136). This, together with earlier NMR evidence (185), led to the idea (136, 185) that histones associate with each other primarily through their COOH-terminal regions (which have "normal" amino acid compositions and which are, therefore, assumed to be roughly globular) and that NH_2-terminal "tails" extend outward to envelop the DNA. The basic tails are then seen as being located outside the DNA in chromatin, hence their cleavage by trypsin.

The simple view that the arginine-rich tetramer is roughly globular, with a surface rich in the side chains of basic amino acid residues (contributed mainly by the NH_2-terminal regions of the polypeptide chains) which is the site of interaction with DNA phosphates has been rejected in two NMR studies (186, 187) in favor of a structure with mobile tails. Proton NMR studies (186) are interpreted as showing tails because of the absence of the line broadening (particularly of resonances from lysine side chains) that would indicate restricted rotation. (The histone spectrum, at 2.5 mg/ml, showed some sharp resonances, whereas all resonances were broadened in serum albumin at 10 mg/ml.) In ^{13}C NMR studies (187), it has been argued that the resonance from glycine can be distinguished and used as an index of the mobility of the polypeptide backbone—in particular, the backbone in the NH_2-terminal region because most of the glycines are located there. Since a spectrum of the tetramer shows much greater mobility than that of lysozyme, a protein known to be truly globular, it is concluded (187) that the tetramer has NH_2-terminal, conformationally free tails. However, whereas the tetramer spectrum was measured at 100 mg/ml (~2 mM), lysozyme was measured at 250 mg/ml (~20 mM) at which concentration intermolecular interactions might well restrict any mobility of the backbone that existed.

Although there is a consensus from the two NMR studies (186; 187) on mobility of the NH_2-terminal regions of the histone chains in the $(H3)_2(H4)_2$ tetramer, this conclusion should at present be regarded as tentative. The complexity of protein NMR spectra does not allow any definite assignment of particular resonances to particular groups in a protein; furthermore, any comparisons of two proteins should be carried out under conditions that are as nearly identical as possible. Other detailed criticisms of conclusions drawn from NMR studies about histone conformations have recently been advanced (58).

The features of histone and chromatin structure for which tails have been invoked could equally well be explained in other ways. A high degree of hydration arising from a high surface density of positive charge could account for the somewhat low sedimentation coefficient of the tetramer (10, 186). A protease-sensitive domain could account for the cleavage by trypsin of 20–30 residues from the NH_2-terminus of the histones. In fact, there is good analogy for the latter in the cleavage of a 20-residue peptide (the S-peptide) from the NH_2-terminus of ribonuclease A by subtilisin (188), and of a 48-residue peptide from the NH_2-terminus of staphylococcal nuclease by trypsin (189); there is no disagreement on the globular nature of these proteins because their three-dimensional structures are known (189, 190). On balance, there seems to be no convincing reason for abandoning the simple idea that the histone tetramer is a fairly typical globular protein, comparable in dimensions and subunit structure to hemoglobin, but with a high density of surface positive charge which in chromatin binds the DNA.

H1 is somewhat different from the other histones, in structure as well as in function. Its hydrodynamic properties (12, 84) and behavior on gel filtration (10) suggest a very asymmetric molecule. The amino acid sequence shows a polar NH_2-terminal region (1–15 acidic, 16–39 basic), a central apolar region (residues 44–111), which is undoubtedly globular as confirmed by NMR studies (191), and a very basic COOH-terminal region (residue 111 to end), which is presumably a major site of interaction with DNA. Chemical cross-linking does not detect interaction of H1 with the other histones, either in solution or in chromatin at physiological ionic strength (25), although H1 self-associations do appear to occur in chromatin as described earlier (see under "Associations of Histones in Chromatin").

Histone Modifications

Histones undergo a variety of sequence-specific, postsynthetic enzymic modifications at particular positions in the amino acid sequence (see refs. 4 and 192). The side chains of lysine residues can be N-acetylated, or N-methylated (mono-, di-, or trisubstituted); serine residues can be O-acetylated or O-phosphorylated; histidine and arginine residues can be N-methylated; and the histone NH_2-terminal residue can be acetylated. There is a striking contrast between the structural variety introduced by these modifications and the structural constraints imposed by conservation of amino acid sequence of the histones, particularly H3 and H4. A possible interpretation is that precise and tight binding of histones to each other and to DNA is structurally critical (hence the conserved amino acid sequences), but that release of these interactions is also essential for particular chromosomal functions (e.g., replication, transcription) and this could be achieved by introducing structural perturbations through enzymic modifications of amino acid side chains.

Different types of modification have different consequences: N-acetylation causes a loss of one positive charge; N-methylation of lysine causes no charge change but an increase in basicity of the nitrogen and increased hydrophobicity; and O-phosphorylation introduces two negative charges. The ϵ-N-acetyl groups and O-phosphate groups turn over rapidly, whereas NH_2-terminal acetyl groups and N-methyl groups are stable. ϵ-N-acetylation of H2A, H2B, H3, and H4 occurs in S-phase of the cell cycle in several tissues as well as in cultured cells (see ref. 192). It had earlier been suggested that cycles of acetylation and deacetylation might be essential for assembly of newly synthesized histones into chromatin and for attainment of the final correct fit (193). All five histones can be O-phosphorylated, and some phosphorylation events seem to be specific to certain stages of the cell cycle, whereas others occur throughout (194) (for more detail, see ref. 192). Some phosphorylations of H1, for example in S-phase, may be related to decondensation of chromatin for replication to proceed, whereas metaphase phosphorylation at a different subset of sites in H1 may promote chromosome condensation (195). Other types of histone modification are discussed more fully in reference 192. In many cases, the positions of modification in the amino acid sequence have been established, but the precise biological significance is still far from being understood.

An additional modification of histones that could have profound implications is poly(ADP-ribosyl)ation (196, 197). Histone H1 in trout testis nuclei is poly(ADP-ribosyl)ated at three sites (192, 198), in each case at the γ-carboxyl group of a glutamic acid residue (probably by a glycosidic ester linkage to the 1'-position of the ribose). One of the residues modified, Glu-2, carries up to 7 residues of ADP-ribose, and the introduction of 7×1.5 negative charges (at pH 7.0) is likely to modify profoundly the binding of the NH_2-terminal region to DNA in chromatin. Much greater alteration of charge can be produced by poly(ADP-ribosyl)ation than by phosphorylation, and the former rather than the latter could turn out to be the biologically important way of modulating the action of H1 in chromatin (for instance, its role in chromatin condensation and decondensation) (192).

Histone Variants

A striking feature of the four main histones is the conservation of their amino acid sequences through evolution, particularly in the case of H3 and H4 (see under "Histones"). Variants of H3, H2A, and H2B have nonetheless been found (variant forms ("subfractions") of H1 within a tissue, in addition to species specific forms (see ref. 4), have been known for some time). In the case of H3 and H2A, these variants are nonallelic and have been preserved in parallel during evolution. There seems to be a correlation between variants and differentiation, because different variants of H2A and H2B, as well as of H1, are synthesized at different stages of embryogenesis in the sea urchin (199) (one set of variants being synthesized until blastula stage and another set thereafter), and different mammalian tissues contain different amounts of variant forms (8). This suggests that tissue-specific modulation of the basic structure of chromatin may be required, possibly for differential gene expression, and it is possible that the presence of histone variants, together with enzymic modification of histones (see above), could provide for such modulation.

Histone variants are often well resolved in acid-urea polyacrylamide gels containing Triton X-100 (8, 200). The amino acid substitutions in the mammalian H3, H2A, and H2B variants are largely conservative (201). An H2A variant has serine replacing threonine at position 16 in the sequence, and methionine replacing leucine at position 51; three H2B variants have serine replacing glycine at position 75 and/or glutamine for glutamic acid at position 76; H3 variants have serine replacing cysteine at position 96, and if serine occupies this position there may also be substitutions at positions 89 and 90. All tissues seem to contain all variants, but in different relative amounts (a particular variant may represent less than 1% or as much as 50% of that histone type (8)). In mammalian nuclei, at least 136 different histone octamers with composition $(H2A)_2(H2B)_2(H3)_2(H4)_2$ could, therefore, be formed, each having a further potential for variation by side chain modification (8). All the substitutions in mammalian variants noted above (except that at position 16 in H2A) are located in hydrophobic regions which are likely to form the contact regions between histones in the octamer in chromatin, so that hybrid nucleosome cores may well differ in stability. Zweidler (8) suggests that "as in the case of isozymes, different populations of histone complexes might be optimal for genome function in the different nuclear milieus of cells in different metabolic states."

NONHISTONE PROTEINS

Nonhistone proteins are those proteins other than histones that are found associated with chromatin prepared by a variety of procedures (see ref. 4 for a survey of the literature up to 1975). The most abundant nonhistone proteins form a pattern of about 100 bands in an SDS-polyacrylamide gel (202). Much greater resolution is possible with the use of the two-dimensional gel system of O'Farrell (203) in which separation in the first dimension is achieved by isoelectric focusing and in the second dimension by electrophoresis in the presence of SDS. For example, about 400 nonhistone proteins were demonstrated in HeLa cell nuclei, most of which were present at less than 10,000 copies per nucleus; as few as 500 copies could be detected (204).

Many nonhistone proteins may be nucleoplasmic (possibly even cytoplasmic) contaminants that bind to chromatin during isolation, as well as membrane proteins, whose attachment to chromosomal material may or may not be structurally significant. The search for specific gene activators among the nonhistone proteins has recently been reviewed (205). Some specificity in in vitro transcription of reconstituted chromatin (see above) suggests (despite the difficulties inherent in such experiments, summarized in refs. 152 and 206) that the nonhistone proteins do include regulatory proteins, at least for the transcription of globin genes (151) and histone genes (207). Another example of a regulatory role for nonhistone proteins may be found in steroid hormone action. The effect of steroid hormones on transcription in target tissues is mediated by a cytoplasmic receptor which moves into the nucleus. Specific binding of the hormone-receptor complex to chromatin seems, in some cases at least, to require particular nonhistone proteins, because a subfraction of nonhistones in chick oviduct is necessary for binding the progesterone receptor (208). Gene activation by hormones is often accompanied by synthesis of specific nuclear proteins (see ref. 209); the function of these newly synthesized proteins is still a matter of conjecture.

Measurement of specific, rather than nonspecific, binding to chromatin presents a problem. Lin and Riggs (210) pointed out, with particular reference to the *lac* repressor in prokaryotes, that any protein with high affinity for a particular DNA sequence must also have a finite, albeit lower, affinity for DNA in general, and they suggested that in eukaryotes there may be numerous nonspecifically bound copies of a given regulatory protein for every specifically bound copy. Indeed, Yamamoto and Alberts (211) have shown that estradiol-receptor complexes bind with a dissociation constant of about 5×10^{-4} M to any DNA (eukaryotic, prokaryotic, or synthetic), thereby explaining earlier observations that a nucleus contains at least 30,000 binding sites for the complex. The number of specific binding sites was calculated to be about 1,000, and ways of detecting them were suggested (211).

A promising approach to the study of nonhistone proteins in polytene chromosomes involves indirect immunofluorescence. Rabbit antibodies, raised against *Drosophila* nonhistone proteins fractionated in SDS gels, are adsorbed to polytene chromosomes and visualized with fluorescent goat antirabbit antibodies

(212). Over a thousand chromatin fibers are lined up in register in a polytene chromosome, so the binding of a single protein to a sequence that occurs in one copy per haploid genome leads to an accumulation of antibody molecules and thus a region of fluorescence at a single locus on the chromosome. So far, the method has demonstrated the localization of some subfractions of total nonhistone proteins in certain bands in *Drosophila* salivary gland chromosomes (213). Another potentially useful approach is the bulk preparation by cloning techniques of defined eukaryotic DNA sequences which may then be used to assay for specific binding of particular nonhistone proteins.

Structural studies on one class of nonhistone proteins are quite far advanced. These are the HMG proteins ("high mobility group," a reference to electrophoretic mobility in polyacrylamide gels) (214). There are 10^5–10^6 copies of each per nucleus, and in isolated nucleosomes they occur at the level of about 1 molecule per 15 nucleosomes (215). It is not clear whether they are uniformly distributed in chromatin or whether they are localized in specific regions of the genome. Isolation procedures have been described for proteins HMG-1, HMG-2 (both MW ca. 27,000) (216), and HMG-17 (MW 9,250) (217). They have a relatively high content of polar amino acids, particularly of lysine (about 22%, 22%, and 24%, respectively). Like the histones, they show an asymmetry in the distribution of charged residues within the amino acid sequence, the NH_2-terminal ends being relatively basic. The complete sequence of HMG-17 (89 residues) has recently been determined (218). The role of this class of nonhistone proteins is not known, but it may be significant that isolated HMG-1 and 2 have recently been shown to interact in solution with certain subfractions of H1 (219). Structural studies on certain other nonhistone proteins, also in search of a role, are in progress, e.g., proteins A24 (220) and D1 (221).[1]

REPLICATION OF CHROMATIN

The size of the Okazaki fragments produced in replicating *Drosophila* DNA and the size of the single strand gaps near the replication forks (222) suggest (16) that DNA might be made available for replication in nucleosome-sized units. There is no evidence as yet on whether histones remain associated with DNA through strand separation and DNA synthesis (see below). The distribution of histones and nucleosomes near the replication fork has, however, been examined (223–227). In the absence of new histone synthesis (cycloheximide block), there is an excess of DNA, and nucleosomes seem to form cooperatively on one or both daughter strands (223, 224). Pulse labeling of DNA at the replication fork and analysis of newly replicated chromatin also lead to the conclusion that nucleosomes are formed cooperatively on a daughter chromosome (226). Two conflicting results have been obtained for normal replication in which histone synthesis is not limiting. One result is that histones are distributed randomly on old and new strands of DNA (145), and the other is that newly synthesized histone associates with newly synthesized DNA (144). The latter has been incorporated into a proposal (124, 143) that the octameric core of the nucleosome splits during replication into two heterotypic tetramers (see under "The Protein Core of the Nucleosome"), one of which remains associated with each parental strand of DNA; after DNA replication, newly synthesized histones are bound to the new DNA and the octamer is reformed.[1]

[1]*NOTE ADDED IN PROOF:*
There have been a number of advances since this chapter was written, many of which were reported at the Cold Spring Harbor Symposium (228) in June 1977. Only a few of those that bear directly on the chapter will be mentioned here.

A major advance has been the x-ray analysis by Finch and Klug and their co-workers (228, 229) of single crystals of (proteolytically cleaved) nucleosome core particles. The particle is a slightly wedge-shaped disc of dimensions $55 \times 55 \times 110$ Å and is consistent with about 1¾ turns of DNA around a histone core, with about 75–82 base pairs per turn. A similar disc-shaped core particle had also been suggested by Richards et al. (230) from low angle x-ray and neutron scattering studies in solution, although such studies, unlike x-ray diffraction by crystals, could not alone prove a model. The actual number of base pairs per turn is taken by Finch et al. (229) to be 80, since sites in the DNA cut with similar frequencies are 80 bases apart on a strand, as shown by Lutter (228); in a flat DNA superhelix, such as that proposed, such sites will share a common environment and be protected to similar extents. The 1¾ turns of DNA per 140 base pair core particle (implying 2½ turns per 200 base pair nucleosome if the whole length is wound superhelically) can be reconciled with 1¼ apparent superhelical turns per nucleosome measured for the DNA in solution [or rather with the measured change in linkage number of the DNA (231)] if there is a decrease in the helical screw rotation angle of free DNA as it binds to histone to form a nucleosome (229). There would be a corresponding change in the number of base pairs per turn of the duplex, and since this is taken to be 10 base pairs for DNA in the nucleosome (from DNase I digestion results) 10–11 base pairs per turn are predicted for B-form DNA in solution (229). The Symposium volume (228) also contains full details of the two studies, by Lutter and by Noll, of DNase I cleavage sites in the core particle to which reference was made in the text (128, 129). Lutter's analysis of the kinetics of digestion is consistent with a dyad axis of symmetry in the particle.

The complete sequence of trout testis H1 has been published (232). The role of the arginine-rich histones in folding DNA has been studied further (233–235) but there is some disagreement on whether the 130–140 base pair particle (see text) contains one or two $(H3)_2(H4)_2$ tetramers. Two papers suggest that in vitro reconstitution of nucleosomes proceeds by assembly of preformed octamers with DNA (Thomas and Butler and Simpson et al. in ref. 228). Weintraub et al. (228) show that histones segregate conservatively as octamers during chromatin replication over 3–4 generations and that the assembly of new octamers is also conservative, new histones not mixing with old. This argues against the model previously proposed (124,143) in which histones segregate semi-conservatively as half-octamers (heterotypic tetramers) during replication. Keller et al. (228) and Oudet et al. (228) confirm that the SV40 minichromosome prepared in 0.15 M rather than 0.4 M NaCl contains H1:the nature of this H1 is still uncertain since it is removed in 0.2–0.3 M NaCl in contrast with typical eukaryotic H1. Mini-chromosomes isolated containing H1 appear compact in the electron microscope but have the same number of superhelical turns in the DNA and show the same micrococcal nuclease digestion pattern as the minichromosome lacking H1 (see text). A study (236) of the arrangement of histones with respect to base sequences in polyoma virus and SV40 suggests that the arrangement may not in fact be completely random (see "The Phase Problem") but that nucleosomes occupy a small number of distinct alternative positions. Rouvière-Yaniv (228) has shown that the *E. coli* DNA binding protein HU is a dimer present in amount sufficient to account for one dimer per 200 base pairs of DNA, so its role in DNA packing may be similar to that of histones in eukaryotes. It binds preferentially to supercoiled DNA.

A striking result for nonhistone proteins is the identification by Laemmli's group (Cheng et al., ref. 228) of a protein "scaffold" containing a relatively small number of nonhistone proteins in human metaphase chromosomes, and the visualization of loops 30,000–90,000 base pairs long emerging from this scaffold. A major trout nonhistone protein seems to be associated with the linker DNA of the nucleosome (237). Histone H5 does not seem to be responsible for the long linker DNA length in chick erythrocyte nuclei since chick erythroid cells which also contain H5 have the same DNA repeat length (about 200 base pairs) as chick liver nuclei (238). Many groups are probably already exploiting the observation (239) that treatment of cells with *n*-butyrate causes extensive acetylation of histone H4 to examine the effect of this modification on nucleosome structure and function.

The state of the chromatin fiber in transcription has been studied further by electron microscopy (Foe, McKnight et al., and Franke et al., in ref. 228) and the evidence has been presented (Bellard et al., ref. 228) for localization of the actively transcribed ovalbumin gene in the series of fragments produced by micrococcal nuclease digestion of chromatin from the oviduct of the laying hen. The problems of interpretation that arise for reiterated ribosomal genes (see text) do not apply in the case of the single copy ovalbumin sequence in a population of cells of which most are actively transcribing the gene. The apparent conflict between some electron microscopy results and nuclease digestion studies on the presence of nucleosomes on actively transcribed genes (see text) may perhaps be resolved in part (Bellard et al., ref. 228) by the observation that a nucleofilament in which the DNA is compacted in nucleosomes may undergo reversible structural transitions to an extended state (a thickened nucleohistone fiber in the electron microscope) (Oudet et al., ref. 228). This would also be expected to reduce the mechanical problems that would arise in transcription of DNA coiled in a nucleosome.

There have been interesting suggestions relating to higher order structures. The "superbead" containing 6–10 nucleosomes, and held together by H1, has been suggested as an alternative to the solenoid as the true superstructural element in native chromatin, on the basis of electron microscopy and nuclease digestion studies (240, 241). Meanwhile, a detailed model has been proposed for the higher order coiling of a nucleofilament into a solenoid (Worcel et al., ref. 228). The metaphase chromosome has been described, on microscopic evidence, as a "hierarchy of helices" comprising a solenoid wound into a further helix. This supersolenoid would have a DNA packing ratio of 1600, and, if further folded to the dimensions of chromatids, would give the additional factor of about 5 necessary to account entirely for the DNA packing ratio in metaphase chromosomes (242).

ACKNOWLEDGMENT

I am most grateful to Dr. Roger Kornberg for many constructive comments on a draft of this chapter and to Dr. Aaron Klug for comments on the final manuscript. I also thank those friends and colleagues who have allowed me to cite their results before publication.

REFERENCES

1. Rouvière-Yaniv, J., and Gros, F. (1975). Proc. Natl. Acad. Sci. USA 72:3428.
2. Griffith, J. (1976). Proc. Natl. Acad. Sci. USA 73:563.
3. Pettijohn, D. E. (1976). Crit. Rev. Biochem. 4:175.
4. Elgin, S. C. R., and Weintraub, H. (1975). Annu. Rev. Biochem. 44:725.
5. Isenberg, I., and Van Holde, K. E. (1975). Accounts Chem. Res. 8:327.
6. Kaye, J. S., and McMaster-Kaye, R. (1974). Chromosoma 46:397.
7. Shires, A., Carpenter, M. P., and Chalkley, R. (1975). Proc. Natl. Acad. Sci. USA 72:2714.
8. Zweidler, A. (1976). In V. G. Allfrey, E. K. F. Bautz, B. J. McCarthy, R. T. Schimke, and A. Tissières (eds.), Organization and Expression of Chromosomes, p. 187. Dahlem Konferenzen, Berlin.
9. van der Westhuyzen, D. R., and von Holt, C. (1971). FEBS Lett. 14:333.
10. Kornberg, R. D., and Thomas, J. O. (1974). Science 184:865.
11. Roark, D. E., Geoghegan, T. E., and Keller, G. H. (1974). Biochem. Biophys. Res. Commun. 59:542.
12. Kelley, R. I. (1973). Biochem. Biophys. Res. Commun. 54:1588.
13. D'Anna, J. A., and Isenberg, I. (1974). Biochemistry 13:4992.
14. Pardon, J. F., and Wilkins, M. H. F. (1972). J. Mol. Biol. 68:115.
15. Hewish, D. R., and Burgoyne, L. A. (1973). Biochem. Biophys. Res. Commun. 52:504.
16. Kornberg, R. D. (1974). Science 184:868.
17. Ris, H., and Kubai, D. F. (1970). Annu. Rev. Genet. 4:263.
18. Oudet, P., Gross-Bellard, M., and Chambon, P. (1975). Cell 4:281.
19. Finch, J. T., and Klug, A. (1976). Proc. Natl. Acad. Sci. USA 73:1897.
20. Woodcock, C. L. F. (1973). J. Cell Biol. 59:368(a).
21. Olins, A. L., and Olins, D. E. (1974). Science 183:330.
22. Woodcock, C. L. F., Safer, J. P., and Stanchfield, J. E. (1976). Exp. Cell Res. 97:101.

23. Franke, W. W., Scheer, U., Trendelenburg, M. F., Spring, H., and Zentgraf, H. (1976). Cytobiologie 13:401.
24. Noll, M. (1974). Nature (Lond.) 251:249.
25. Thomas, J. O., and Kornberg, R. D. (1975). Proc. Natl. Acad. Sci. USA 72:2626.
26. Griffith, J. D. (1975). Science 187:1202.
27. Finch, J. T., Noll, M., and Kornberg, R. D. (1975). Proc. Natl. Acad. Sci. USA 72:3320.
28. Van Holde, K. E., Sahasrabuddhe, C. G., Shaw, B. R., Van Bruggen, E. F. J., and Arnberg, A. C. (1974). Biochem. Biophys. Res. Commun. 60:1365.
29. Sperling, L., and Tardieu, A. (1976). FEBS Lett. 64:89.
30. Noll, M. (1974). Nucleic Acids Res. 1:1573.
31. Pardon, J. F., Worcester, D. L., Wooley, J. C., Tatchell, K., Van Holde, K. E., and Richards, B. M. (1975). Nucleic Acids Res. 2:2163.
32. Hjelm, R. P., Kneale, G. G., Suau, P., Baldwin, J. P., Bradbury, E. M., and Ibel, K. (1977). Cell 10:139.
33. Burgoyne, L. A., Hewish, D. R., and Mobbs, J. (1974). Biochem. J. 143:67.
34. Clark, R. J., and Felsenfeld, G. (1971). Nature (New Biol.) 229:101.
35. Noll, M., and Kornberg, R. D. (1977). J. Mol. Biol. 109:393.
36. Bokhon'ko, A., and Reeder, R. H. (1976). Biochem. Biophys. Res. Commun. 70:146.
37. Thomas, J. O., unpublished results.
38. Thomas, J. O., and Furber, V. (1976). FEBS Lett. 66:274.
39. Thomas, J. O., and Thompson, R. J. (1977). Cell 10:633.
40. Shaw, B. R., Herman, T. M., Kovacic, R. T., Beaudreau, G. S., and Van Holde, K. E. (1976). Proc. Natl. Acad. Sci. USA 73:505.
41. Spadafora, C., Bellard, M., Compton, J. L., and Chambon, P. (1976). FEBS Lett. 69:281.
42. Compton, J. L., Bellard, M., and Chambon, P. (1976). Proc. Natl. Acad. Sci. USA 73:4382.
43. Noll, M. (1976). Cell 8:349.
44. Lohr, D., and Van Holde, K. E. (1975). Science 188:165.
45. Southern, E. M. (1975). J. Mol. Biol. 94:51.
46. Bellard, M., Oudet, P., Germond, J.-E., and Chambon, P. (1976). Eur. J. Biochem. 70:543.
47. Whitlock, J. P., and Simpson, R. T. (1976). Nucleic Acids Res. 3:2255.
48. Sahasrabuddhe, C. G., and Van Holde, K. E. (1974). J. Biol. Chem. 249:152.
49. Morris, N. R. (1976). Cell 9:627.
50. Thomas, J. O., Mechler, B., Cantin, E., and Northcote, D. H., unpublished results.
51. Compton, J. L., Hancock, R., Oudet, P., and Chambon, P. (1976). Eur. J. Biochem. 70:555.
52. Lohr, D., Corden, J., Tatchell, K., Kovacic, R. T., and Van Holde, K. E. (1977). Proc. Natl. Acad. Sci. USA 74:79.
53. Morris, N. R. (1976). Cell 8:357.
54. Johnson, E. M., Littau, V. C., Allfrey, V. G., Bradbury, E. M., and Matthews, H. R. (1976). Nucleic Acids Res. 3:3313.
55. Lipps, H. J., and Morris, N. R. (1977). Biochem. Biophys. Res. Commun. 74:230.
56. Palau, J., Ruiz-Carrillo, A., and Subirana, J. A. (1969). Eur. J. Biochem. 7:209.
57. Sarkander, H. I., Fleishcher-Lambropoulos, H., and Brade, W. P. (1975). FEBS Lett. 52:40.
58. Kornberg, R. D. (1977). Annu. Rev. Biochem. 46:931.
59. Prunell, A., and Kornberg, R. D. Phil. Trans. R. Soc. Lond. In press.

60. Noll, M., Thomas, J. O., and Kornberg, R. D. (1975). Science 187:1203.
61. Thomas, J. O., and Kornberg, R. D. (1975). FEBS Lett. 58:353.
62. Sperling, L., and Klug, A. (1977). J. Mol. Biol. 112:253.
63. Rees, A. W., Debuysere, M. S., and Lewis, E. A. (1974). Biochim. Biophys. Acta 361:97.
64. Gottesfeld, J. M., Garrard, W. T., Bagi, G., Wilson, R. F., and Bonner, J. (1974). Proc. Natl. Acad. Sci. USA 71:2193.
65. Woodhead, L., and Johns, E. W. (1976). FEBS Lett. 62:115.
66. Davies, G. E., and Stark, G. R. (1970). Proc. Natl. Acad. Sci. USA 66:651.
67. Thomas, J. O., and Kornberg, R. D. (1977). Methods Cell Biol. 18:429.
68. Hardison, R. C., Eichner, M. E., and Chalkley, R. (1975). Nucleic Acids Res. 2:1751.
69. Hand, E. S., and Jencks, W. P. (1962). J. Am. Chem. Soc. 84:3505.
70. Hunter, M. J., and Ludwig, M. L. (1962). J. Am. Chem. Soc. 84:3491.
71. Brutlag, D., Schlehuber, C., and Bonner, J. (1969). Biochemistry 8:3214.
72. Olins, D. E., and Wright, E. B. (1973). J. Cell Biol. 59:304.
73. Jackson, V. J., and Chalkley, R. (1974). Biochemistry 13:3952.
74. Chalkley, R., and Hunter, C. (1975). Proc. Natl. Acad. Sci. USA 72:1304.
75. Hyde, J. E., and Walker, I. O. (1975). FEBS Lett. 50:150.
76. Van Lente, F., Jackson, J. F., and Weintraub, H. (1975). Cell 5:45.
77. Bonner, W. M., and Pollard, H. B. (1975). Biochem. Biophys. Res. Commun. 64:282.
78. Martinson, H. G., and McCarthy, B. J. (1975). Biochemistry 14:1073.
79. Martinson, H. G., Shetlar, M. D., and McCarthy, B. J. (1976). Biochemistry 15:2002.
80. Martinson, H. G., and McCarthy, B. J. (1976). Biochemistry 15:4126.
81. Richards, B. M., and Pardon, J. F. (1970). Exp. Cell Res. 62:184.
82. Thomas, J. O. (1977). In P. O. P. Ts'o (ed.), Molecular Biology of the Mammalian Genetic Apparatus, Vol. 1., p.199. Elsevier/North Holland Biomedical Press, Amsterdam.
83. Olins, A. L., Carlson, R. D., Wright, E. B., and Olins, D. E. (1976). Nucleic Acids Res. 3:3271.
84. Edwards, P. A., and Shooter, K. V. (1969). Biochem. J. 114:227.
85. Varshavsky, A. J., Bakayev, V. V., and Georgiev, G. P. (1976). Nucleic Acids Res. 3:477.
86. Ilyin, Y. V., Varshavsky, A. Y., Mickelsaar, U. N., and Georgiev, G. P. (1971). Eur. J. Biochem. 22:235.
87. Finch, J., Kornberg, R. D., and Noll, M., unpublished results.
88. Varshavsky, A. J., Bakayev, V. V., Chumackov, P. M., and Georgiev, G. P. (1976). Nucleic Acids Res. 3:2101.
89. Christiansen, G., Landers, T., Griffith, J., and Berg, P. In press.
90. Crémisi, C., Pignatti, P. F., Croissant, O., and Yaniv, M. (1976). J. Virol. 17:204.
91. Laskey, R. A., Mills, A. D., and Morris, N. R. (1977). Cell 10:237.
92. Germond, J. E., Hirt, B., Oudet, P., Gross-Bellard, M., and Chambon, P. (1975). Proc. Natl. Acad. Sci. USA 72:1843.
93. Keller, W. (1975). Proc. Natl. Acad. Sci. USA 72:4876.
94. Bauer, W., and Vinograd, J. (1968). J. Mol. Biol. 33:141.
95. Vogel, T., and Singer, M. (1975). J. Biol. Chem. 250:796.
96. Böttger, M., Scherneck, S., and Fenske, H. (1976). Nucleic Acids Res. 3:419.
97. Davies, H. G., Murray, A. B., and Walmsley, M. E. (1974). J. Cell Sci. 16:261.

98. Davies, H. G., and Haynes, M. E. (1976). J. Cell Sci. 21:315.
99. Crick, F. H. C. (1976). *In* V. G. Allfrey, E. K. F. Bautz, B. J. McCarthy, R. T. Schimke, and A. Tissières (eds.), Organization and Expression of Chromosomes, p. 71. Dahlem Konferenzen, Berlin.
100. Varshavsky, A. J., Ilyin, Y. V., and Georgiev, G. P. (1974). Nature (Lond.) 250:602.
101. Beerman, W. (1972). *In* W. Beermann (ed.), Results and Problems in Cell Differentiation, Vol. 4, p. 1. Springer-Verlag, Berlin.
102. Worcel, A., and Burgi, E. (1972). J. Mol. Biol. 71:127.
103. Cook, P. R., and Brazell, I. A. (1975). J. Cell Sci. 19:261.
104. Ide, T., Nakane, M., Anzai, K., and Andoh, T. (1975). Nature (Lond.) 258:445.
105. Benyajati, C., and Worcel, A. (1976). Cell 9:393.
106. Wilkins, M. H. F., Zubay, G., and Wilson, H. R. (1959). J. Mol. Biol. 1:179.
107. Luzzati, V., and Nicolaieff, A. (1963). J. Mol. Biol. 7:142.
108. Kornberg, R. D. (1975). *In* W. J. Peacock and R. D. Brock (eds.), The Eukaryote Chromosome, p. 245. Australian National University Press, Canberra.
109. Kornberg, R. D., Klug, A., and Sperling, L., unpublished results.
110. Carpenter, B. G., Baldwin, J. P., Bradbury, E. M., and Ibel, K. (1976). Nucleic Acids Res. 3:1739.
111. Carlson, R. D., and Olins, D. E. (1976). Nucleic Acids Res. 3:89.
112. Harrison, S. C., and Kornberg, R. D. (1976). *In* D. P. Nierlich, W. N. Rutter, and C. F. Fox (eds.), Molecular Mechanisms in the Control of Gene Expression, p. 7. Academic Press, New York.
113. Baldwin, J. P., Boseley, P. G., Bradbury, E. M., and Ibel, K. (1975). Nature (Lond.) 253:245.
114. Olins, A. L., Carlson, R. D., Wright, E. B., and Olins, D. E. (1976). Nucleic Acids Res. 3:3271.
115. Bakayev, V. V., Melnickov, A. A., Osicka, V. D., and Varshavsky, A. J. (1975). Nucleic Acids Res. 2:1401.
116. Shaw, B. R., Corden, J. L., Sahasrabuddhe, C. G., and Van Holde, K. E. (1974). Biochem. Biophys. Res. Commun. 61:1193.
117. Van Holde, K. E., Shaw, B. R., Lohr, D., Herman, T. M., and Kovacic, R. T. (1975). *In* G. Bernardi and F. Gros (eds.), Organization and Expression of the Eukaryotic Genome. FEBS Symposia, Vol. 38, p. 57. North-Holland/American Elsevier, Amsterdam.
118. Loening, U. E. (1967). Biochem. J. 102:251.
119. Staynov, D. Z., Pinder, J. C., and Gratzer, W. B. (1972). Nature (New Biol.) 235:108.
120. Maniatis, T., Jeffrey, A., and van deSande, H. (1975). Biochemistry 14:3787.
121. Finch, J. T., Klug, A., Lutter, L., Rhodes, D., and Rushton, B., unpublished results.
122. Simpson, R. T., and Bustin, M. (1976). Biochemistry 15:4305.
123. Stuhrmann, H. B. (1974). J. Appl. Cryst. 7:173.
124. Weintraub, H., Worcel, A., and Alberts, B. (1976). Cell 9:409.
125. Crick, F. H. C., and Klug, A. (1975). Nature (Lond.) 255:530.
126. Sobell, H. M., Tsai, C.-C., Gilbert, S. G., Jain, S. C., and Sakore, T. D. (1976). Proc. Natl. Acad. Sci. USA 73:3068.
127. Simpson, R. T., and Whitlock, J. P. (1976). Cell 9:347.
128. Noll, M., manuscript in preparation.
129. Lutter, L., manuscript in preparation.
130. Sollner-Webb, B., and Felsenfeld, G. (1977). Cell 10:537.

131. Altenburger, W., Hörz, W., and Zachau, H. G. (1976). Nature (Lond.) 264:517.
132. Greil, W., Igo-Kemenes, T., and Zachau, H. G. (1976). Nucleic Acids Res. 3:2633.
133. Oosterhof, D. K., Hozier, J. C., and Rill, R. L. (1975). Proc. Natl. Acad. Sci. USA 72:633.
134. Oudet, P., Germond, J. E., Bellard, M., Spadafora, C., and Chambon, P. Phil. Trans. R. Soc. Lond. In press.
135. Axel, R., Melchior, W., Sollner-Webb, B., and Felsenfeld, G. (1974). Proc. Natl. Acad. Sci. USA 71:4101.
136. Weintraub, H., and Van Lente, F. (1974). Proc. Natl. Acad. Sci. USA 71:4249.
136a. Cantor, C. R. (1976). Proc. Natl. Acad. Sci. USA 73:3391.
137. Weintraub, H. (1975). Proc. Natl. Acad. Sci. USA 72:1212.
138. Simpson, R. T. (1976). Proc. Natl. Acad. Sci. USA 73:4400.
139. Levina, E. S., and Mirzabekov, A. D. (1975). Dokl. Akad. Nauk. SSSR 221:1222.
140. Weintraub, H., Palter, K., and Van Lente, F. (1975). Cell 5:85.
141. Thomas, J. O., and Butler, P. J. G. (1977). J. Mol. Biol. 116:769–781.
142. Campbell, A. M., and Cotter, R. I. (1976). FEBS Lett. 70:209.
143. Alberts, B., Worcel, A., and Weintraub, H. (1977). In E. M. Bradbury and K. Javaherian (eds.), The Organization and Expression of the Eukaryotic Genome, p. 165. Academic Press, London.
144. Tsanev, R., and Russev, G. (1974). Eur. J. Biochem. 43:257.
145. Jackson, V., Granner, D. K., and Chalkley, R. (1975). Proc. Natl. Acad. Sci. USA 72:4440.
146. Paul, J., and Gilmour, R. S. (1968). J. Mol. Biol. 34:305.
147. Bekhor, I., Kung, M., and Bonner, J. (1969). J. Mol. Biol. 39:351.
148. Gilmour, R. S., and Paul, J. (1970). FEBS Lett. 9:242.
149. Spelsberg, T. C., Hnilica, L. A., and Ansevin, A. T. (1971). Biochim. Biophys. Acta 228:550.
150. Paul, J., Gilmour, R. S., Affara, N., Birnie, G. D., Harrison, P. R., Hell, A., Humphries, S., Windass, J. D., and Young, B. D. (1973). Cold Spring Harbor Symp. Quant. Biol. 38:885.
151. Gilmour, R. S., Sindass, J. D., Affara, N., and Paul, J. (1975). J. Cell Physiol. 85:449.
152. Paul, J. (1976). In V. G. Allfrey, E. K. F. Bautz, B. J. McCarthy, R. T. Schimke, and A. Tissières (eds.), Organization and Expression of Chromosomes, p. 317. Dahlem Konferenzen, Berlin.
153. Gadski, R. A., and Chae, C.-B. (1976). Biochemistry 15:3812.
154. Boseley, P. G., Bradbury, E. M., Butler-Browne, G. S., Carpenter, B. G., and Stephens, R. M. (1976). Eur. J. Biochem. 62:21.
155. Yaneva, M., Tasheva, B., and Dessev, G. (1976). FEBS Lett. 70:67.
156. Steinmetz, M., Streeck, R. E., and Zachau, H. G. Phil. Trans. R. Soc. Lond. In press.
157. Thomas, J. O., and Butler, P. J. G. (1977). Cold Spring Harbor Symp. Quant. Biol. 42. In press.
158. M. Nomura, A. Tissières, and P. Lengyel (eds.). (1974). Ribosomes. Cold Spring Harbor Laboratory, Cold Spring Harbor, New York.
159. Camerini-Otero, R. D., Sollner-Webb, B., and Felsenfeld, G. (1976). Cell 8:333.
160. Sollner-Webb, B., Camerini-Otero, R. D., and Felsenfeld, G. (1976). Cell 9:179.
161. Axel, R., Cedar, H., and Felsenfeld, G. (1975). Biochemistry 14:2489.
162. Lacy, E., and Axel, R. (1975). Proc. Natl. Acad. Sci. USA 72:3978.
163. Kuo, M. T., Sahasrabuddhe, C. G., and Saunders, G. F. (1976). Proc. Natl. Acad. Sci. USA 73:1572.

164. Gottesfeld, J. M., Murphy, R. F., and Bonner, J. (1975). Proc. Natl. Acad. Sci. USA 72:4404.
165. Gottesfeld, J. M. (1977). Phil. Trans. R. Soc. Lond. In press.
166. Piper, P. W., Celis, J., Kaltoft, K., Leer, J. C., Nielsen, O. F., and Westergaard, O. (1976). Nucleic Acids Res. 3:493.
167. Mathis, D. J., and Gorovsky, M. A. (1976). Biochemistry 15:750.
168. Reeves, R., and Jones, A. (1976). Nature (Lond.) 260:495.
169. Weintraub, H., and Groudine, M. (1976). Science 193:848.
170. Garel, A., and Axel, R. (1976). Proc. Natl. Acad. Sci. USA 73:3966.
171. Miller, O. L., and Beatty, B. R. (1969). Genet. Suppl. 61:134.
172. Laird, C. D., Wilkinson, L. E., Foe, V. E., and Chooi, W. Y. (1976). Chromosoma (Berlin) 58:169.
173. Bostock, C. J., Christie, S., and Hatch, F. T. (1976). Nature (Lond.) 262:516.
174. Hörz, W., Igo-Kemenes, T., Pfeiffer, W., and Zachau, H. G. (1976). Nucleic Acids Res. 3:3213.
175. Lipchitz, L., and Axel, R. (1976). Cell 9:355.
176. Polisky, B., and McCarthy, B. (1975). Proc. Natl. Acad. Sci. USA 72:2895.
177. Crémisi, C., Pignatti, P. F., and Yaniv, M. (1976). Biochem. Biophys. Res. Commun. 73:548.
178. D'Anna, J. A., and Isenberg, I. (1973). Biochemistry 12:1035.
179. D'Anna, J. A., and Isenberg, I. (1974). Biochemistry 13:2098.
180. Sperling, R., and Bustin, M. (1974). Proc. Natl. Acad. Sci. USA 71:4625.
181. Sperling, R., and Bustin, M. (1975). Biochemistry 14:3322.
182. Sperling, R., and Bustin, M. (1976). Nucleic Acids Res. 3:1263.
183. Skandrani, E., Mizon, J., Sautière, P., and Biserte, G. (1972). Biochimie 54:1267.
184. Rubin, R. L., and Moudrianakis, E. N. (1975). Biochemistry 14:1718.
185. Bradbury, E. M., and Rattle, H. W. E. (1972). Eur. J. Biochem. 27:270.
186. Moss, T., Crane-Robinson, C., and Bradbury, E. M. (1976). Biochemistry 15:2261.
187. Lilley, D. M. J., Howarth, O. W., Clark, V. M., Pardon, J. F. and Richards, B. M. (1976). FEBS Lett. 62:7.
188. Richards, F. M., and Vithayathil, P. J. (1959). J. Biol. Chem. 234:1459.
189. Anfinsen, C. B. (1973). Science 181:223.
190. Kartha, G., Bello, J., and Harkar, D. (1967). Nature (Lond.) 213:862.
191. Chapman, G. E., Hartman, P. G., and Bradbury, E. M. (1976). Eur. J. Biochem. 61:69.
192. Dixon, G. H. (1976). In V. G. Allfrey, E. K. F. Bautz, B. J. McCarthy, R. T. Schimke, and A. Tissières (eds.), Organization and Expression of Chromosomes, p. 197. Dahlem Konferenzen, Berlin.
193. Louie, A. J., and Dixon, G. H. (1972). Proc. Natl. Acad. Sci. USA 69:1975.
194. Gurley, L. R., Wallers, R. A., and Tobey, R. A. (1975). J. Biol. Chem. 250:3936.
195. Bradbury, E. M., Inglis, R. J., and Matthews, H. R. (1974). Nature (Lond.) 247:257.
196. Nishizuka, Y., Ueda, K., Yoshibara, K., Yamamura, H., Takeda, M., and Hayaishi, O. (1969). Cold Spring Harbor Symp. Quant. Biol. 34:781.
197. Ueda, K., Omachi, A., Kawaichi, M., and Hayaishi, O. (1975). Proc. Natl. Acad. Sci. USA 72:205.
198. Dixon, G. H., Wong, N., and Poirier, G. G. (1976). Fed. Proc. 35:1623.
199. Cohen, L. H., Newrock, K. M., and Zweidler, A. (1975). Science 190:994.
200. Alfageme, C. R., Zweidler, A., Mahowald, A., and Cohen, L. H. (1974). J. Biol. Chem. 249:3729.
201. Franklin, S. G., and Zweidler, A. (1977). Nature (Lond.) 266:273.

202. Elgin, S. C. R., and Bonner, J. (1970). Biochemistry 9:4440.
203. O'Farrell, P. H. (1975). J. Biol. Chem. 250:4007.
204. Peterson, J. L., and McConkey, E. H. (1976). J. Biol. Chem. 251:548.
205. McConkey, E. H. (1976). In: V. G. Allfrey, E. K. F. Bautz, B. J. McCarthy, R. T. Schimke, and A. Tissières (eds.), Organization and Expression of Chromosomes, p. 273. Dahlem Konferenzen, Berlin.
206. Roeder, R. G. (1976). In V. G. Allfrey, E. K. F. Bautz, B. J. McCarthy, R. T. Schimke, and A. Tissières (eds.), Organization and Expression of Chromosomes, p. 285. Dahlem Konferenzen, Berlin.
207. Stein, G., Park, W., Thrall, C., Mans, R., and Stein, J. (1975). Nature (Lond.) 257:764.
208. Spelsberg, T. C., Steggles, A. W., Chytil, F., and O'Malley, B. W. (1972). J. Biol. Chem. 247:1368.
209. Allfrey, V. G., Inoue, A., Karn, J., Johnson, E. M., Good, R. A., and Hadden, J. W. (1975). CIBA Found. Symp. 28:199.
210. Lin, S.-Y., and Riggs, A. D. (1975). Cell 4:107.
211. Yamamoto, K. R., and Alberts, B. (1975). Cell 4:301.
212. Silver, L. M., and Elgin, S. C. R. (1976). Proc. Natl. Acad. Sci. USA 73:423.
213. Elgin, S. C. R., Silver, L. M., and Wu, C. E. C. In P. O. P. Ts'o (ed.), Molecular Biology of the Mammalian Genetic Apparatus, Vol. 1, p. 127. Elsevier/North Holland Biomedical Press, Amsterdam.
214. Johns, E. W., Goodwin, G. H., Walker, J. M., and Sanders, C. (1975). CIBA Found. Symp. 28:95.
215. Goodwin, G. H., Woodhead, J. L., and Johns, E. W. (1977). FEBS Lett. 73:85.
216. Goodwin, G. H., and Johns, E. W. (1973). Eur. J. Biochem. 40:215.
217. Goodwin, G. H., Nicolas, R. H., and Johns, E. W. (1975). Biochim. Biophys. Acta 405:280.
218. Walker, J. M., Hastings, J. R. B., and Johns, E. W. (1977). Eur. J. Biochem. 76:461.
219. Smerdon, M. J., and Isenberg, I. (1976). Biochemistry 15:4242.
220. Goldkopf, I. L., Taylor, C. W., Baum, R. M., Yeoman, L. C., Olson, M. O. J., Prestayko, A. W., and Busch, H. (1975). J. Biol. Chem. 250:7182.
221. Cohen, L. H., et al., unpublished observations. Cited in Science 190:994.
222. Kriegstein, H. J., and Hogness, D. S. (1974). Proc. Natl. Acad. Sci. USA 71:135.
223. Weintraub, H. (1972). Nature (Lond.) 240:449.
224. Weintraub, H. (1973). Cold Spring Harbor Symp. Quant. Biol. 38:247.
225. Seale, R. L., and Simpson, R. T. (1975). J. Mol. Biol. 94:479.
226. Seale, R. L. (1976). Cell 9:423.
227. Weintraub, H. (1976). Cell 9:419.
228. Cold Spring Harbor Symp. Quant. Biol. (1977). 42. In press.
229. Finch, J. T., Lutter, L. C., Rhodes, D., Brown, R. S., Rushton, B., Levitt, M., and Klug, A. (1977). Nature (Lond.) 269:29.
230. Richards, B., Pardon, J., Lilley, D., Cotter, R., Wooley, J., and Worcester, D. (1977). Cell Biol. Internat. Rep. 1:107.
231. Crick, F. H. C. (1976). Proc. Natl. Acad. Sci. USA 73:2639.
232. MacLeod, A. R., Wong, N. C. W., and Dixon, G. H. (1977). Eur. J. Biochem. 78:281.
233. Bina-Stein, M., and Simpson, R. T. (1977). Cell 11:609.
234. Camerini-Otero, R. D., and Felsenfeld, G. (1977). Nucleic Acids Res. 4:1159.
235. Moss, T., Stephens, R. M., Crane-Robinson, C., and Bradbury, E. M. (1977). Nucleic Acids Res. 4:2477.
236. Ponder, B. A. J., and Crawford, L. V. (1977). Cell 11:35.

237. Levy, W., B., Wong, N. C. W., and Dixon, G. (1977). Proc. Natl. Acad. Sci. USA 74:2810.
238. Wilhelm, M. L., Mazen, A., and Wilhelm, F. X. (1977). FEBS Lett. 79:404.
239. Riggs, M. G., Whittaker, R. G., Neumann, J. R., and Ingram, V. M. (1977). Nature (Lond.) 258:462.
240. Renz, M., Nehls, P., and Hozier, J. (1977). Proc. Natl. Acad. Sci. USA 74:1879.
241. Hozier, J., Renz, M., and Nehls, P. (1977). Chromosoma (Berl.) 62:301.
242. Bak, A. L., Zeuthen, J., and Crick, F. H. C. (1977). Proc. Natl. Acad. Sci. USA 74:1595.

International Review of Biochemistry
Biochemistry of Nucleic Acids II, Volume 17
Edited by B. F. C. Clark
Copyright 1978   University Park Press   Baltimore

7
Transcriptional Control Mechanisms

A. A. TRAVERS
MRC Laboratory of Molecular Biology, Cambridge, England

DNA transcription is the copying of one strand of a DNA template by DNA-dependent RNA polymerase to yield an RNA molecule of complementary nucleotide sequence. The qualitative and quantitative regulation of the sequences transcribed from DNA genomes is a major component of the overall control of gene expression in prokaryotes (1, 2). In eukaryotes, only a few examples of control of transcription per se—as distinct from post-transcriptional modification—have been documented (3). Nevertheless, it seems probable that in this class of organisms as well transcriptional regulation is of comparable significance.

In principle, the selection of a particular DNA sequence for transcription requires that the RNA synthesis be initiated and terminated at precise locations on the template. Such precision implies the specific recognition of regulatory nucleotide sequences by other components of the transcription apparatus. Thus, initiation at a promoter site may be controlled either by an alteration of the recognition capacity of RNA polymerase itself or by a change in the availability of the promoter site. Such changes can be mediated by operon-specific regulatory proteins binding in the promoter region or by gross changes in the conformation of the DNA template. In addition to qualitative controls of this type, other quantitative mechanisms contribute to the overall pattern of RNA species transcribed. For example, a requirement for a high rate of ribosomal RNA (rRNA) synthesis is often met by amplifying the number of copies of rRNA cistrons present (4-6). Similarly, the amount and specific activity of RNA polymerase in a cell are also regulated.

COMPONENTS OF TRANSCRIPTION MACHINERY OF PROKARYOTES

RNA Polymerase

In bacteria, the synthesis of all major bacterial RNA species is thought to be mediated by a single RNA polymerase. This enzyme, which is structurally one of the most complex enzymes in the bacterial cell, is an allosteric heteromultimer which can interact with an array of other regulatory macromolecules and low molecular weight effectors. One consequence of this complexity is that within the cell RNA polymerase molecules may exist as a population of different assemblies of polypeptide chains. Conventional enzyme purification procedures applied to such a mixture yield, in general, the most stable assembly. Therefore, the preferential purification of a particular assembly is not necessarily a measure of its relative abundance within the cell. Consequently, both genetic and structural criteria are required to define the polypeptide composition of RNA polymerase.

The smallest assembly of *Escherichia coli* RNA polymerase known to be enzymically active in the elongation of RNA chains is core RNA polymerase (E) (7), a zinc metalloenzyme (8) consisting of four polypeptide chains, $\alpha_2\beta\beta'$ (9), with an overall protomeric molecular weight of 400,000. The individual α, β, β' polypeptides have molecular weights of 41,000, 155,000, and 165,000, respectively (9). The genes coding for β and β' are contiguous and map at 88.6' on the *E. coli* chromosome (10), whereas that for α maps at 72' (11). In both cases, the genes for the RNA polymerase subunits are adjacent to genes for ribosomal proteins (12). Mutations in the β subunit conferring resistance to the antibiotic rifampicin affect the entire cellular population of polymerase molecules (13), while temperature-sensitive mutants of the β' subunit are similarly effective (14). Thus, both β and β' must be essential components of RNA polymerase. It seems likely that this conclusion is also true for α, although the genetic evidence in this case is less compelling.

By itself, core polymerase lacks the ability to initiate RNA synthesis at promoters but can assume this capability on association with an initiation factor. Two such factors, σ (7) and σ' (15), with molecular weights of 86,000 and 56,000, respectively, have been identified in *E. coli*. A gene coding for σ has been provisionally located at 66' on the *E. coli* chromosome, raising the possibility that σ may be identical with the *dna*G product (16). The σ and σ' polypeptides bind to the core enzyme to form holoenzyme I (Eσ) and holoenzyme II (Eσ'). The complex between core and σ is more stable than that between core and σ'; consequently, holoenzyme I is the form of the polymerase which is normally purified. Nevertheless, in *E. coli, σ'* is a more abundant protein than σ (17) and in crude extracts is normally found in association with ribosomes.

The structure of core polymerase and holoenzyme in other bacteria and cyanophytes is generally similar to that in *E. coli*. Thus, in the vegetative state, *Bacillus subtilis* yields a core polymerase of structure $\alpha_2\beta\beta'$ in which the molecular weights of the individual polypeptides are 45,000, 155,000, and 165,000 (18), respectively. However, the holoenzyme normally isolated from *B. subtilis* differs from the corresponding *E. coli* enzyme in containing a σ subunit of 57,000 daltons (18). As yet it is unclear whether this polypeptide is homologous with the σ' polypeptide of *E. coli*. Holoenzyme from *Lactobacillus curvatus* contains an even smaller σ polypeptide of molecular weight 44,000, which in qualitative properties seems to be functionally equivalent to the σ subunit of *E. coli* polymerase (19). In addition, another RNA polymerase assembly in which core polymerase forms a complex with a polypeptide, y, of 84,000 daltons can be isolated from *L. curvatus*. The polypeptide y, like σ, binds to the β' subunit and causes a low but significant increase in the activity of core polymerase on double-stranded DNA in the absence of σ. However, σ is clearly required for additional activity of Ey. Although y and σ appear to exclude each other on core polymerase, they act synergistically, suggesting that they may affect different steps of the transcription process. A protein x, of 110,000 molecular weight, with some properties analogous to those of y, has also been isolated from *E. coli* and may be identical with the previously reported τ protein (7) and "inactive σ" (20). The complexity of these prokaryotic enzymes should be contrasted with the apparent simplicity of *Halobacterium cutirubrum* RNA polymerase, which has been reported to consist of two polypeptide chains each with a molecular weight of 25,000 (21). Other RNA polymerases of simple structure are coded by phages T3 and T7 (22, 23). These enzymes are single polypeptide chains with a molecular weight of $\sim$105,000.

A higher level of structural complexity is represented by assemblies in which other bacterial macromolecules are bound to polymerase holoenzyme. Methods which have been used to demonstrate such binding are copurification, retention by an RNA polymerase-agarose column (24) and precipitation by antibody to RNA polymerase core or holoenzyme (25). Of these methods, the last is probably the least specific (24), but in all cases an indication of an interaction should be regarded with circumspection until a functional role for the putative transcription factor has been established.

Among the polymerase binding proteins identified by copurification is the δ protein of *B. subtilis* (26), a polypeptide of 21,500 daltons which copurifies with holoenzyme but not with core polymerase isolated by chromatography on phosphocellulose. A functional role for this protein is not yet apparent in its complex with vegetative holoenzyme, but it does stimulate asymmetric transcription by a modified RNA polymerase isolated from cells infected with phage SP01. This stimulation could be a consequence of suppression of incorrect transcription or of enhancement of accurate initiation. A second polymerase binding protein which copurifies with core and holoenzyme from *E. coli* and *B. subtilis* is the ω polypeptide (9). This protein has a molecular weight of $\sim 11,000$ and is present in a stoichiometric ratio of 1–2 mole/mole of β. No function has been established for this protein.

Although a variety of bacterial polypeptides bind to RNA polymerase-agarose columns, in no case has such a polypeptide been identified or assigned a functional role (24). However, included among the abundance of *E. coli* proteins which are precipitated from cell extracts by antiserum to RNA polymerase holoenzyme or core polymerase are the protein biosynthesis factors EF-Ts, EF-Tu, and EF-G, ribosomal protein S1, and stringent starvation protein (17, 25). Of these proteins, the first four have a direct effect on transcription in vitro.

A number of other putative polymerase-binding macromolecules have been identified on the basis of a functional effect on polymerase activity and a demonstration of specific interaction between the polymerase and the macromolecule. The *E. coli* protein biosynthesis components IF-2 (27) and fMet-tRNA$_f^{Met}$ (28) are two such macromolecules. The specific binding of fMet-tRNA$_f^{Met}$ depends on the presence of both formyl group and the initiator tRNA moiety and changes the pattern of in vitro RNA synthesis, stimulating in particular transcription from λ *plac* DNA. Another transcription factor which binds to polymerase is M factor, a preparation which stimulates the initiation of RNA synthesis on a variety of DNA templates, most notably λ DNA and *E. coli* DNA (29, 30). The most highly purified preparations of this factor, which is found in association with ribosomes, consist of nonequimolar mixtures of two polypeptide chains of molecular weights of $\sim 35,000$ and $\sim 55,000$, respectively (30). The possible identity of the 55,000-dalton polypeptide and σ' remains to be established. Yet another protein which may bind to RNA polymerase is a transcription inhibitor, a polypeptide of $\sim 70,000$ daltons, which apparently allows the first round of RNA synthesis to occur normally and then prevents further RNA synthesis (31). The stoichiometry of the inhibitory effect and the lack of specificity for template suggest that the target of the inhibitor is RNA polymerase. One possibility is that in vivo this polypeptide has a role in the regulation of the amount of active RNA polymerase.

Accessory Components of Transcription Machinery

Among the bacterial proteins reported to influence transcription by holoenzyme is a class of low molecular weight, thermostable DNA binding proteins. These include H_1 protein (32), normally isolated as a tetramer with a molecular weight of $\sim 30,000$, and D (33) and H_2 proteins (32), both basic proteins of $\sim 10,000$ molecular weight. Although H_1 and H_2 have been identified by immunological and structural criteria as components of highly purified RNA polymerase preparations, it is as yet uncertain whether this copurification is a consequence of direct interaction with RNA polymerase or whether these small proteins are merely binding to contaminating DNA fragments in the RNA polymerase preparations. These proteins alter the initiation specificity of transcription. Thus, with λb_2 DNA as template, D protein stimulates RNA synthesis only from early λ genes, at the same time increasing the proportion of λ RNA synthesis that can be inhibited by λ repressor (34). Thus, D protein must stimulate RNA synthesis from specific sites. H_1 protein similarly stimulates λ transcription and also markedly increases *lac* mRNA synthesis in vitro (35). In contrast, this protein strongly depresses transcription from rRNA promoters (36). Both D and H proteins increase the T_m of bulk DNA, but, in a manner analogous to the arginine-rich histones, H protein stabilizes GC-rich regions to a far greater extent than AT-rich regions. Thus, these proteins may influence the specificity of transcription by imposing structural constraints on phage and bacterial chromosomes.

In addition to proteins whose primary action affects the initiation step, two proteins, ρ (37) and κ (38), influence RNA chain termination in vitro. In the presence of ρ, normally isolated as a tetramer of $\sim 200,000$ daltons, RNA synthesis terminates at specific sites on the DNA template, and the completed RNA chains are released (37). There is now strong genetic evidence that ρ also acts as a terminator protein in vivo and is the product of the *su*A gene (39–41). Recessive mutations in this gene relieve the transcriptional polarity induced by nonsense and frameshift mutants (42). In contrast, ρ function in vitro induces polarity in the *gal* operon (43). This parallel between the in vitro effect of ρ and the in vivo effect of *su*A mutants is supported by the observation that ρ factor purified from a strain containing the *su*78 mutation in the *su*A gene has altered termination function (39). This ρ factor fails to terminate transcription at some sites which are efficiently recognized by ρ factor purified from an isogenic strain. Furthermore, affinity chromatography of ρ protein from different *E. coli* strains reveals that different *su*A mutations affect ρ in disparate ways (40). One destabilizes ρ, a second increases the molecular weight of the polypeptide, whereas others, presumed to be missense mutations, overproduce ρ. Finally, by making use of a molecular weight difference between ρ isolated from different *E. coli* strains, tight linkage of the ρ gene to the *su*A locus has been demonstrated.

Together these experiments make it almost certain that the *su*A gene defines the structural gene for ρ and not a modifier of ρ protein.

In addition to its transcriptional termination function, ρ possesses a poly(C)-dependent ATPase activity (44). The ρ protein purified from strains carrying a *ts* mutation at the *su*A locus is temperature-sensitive for this activity (45). Such strains are defective in many aspects of bacterial physiology including UV repair, recombination, and energy metabolism. Consequently, it has been suggested that ρ participates in several different cellular processes that require an ATPase. These might include transcriptional termination, *rec*BC nuclease function, and the Mg^{2+}-Ca^{2+}-dependent membrane ATPase.

Operon-specific DNA Binding Proteins

In addition to proteins which appear to influence the overall pattern of transcription in vivo and in vitro, there is a class of highly specific regulatory proteins each of which controls the initiation of RNA synthesis from only one or a few transcription units. Among such proteins are the *lac* (46) and λ (47) repressors which act as negative regulators, the *ara*C protein, which can act either as a positive or a negative regulator (48), and the *crp* protein, which acts exclusively as a positive regulator. In all cases, these proteins bind to a specific DNA site within the promoter region. The repressor and polymerase binding sites overlap and thus the inhibitory activity of repressors can be ascribed to polymerase exclusion. By contrast the binding sites of *crp* protein and polymerase are contiguous (49). Thus, *crp* could facilitate promoter opening either by directly influencing the DNA structure of the promoter or by binding to RNA polymerase holoenzyme (50, 51), thereby increasing its affinity for the promoter.

INITIATION OF RNA SYNTHESIS

Mechanism of Initiation

The production of a defined RNA molecule from a DNA template requires that synthesis of the RNA be initiated at a specific site on the DNA template, then that the coding DNA strand be copied in accordance with the Watson-Crick rules of base pairing, and finally that RNA synthesis be terminated at a second specific site on the DNA template. Thus, one crucial determinant of the quality of the DNA transcript is the accuracy of initiation. The sequence of molecular events leading up to the initiation of an RNA chain was first formulated by Zillig and his collaborators (52) and subsequently confirmed and elaborated (53, 54) (Figure 1). After unproductive random interactions with the DNA template, RNA polymerase holoenzyme "recognizes" a specific structure or sequence within the promoter region. This recognition, which requires the presence of the σ polypeptide, enables polymerase holoenzyme to bind to the DNA in this region at least an order of magnitude more tightly than in nonspecific interactions. After the formation of this I complex there is a change in the conformational state of the DNA in

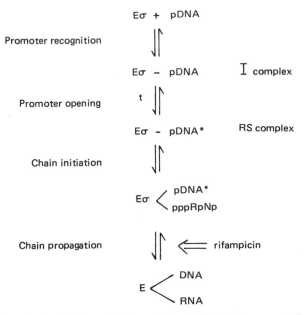

Figure 1. Scheme for the initiation of RNA synthesis. pDNA indicates promoter DNA; R, purine nucleoside; N, any standard nucleoside.

the promoter region which results in the formation of a relatively stable polymerase-promoter complex, the RS complex (55). Once this transition has occurred, the polymerase is poised to initiate the synthesis of an RNA molecule very rapidly (55).

The transition between the DNA conformational states after the initial binding of RNA polymerase to a promoter is cooperative and reversible and often occurs over a narrow temperature range, the midpoint of which defines a transition temperature, t. Below t, the promoter is closed; above t the promoter is open (54, 56). These words are used strictly in terms of the experimentally observed transition and do not imply any particular physical description of the relevant DNA conformations. The strong dependence of promoter opening on temperature indicates a positive enthalpy for the process. Thus, the observed transition temperature is, to a first approximation, a measure of the activation energy for opening. This energy is presumably made available by the tight binding consequent upon promoter recognition. The probability of opening and hence of initiation could be altered either by directly changing the activation energy required for the DNA conformation change or by changing the initial affinity of polymerase holoenzyme for the promoter site. Changes in either of these parameters would consequently be reflected in an increase or decrease in t for a particular promoter.

What is the nature of the conformational change in the promoter region? The binding of RNA polymerase to superhelical λ DNA results in an unwinding of the double helix of 120° per bound polymerase molecule (with an uncertainty factor of 2) (57). This unwinding is observed at 37°C and not at 0°C, and its extent is unaffected by subsequent RNA synthesis. This phenomenon would be consistent with either a local melting of the double helix over a region of 4–8 base pairs or a conversion of DNA from the B to the A form over a region of 30–60 base pairs. The former possibility is supported by the observation that the 3'-end of a growing RNA chain is hybridized to the complementary DNA strand, suggesting that at least during RNA synthesis DNA strand separation does occur over a limited region (58). Because promoter opening places the polymerase in a state in which it can initiate RNA synthesis very rapidly, the simplest model is that promoter opening can be equated with a limited melting of DNA in the promoter region. Is this interpretation compatible with other evidence? The enthalpy of the conformational change at the T7 early promoter is ~ 58 kcal/mol (59), a figure which would be consistent with a melting of about eight base pairs. This, in turn, agrees well with the extent of unwinding induced by bound polymerase on supercoiled DNA.

Evidence that changes in DNA structure affect t for promoter opening is based largely on comparisons of the template activities of the relaxed and supercoiled forms of circular DNA species. Thus, the formation of polymerase promoter complexes is 20-fold greater on the supercoiled form of ϕX174 RF DNA than on the corresponding relaxed form (60). A similar pattern is observed for the two forms of PM2 DNA (61). In this case, supercoiling both lowers the average transition temperature and increases the number of promoter sites available to the enzyme. These studies measured the overall pattern of RNA synthesis from the templates tested, and consequently it is not rigorously established that supercoiling alters t for particular promoters, as opposed to altering the total pattern of available promoters.

The transition temperature for opening may also be changed by the addition of glycerol or dimethylsulfoxide in vitro (62, 63). A solution of glycerol (20% v/v) reduces t for rRNA synthesis from *E. coli* DNA by 6°C; 10% glycerol reduces t for T4 RNA synthesis by 16°C (63). In addition, glycerol activates the normally inactive λ *sex* promoter and releases the *gal* promoter from the constraints necessary for positive control by *crp* protein (62). Dimethylsulfoxide has a very similar effect and is equally or even more effective on a molar basis than glycerol. The target of these compounds is believed to be the DNA template, whose conformation they could influence by altering the disordered water structure around the double helix. Such disruption would result in a lowering of the average T_m of DNA (64).

Another factor which affects the transition temperature is the monovalent cation concentration. By raising the KCl concentration, t for the *gal* promoter (62) and t for the principal early promoter of phage T7 (65) are raised. In this case, it seems likely that increasing ionic strength antagonizes the electrostatic interaction between RNA polymerase and the DNA template and so decreases the energy available for opening. However, the effect of change in KCl concentration is in general more complex (see under "Promoter Selection").

Variations in the structure of RNA polymerase modifying the affinity of the enzyme for promoter sites also influence the transition temperature. A good example is provided by the comparative properties of polymerase holoenzymes isolated from *E. coli* and from *Bacillus stearothermophilus* (66). The optimum growth temperatures for these organisms are 35–40°C and 60–65°C, respectively. Similarly, *t* for the respective holoenzymes on a T4 DNA template differs by about 20°C, that for the *E. coli* enzyme being 16.5°C and that for the *B. stearothermophilus* enzyme 35–45°C. Structural modifications after T4 infection also increase *t* for initiation at T4 promoters (67) and the polymerase effectors ppGpp (68) and EF-TuTs (56) increase and decrease *t* for rRNA promoter opening, respectively.

Conditions which favor the formation of the open promoter substantially increase the stability of the polymerase-promoter complex so that at 37°C the half-life of this complex is of the order of tens of minutes (69). Early experiments suggesting a half-life of 60 hr for such complexes on T7 DNA (65) depended on the assumption of a single polymerase binding site on that template. In fact, 3–4 polymerase holoenzyme molecules bind per DNA template. The stability of the open promoter complex allows the enzyme to initiate an RNA chain very rapidly, i.e., within 0.2 s, when presented with the substrate triphosphates (55).

This scheme for the mechanism of initiation heavily depends on results from in vitro experiments with the use of inhibitors of RNA polymerase that act prior to or immediately after chain initiation and which do not, in general, block subsequent chain extension. These inhibitors include the rifamycin group of antibiotics and polyanions such as heparin and poly(rI). The rifamycins bind to the β subunit (70) of polymerase and inhibit RNA synthesis both in vivo and in vitro. Initial studies on the mode of action of the antibiotic suggested that once RNA polymerase has initiated an RNA chain addition of rifamycin has little or no immediate effect on subsequent chain extension (71, 72). Yet in the presence of rifamycin, RNA polymerase catalyzes the DNA-dependent, promoter-specific synthesis of large amounts of dinucleoside tetraphosphate of identical sequence to the 5'-terminus of the RNA molecule normally synthesized (73, 74). This "abortive initiation" is under most conditions a relatively insignificant aspect of polymerase function. It remains unclear whether the process involves an intermediate in the normal initiation process or arises from an aberrant pathway. The former possibility, if correct, would suggest that rifampicin acts as an inhibitor of chain elongation after the formation of the first phosphodiester bond. The presence of DNA, and possibly other components of the reaction, substantially reduces the rate of binding of antibiotic to the enzyme (75, 76); consequently, at low concentrations ($\sim$10 μg/ml) rifampicin acts much more slowly on RNA polymerase engaged in catalytic activity than on free enzyme. At very high concentrations (500 μg/ml), rifampicin does, however, inhibit all chain elongation (77).

Most mechanistic studies employing rifampicin use the "rifampicin challenge" technique in which rifampicin and nucleoside triphosphates are added to a preformed polymerase-DNA complex (55). The antibiotic and substrate then effectively compete for the enzyme, the former inhibiting and the latter promoting RNA synthesis. The conclusions from this type of experiment do not depend on rifampicin inhibition at a particular step in the sequence of initiation events. However, conclusions concerning the order of binding of the first two triphosphates are so dependent (78).

Promoter Selection

The relative characteristics of the transcription of different DNA templates by *E. coli* holoenzyme I are largely determined by the interaction of polymerase with promoter sites. The simplistic view that the extent of promoter opening is a function of temperature and ionic strength so that opening is facilitated by increasing temperature and/or decreasing ionic strength (54) is almost certainly an understatement of the actual complexity of the process of promoter selection. Thus, there is considerable evidence that the response of the synthesis of different RNA species to variations in ionic strength is not strictly correlated with transition temperature. For example, at 0.075 M KCl, t for the opening of rRNA and ϕ80 promoters is approximately the same, yet rRNA synthesis declines much more rapidly with increasing ionic strength than does ϕ80 RNA synthesis (68). Similarly, substitution of bromouracil residues for thymine in T4 DNA increases t by 10–15°C at very low ionic strength and decreases t at high ionic strength (79). This is the opposite pattern to what would be predicted from a simple effect of BrdU substitution on the affinity of RNA polymerase for promoter sites.

The response of transcription to variation in monovalent cation concentration is strongly promoter-dependent. Thus, under conditions of enzyme excess, the optimal KCl concentration for rRNA and su^+_{III} tRNA synthesis is <0.1 M (68, 80, 81), that for λ RNA synthesis is $\sim$0.1 M (82), and that for T4 and T7 RNA synthesis is $\sim$0.2 M (82). These optimal concentrations are also highly dependent on the ratio of polymerase to DNA. A similar dependence on magnesium ion concentration is observed, with rRNA and su^+_{III} tRNA synthesis being favored by low magnesium concentrations ($\leq$5 mM) and T4 RNA synthesis by higher magnesium concentrations ($\sim$10 mM) (83).

The relative rates of transcription of different RNA species also vary with temperature in a manner which is not wholly consistent with the pattern predicted by simple promoter opening. A particularly striking example is the halving of the rate of su^+_{III} tRNA synthesis over the temperature range of 35–45°C (81). A similar but less marked decline is apparent in the case of rRNA synthesis, but little or no diminution over this temperature range is observed for ϕ80 RNA synthesis (84). However, no decline is observed for either rRNA or su^+_{III} tRNA synthesis in the presence of the polymerase effector ppGpp (81, 84) (see under "Modulation of RNA Polymerase Initiation Specificity").

These changes in the patterns of promoter selection could be consequent upon structural changes in RNA polymerase, in the promoter sites, or in both. There is little if any evidence for ionic strength dependent direct changes in the structure of promoters. The effects of raising monovalent cation concentration are apparent in the presence and absence of 10 mM magnesium ion (62). The latter stabilizes DNA structure 100 times more effectively than monovalent cations on a molar basis (85), and thus it appears unlikely that the ionic strength variations have any gross effect on promoter structure. Nevertheless, it is not excluded that such variations may exert more subtle influences by, for example, altering the apparent charge of groups which protrude in the grooves of the DNA double helix. By contrast, there is ample documentation that the structure of RNA polymerase depends on both ionic strength and temperature. Thus, below 0.12 M KCl, holoenzyme is an equilibrium mixture of protomeric and dimeric forms, the dimeric form predominating as the KCl concentration is decreased (86). Other temperature- and ionic strength-dependent structural transitions in the enzyme are suggested by abrupt changes in the emission of a reporter molecule bound to polymerase holoenzyme (87). These changes occur over a narrow range of temperature ($\sim2°C$), the precise temperature at which the transition occurs depending on the monovalent cation concentration so that the break in the emission curve is observed at lower temperatures with increasing ionic strength. Similar subtle structural transitions in the enzyme have also been deduced from measurements of the kinetics of the partial quenching of the tryptophan emission of the protein induced by the binding of rifampicin to RNA polymerase (75). The kinetics of this quenching is at least biphasic and is consistent with the presence of two conformers of the enzyme in isomeric equilibrium with different reactivities for rifampicin. The relative contributions of the two phases to the total quenching are variable in that the fast phase is favored by the absence of magnesium and by high temperature. Together, these studies suggest the existence of conformational isomers of RNA polymerase. There is also some correlation between the inferred structural changes and alterations of template specificity. For example, the preferential synthesis of rRNA and su^+_{III} tRNA correlates with a predominance of the fast phase of rifampicin binding. Nevertheless, further studies are required to establish whether this correlation is general. The existence of temperature-dependent polymerase transitions has led to the suggestion that the transition temperature for promoter opening is a manifestation of this phenomenon (87). This seems unlikely in view of the dependence of t on DNA structure. Nevertheless, such temperature-dependent variations in enzyme structure may well be reflected in alterations in promoter selection by the enzyme.

The changes in the pattern of promoter selection consequent upon variations of ionic strength and temperature, coupled with the changes in initiation specificity induced by the nucleotide derivative ppGpp (68), strongly suggest that

RNA polymerase holoenzyme can discriminate between different promoter sites. What is the mechanism of this discrimination? Theoretically, the enzyme could exist either as a single form with a uniquely defined broad specificity or in a small number of conformational states, each of which would have a preferred affinity for a particular class of promoter sites. In the first model, discrimination would depend solely on the relative affinity of the enzyme for particular promoters, whereas in the second, discrimination would also depend on the distribution of the enzyme between the different states.

These formal possibilities can be distinguished experimentally by in vitro template competition experiments. First, an in vitro system is characterized in which RNA synthesis proceeds simultaneously from two types of promoter, e.g., phage and rRNA promoters. This system is then perturbed by the addition of increasing quantities of another DNA template which causes a redistribution of polymerase molecules by competing for the enzyme. If polymerase exists as a single form with invariant specificity, competition of RNA synthesized from different promoters should be independent of the competing template and depend on only the relative affinity of the enzyme for the respective promoters. In contrast, if the enzyme exists as a mixture of multiple forms, each with a different initiation specificity, the character of the competition observed should depend on the nature of the competing template. In a mixed template system containing $\phi 80$ d_3 $rrnB^+$ ilv^+ su^+7 DNA and T2 DNA, addition of T7 DNA results in a reduction of T2 RNA synthesis and a stimulation of rRNA synthesis (88). Contrastingly, addition of E. coli DNA reduces the synthesis of both rRNA and T2 RNA but competes with rRNA synthesis to a greater degree. Therefore, the character of the competition observed depends on the nature of the competing template.

The phenomenon of interference is another consequence of models in which one form of the enzyme has a high affinity for a particular promoter and other forms have a lower affinity. All forms of the enzyme bind to a particular promoter because variations in affinity are not absolute. However, the low affinity form(s) of the enzyme may have a slower rate of chain initiation than the high affinity form because a lower proportion of the initial polymerase-promoter binding events results in a productive open polymerase-promoter complex in the case of the low affinity form(s) of the enzyme. Thus, in template competition experiments, the removal of enzyme with a high affinity for T7 promoters by the addition of T7 DNA results in an increase in the rate of rRNA synthesis from $\phi 80$ d_3 $rrnB^+ilv^+su^+7$ DNA (88). This result suggests that a fraction of the population of RNA polymerase molecules blocks optimum initiation from rRNA promoters. Similarly, under conditions of low ionic strength, i.e., conditions which favor rRNA synthesis, transcription from T4 DNA and T7 DNA with enzyme excess is substantially below the optimum rate. This phenomenon is no longer apparent with template excess (89). Again it must be inferred that at low ionic strength a subpopulation of polymerase molecules decreases the rate of initiation from T4 and T7 DNA. Because this interference strongly depends on the type of template and on the parameters of temperature and ionic strength, a correlation with initiation specificity is clearly indicated.

These results are consistent with the hypothesis (56) that RNA polymerase holoenzyme I exists as a mixture of functionally distinct forms, each of which preferentially initiates at a particular class of promoter sites. The relative abundance of these forms would depend on ionic strength and temperature and would constitute a major determinant in the specificity of promoter selection.

REGULATION OF BACTERIAL RNA POLYMERASE

Ostensibly all the DNA-dependent RNA polymerases involved in transcription perform the same function in the synthesis of RNA. Yet in mitochondria, a polymerase consisting of a single polypeptide chain of $\sim$60,000 daltons suffices for the actual process of transcription (90). The relative complexity of the $E.$ $coli$ enzyme must, therefore, be concerned with functions other than the purely enzymic synthesis of RNA. An obvious possibility is that the heteromultimeric structure of $E.$ $coli$ RNA is required for regulatory purposes. Indeed, biochemical and genetic evidence now strongly suggests that RNA polymerase plays a central role in the overall control of RNA synthesis during normal growth. This mode of transcriptional control thus complements the operation of operon-specific DNA binding proteins. Together, these mechanisms ensure that the production of the major cellular components is balanced and geared to the environmental constraints imposed on the cell.

Modulation of RNA Polymerase Initiation Specificity

Aspects of polymerase function which are subject to regulation include initiation specificity, termination specificity, and specific activity. Two modes of modulation of initiation specificity can be distinguished. During normal growth, holoenzyme I, and probably holoenzyme II as well, acts as an allosteric enzyme with a variable initiation specificity (2). The enzyme can discriminate between different promoters; this discrimination is controlled by a variety of effectors which bind to holoenzyme. Among these effectors are the nucleotide derivative ppGpp, fMet-tRNA$_f^{Met}$, and the protein synthesis factors IF-2 and EF-TuTs. A second mode of modulating initiation specificity is observed during the development of $B.$ $subtilis$ phage SP01. Here, the σ subunit of the host polymerase is replaced by a series of small phage-coded polypeptides which sequentially direct the synthesis of different classes of phage RNA.

In vivo, any alteration in the initiation specificity of RNA polymerase might be expected to result in gross changes in the pattern of RNA species synthesized. One such change occurs on starvation of certain bacterial strains for a required amino acid. In this stringent response, the initiation of stable RNA species, both rRNA and tRNA, is severely restricted, whereas in contrast, the production of certain mRNA species, for example ϕ80 mRNA (91) and lac mRNA, is relatively unaffected. Another rapid response to amino acid starvation is the intracellular accumulation of the guanine nucleotide derivatives ppGpp and pppGpp to millimolar levels (92). Mutants unable to accumulate these nucleotide species fail to shut off rRNA synthesis and are said to show relaxed control, suggesting that one or both of these nucleotides may effect the preferential shut-off of stable RNA species (93). Strains of *E. coli* which fail to accumulate pppGpp after amino acid starvation but still accumulate ppGpp normally maintain the stringent control of RNA synthesis, suggesting that in vivo ppGpp may be the main motivity of stringency (93, 94).

In vitro ppGpp alters the pattern of RNA species synthesized both in crude extracts (95–97) and by highly purified RNA polymerase holoenzyme I (68, 98). With the DNA of the transducing phages $\phi80$ d_3 $rrnB^+ilv^+su^+7$ (68) or λ d_5 rrn ilv^+ (97, 98) as templates, physiological concentrations of ppGpp preferentially inhibit the synthesis of rRNA relative to either $\phi80$ or λ RNA in both crude and highly purified in vitro systems. Of other types of guanine nucleotides, pppGpp has a similar but less pronounced effect (98), whereas ppG does not change the pattern of transcription (68, 98). It seems that ppGpp acts by inhibiting the formation of stable polymerase-rRNA promoter complexes and is without significant effect on the subsequent extent of rRNA synthesis once such complexes have been formed (68). Thus, ppGpp alters the initiation specificity of RNA polymerase. The 'nucleotide' increases t for opening the rRNA promoter at 0.075 M KCl by 10–15°C from $\sim$25°C to $\sim$35–40°C (68). By contrast, ppGpp has little effect on t for opening $\phi80$ promoters. This differential response of the synthesis of RNA species to ppGpp establishes that RNA polymerase by itself can discriminate between different types of promoter and that this discrimination can be regulated in vitro.

The K_i value for the preferential inhibition of rRNA synthesis by ppGpp is 100–150 μM (56, 68, 95–98), a value which agrees well with the in vivo concentration of ppGpp at which the half-maximal inhibition of rRNA accumulation is observed (99). In a similar highly purified in vitro system, ppGpp also selectively inhibits the synthesis of a second RNA species su^+_{III} tRNA(81) whose accumulation in vivo is stringently controlled (91). However, in contrast to rRNA synthesis the K_i value for this selective inhibition is only $\sim$4 μM (81), a value close to that observed for the inhibition of poly(r(G-C)) synthesis from poly(d(I-C)) (100). The inhibition of the synthesis of rRNA, su^+_{III} tRNA and poly(r(G-C)) by ppGpp contrasts with the response of the synthesis of other RNA species. Thus, 100 μM ppGpp has no effect on the synthesis of poly(r(A-U)), whereas T4 RNA synthesis is stimulated by low concentrations ($\sim$5–20 μM) of ppGpp. The synthesis of *lac* mRNA initiated at the promoter containing the UV$_5$ mutation is substantially increased by higher concentrations ($\sim$200 μM) of the nucleotide derivative.

The target of ppGpp is almost certainly RNA polymerase itself. On incubation of the enzyme and nucleotide derivative together at 0°C in the absence of substrates or template, the transcriptional specificity of RNA polymerase slowly changes (68). When the enzyme is assayed immediately after mixing with ppGpp, an initial inhibition of all types of RNA synthesis is observed. Thereafter, on further prior incubation, the initiation specificity of the enzyme slowly changes so that the enzyme recovers virtually all of its ability to transcribe $\phi80$ and T2 DNA, but its ability to synthesize rRNA diminishes further, plateauing at $\approx\curvearrowright 0\%$ of the control value. Functional removal of ppGpp by dilution reverses the change in specificity with approximately the same kinetics. Other guanosine nucleotide derivatives, for example ppG, do not produce this effect. The specificity change induced by ppGpp is thus reversible and nucleotide derivative-specific. Although the rate of change of initiation specificity is slow by comparison with other protein isomerization constants, the immediate inhibition of polymerase activity on mixing enzyme and the nucleotide species suggests that the rate is not a consequence of a slow, simple association of two components. Rather, the data are compatible with an initial fast association followed either by a slow structural change in the enzyme or by a second much slower association. This interpretation is compatible with the kinetics of quenching of the fluorescence of RNA polymerase by ppGpp. These kinetics are at least biphasic at 25°C, the two modes having half-times of approximately 0.5 s and 20 s (Jovin and Travers, unpublished observations).

To what extent can the existence of conformational isomers as deduced from functional and physical data be related to observed changes in the initiation specificity of RNA polymerase holoenzyme? In the presence of 200 μM ppGpp, the relative competition of T2 RNA and rRNA synthesis from T2 DNA and $\phi80$ d_3 DNA, respectively, by other DNA species is virtually independent of the nature of the competing template (88). Thus, at this ppGpp concentration, in contrast to the situation in the absence of this component, there is no evidence for specificity isomers of the enzyme. Furthermore, the characterization of two distinct K_i values for the inhibition of rRNA synthesis by ppGpp strongly suggests the existence of at least three states of RNA polymerase, states which may or may not be related to the conformational isomers already discussed. The synthesis of su^+_{III} tRNA and poly(r(G-C)) is selectively shut off at 10 μM ppGpp, whereas rRNA synthesis is likewise inhibited at 200 μM ppGpp, at which concentration a substantial increase in *lac* mRNA synthesis is observed (Figure 2). This type of data may be interpreted at present either in terms of the existence of discrete conformations of the enzyme which are stable over a particular range of ppGpp concentrations or in terms of a ppGpp concentration-dependent shift in equilibrium between a mixture of isomers. A further possibility is that the nucleotide derivative directly inactivates one class of polymerase molecules, thereby reducing the total amount of enzyme available. Although the kinetics of the change in specificity induced by ppGpp is more compatible with a conversion of polymerase specificity rather than with inactivation, the latter cannot be excluded solely on these grounds.

Are there additional low molecular weight effectors of polymerase activity or specificity? The purine nucleotide derivatives, ATP and GTP, bind to the free enzyme with a binding constant of 1.5×10^{-5} M (101). This binding, like that of

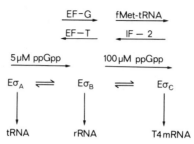

Figure 2. The control of polymerase specificity by modulation of the equilibrium between polymerase conformations. The *arrows* indicate the direction in which the effectors shift the equilibrium.

ppGpp, is strongly competed for by rifampicin. Because the concentration of purine nucleoside triphosphates required for the initiation of transcription is of a similar magnitude, it has long been assumed that the triphosphates bound to the free enzyme were the same triphosphates that were incorporated into the 5'-terminal position of the nascent RNA chain. Yet rifampicin does not significantly inhibit the rate of dinucleoside tetraphosphate synthesis in the abortive initiation reaction even though the requirement for a purine nucleoside triphosphate in the 5'-terminal position is retained (73, 74). Thus the possibility remains open that triphosphates binding to free holoenzyme perform another function, possibly analogous to that of ppGpp.

Coupling of Transcription and Translation

The polymerase effector ppGpp controls the synthesis of the major macromolecular components of the cell that are required for the process of translation and is itself a product of the translation machinery. Its rapid synthesis by an idling reaction on the ribosome is triggered by the codon-dependent binding of unchanged tRNA to the acceptor site (102, 103). The steady state concentration of the nucleotide derivative is thus coupled to and requires a functioning ribosome. In addition, the rate of degradation of ppGpp depends on the energy input of the cell so that breakdown is facilitated when available energy is high (104, 105). In this way, the level of ppGpp in the cell, and hence its effect on transcriptional specificity, is geared to both protein synthesis and energy metabolism.

In vivo evidence shows that ppGpp, although a major effector of transcriptional specificity, may not be the sole controlling element. The close involvement of transcription with translation is suggested by the suppression of ribosomal protein mutants by mutation of the β subunit of RNA polymerase (106). Furthermore, the antibiotics neomycin and spectinomycin, which inhibit polypeptide chain elongation, relieve the inhibitory effect of ppGpp on rRNA (107).

In vitro several components of the translation machinery have been shown to affect RNA polymerase holoenzyme specificity and in every case appear to act by modulating the initiation specificity of the enzyme. Thus, the charged tRNA, fMet-tRNA$_f^{Met}$, stimulates transcription of *lac* mRNA (28) while concomitantly reducing initiation at λ and rRNA promoters. By contrast, the initiation factor IF-2 stimulates rRNA synthesis from λ d$_5$ *ilv* DNA by 3–5-fold and at the same time reduces transcription of λ RNA sequences by a comparable amount (27). Although the activity of these effectors in vitro is relatively insensitive to the conditions of the reaction, binding of fMet-tRNA$_f^{Met}$ to RNA polymerase strongly depends on Mg^{2+} concentration, optimum binding being observed at 5 mM Mg^{2+} (28). The protein synthesis elongation factors EF-TuTs and EF-G also modulate holoenzyme activity. With *E. coli* DNA as template, EF-TuTs decreases the transition temperature for rRNA synthesis while simultaneously increasing the sensitivity of transcription to ppGpp (56). With ϕ80 d$_3$ DNA as template, the response of rRNA synthesis to the addition of EF-TuTs is more complex. At low Mg^{2+} or K$^+$ concentrations, the factor reduces the absolute and proportional rate of rRNA synthesis, whereas at higher concentrations the reverse effect is observed. Synthesis of su^+_{III} tRNA responds in a very similar manner (108). In general, both EF-Tu and EF-Ts are required for the maximum effect on transcription, although a small effect is often apparent with EF-Tu alone. EF-G acts as an antagonist to EF-TuTs in its effect on rRNA synthesis, stimulating rRNA transcription at low ionic strength and depressing it at high ionic strength (108). Although the available evidence indicates that many effectors bind to RNA polymerase, it is not necessary to postulate a comparable number of discrete binding sites. For example, IF-2, EF-G, and EF-Tu all bind to the same site on the 50 S subunit of the ribosome; therefore, their interaction with RNA polymerase could occur at a single site analogous to the ribosomal binding site (109).

The modulation of RNA polymerase specificity by at least four components of the translation machinery suggests that in vivo transcription and translation during normal growth may be tightly coupled by such interactions. The crucial question is whether the variations of initiation specificity observed in vitro are indeed biologically relevant. However, in only one case have mutants of a putative effector been isolated. These are thermosensitive mutants of EF-G (110). Normally such mutants are relaxed for rRNA synthesis at the restrictive temperature; i.e., they synthesize rRNA when protein synthesis is inhibited. However, because EF-G is involved in the conversion of pppGpp to ppGpp (111), these mutants in general also fail to accumulate ppGpp at the restrictive temperature. Nevertheless, one *B. subtilis* mutant with a *ts* EF-G accumulates both ppGpp and rRNA at the restrictive temperature. This phenotype would be consistent with the in vitro effects of EF-G on rRNA synthesis.

Logic of Bacterial Growth

The number of ribosomes a bacterium produces is in general approximately adjusted to the environment it grows in so that during exponential growth no more ribosomes are synthesized than can be engaged in protein synthesis (112).

This correlation has led to the view that the level of ribosomes limits protein synthesis and thus regulation of ribosome production is a major facet of the control of bacterial growth. Three general models exist for the control of ribosome production at the level of transcription: 1) a specific regulation of the Jacob-Monod type governing the initiation of ribosomal protein mRNA in particular; 2) a constitutive formation of r-protein mRNA in which the frequency of transcription is determined passively by a general competition for polymerase molecules among available promoters; and 3) a specific regulation of r-protein mRNA and rRNA synthesis in which the distribution of polymerase molecules between different transcription units is determined actively by polymerase effectors. These models are not mutually exclusive, but the extent to which any one acts as a major determinant of ribosome production remains to be ascertained. One prediction of the second model is that an increase of all constitutive gene products proportional to ribosomal gene transcription should be observed with increasing growth rate. Analysis of the rates of transcription of constitutive *trp* and *lac* operons has not verified this prediction (113, 114). By contrast, there is considerable evidence that ppGpp controls r-protein production in crude in vitro systems in which competition with other promoters is minimal (97). Moreover, the patterns of control observed in vitro correlate well with the in vivo patterns recorded under conditions of intermediate ppGpp concentrations associated with partial amino acid starvation (17). Thus, assuming that the major effects of ppGpp are at the level of transcription, it must be inferred that the frequency of initiation on ribosomal genes is, at least partially, under active control. If indeed the modulation of polymerase specificity is a major determinant of the control of overall transcription patterns, how can the activity of the modulating effectors be integrated to produce a homeostat? Any molecular model for such a control scheme is at present necessarily conjectural but must consider the logic of the control, the nature of the effectors, and the dynamic balance between them.

One element of such a scheme is ppGpp acting as an effector of polymerase specificity in a manner similar to its in vitro function. If the "nucleotide" acts by altering the position of an equilibrium between specificity isomers, the actual effect of ppGpp on transcription in vivo depends on the relative free concentrations of ppGpp and of any effector which shifts the equilibrium in the reverse direction. Thus, the mere accumulation of ppGpp need not invariably result in a shut-off of stable RNA synthesis. The role of ppGpp during normal growth seems in some situations not to be determinative, for example, during shift down from a rich to a poorer medium. More probably, one of the major functions of the nucleotide is to effect a shutdown of ribosome biosynthesis as a response to the imposition of conditions inimical to growth.

A second element is the control of the distribution between specificity forms of RNA polymerase by the cycling factors of translation. These effectors would signal the need for rRNA or tRNA or for particular classes of mRNA species depending on which component of the translation machinery was momentarily limiting. For example, the initiation factor IF-2 is released on initiation of a polypeptide chain. This release would constitute a signal that a ribosome had been engaged in protein synthesis. In the presence of an excess of ribosomes, the factor would once more be engaged in the initiation of protein synthesis. However, if ribosomes were limiting, the factor would be available to direct the polymerase to synthesize rRNA. fMet-tRNA$_f^{Met}$, which is consumed by the act of initiation, alters polymerase specificity in the reverse direction. These two components of the translation initiation machinery, which themselves mutually interact, thus act as a balancing couple in controlling transcription. Similar considerations apply to the protein synthesis elongation factors EF-TuTs and EF-G, which again seem to act as a balancing couple. Interpretation of their in vitro roles is complicated by the strong dependence of the magnitude and direction of the observed specificity changes on the precise reaction conditions. However, the continued synthesis of rRNA in the presence of ppGpp in a *B. subtilis* strain containing a thermolabile EF-G suggests that EF-G in vivo may switch RNA polymerase specificity in the same direction as ppGpp, i.e., away from rRNA synthesis. This argument would indicate that EF-TuTs should switch on rRNA and tRNA synthesis. It is perhaps surprising that EF-Tu, which is the most abundant protein in *E. coli* (114), should be implicated as a regulatory protein. However, a second paradox is apparent in the observation that the concentration of ppGpp required in vitro to switch off su^+_{III} tRNA synthesis is well below the normal in vivo basal levels. Thus, in the bacterial cell, the relatively high concentrations of EF-TuTs might be sufficient to override the low K_i for ppGpp inhibition of su^+_{III} tRNA synthesis with the result that the equilibria between different forms of RNA polymerase would be in a sensitive position for the control of tRNA synthesis.

A third necessary element in the logic of any control scheme is the mode by which the amounts of the regulating effectors are themselves controlled. Any homeostatic model requires that the level of an effector within the cell be regulated in a mode opposite to that of the macromolecules it itself controls. In the bacterial cell, the concentration of ribosomes increases approximately in proportion to the growth rate of the cell (112). Consequently, the rate of synthesis of the major macromolecular components of the ribosome increases approximately in proportion to the second power of the growth rate. Of the putative transcriptional effectors, the variation of the level of IF-2 in the cell with increasing growth rate is proportioned to total protein (115), that of EF-G to ribosomal protein (116). These patterns would be consistent with a homeostatic model. The data discussed so far relate to polymerase holoenzyme I. The relative roles of σ and σ' in vivo are at present unknown, and therefore, the possibility that the regulation of template selection by the σ'-containing enzyme differs from that of the σ-containing enzyme remains. Furthermore, the purification of σ' from ribosomes suggests that the availability of σ' to combine with core enzyme may be directly coupled to protein synthesis.

Bacteriophage Development

The production of a bacteriophage by an infected cell is an extremely specialized, albeit fatal, form of development. The infecting virus commits the host cell to produce more virus and in so doing alters the transcription machinery of the cell so that phage genes are preferentially expressed.

The small coliphages T3 and T7 provide the simplest examples of this type of programmed transcription. The genome of these phages can be conveniently divided into two groups of genes. One group, the early genes comprising 18% of the genome, is expressed immediately after infection and is transcribed by the host polymerase. Subsequently, a second group of genes, the late genes, is turned on while concomitantly early gene and host transcription are switched off. In phage T7, an early gene product, the 0.7 protein, is required for the shut-off of host transcription (117). The product of this gene is a protein kinase which phosphorylates the β and β' subunits of *E. coli* RNA polymerase. This phosphorylation abolishes initiation on the host genome and reduces initiation on T7 DNA. An additional consequence is the failure of the modified enzyme to recognize a termination signal on T7 DNA. A second T7-coded protein, the 0.3 protein, may functionally complement the 0.7 protein in effecting this block (118). This smaller protein, whose analogue is absent in T3-infected cells, binds to RNA polymerase, as judged by the criteria of coprecipitation by antibody to RNA polymerase and of retention by a polymerase-agarose column (24). Another postulated role for the 0.3 protein is an involvement in the mediation of the efflux of potassium ions that occurs soon after T7 infection (119). This efflux, with the resultant alteration in intracellular ionic conditions, has been implicated as a direct causative factor in the switch from early to late T7 transcription. A further T7 protein, a late gene product, also effects the shut-off of host and early T7 RNA synthesis. This protein binds specifically to the host polymerase, inhibiting further initiation of RNA chains.

The expression of T7 and T3 late genes depends on phage-specific protein synthesis and in particular on the function of phage gene 1 (120), which codes for an RNA polymerase which is a single polypeptide chain of 105,000 daltons (23). Unlike the polymerase of the host cell, the phage-coded polymerase is not inhibited by rifampicin, and as a consequence phage production becomes insensitive to this antibiotic early in the lytic cycle. In vitro, the phage polymerases transcribe the homologous phage DNA, asymmetrically synthesizing primarily late phage RNA (121). The initiation specificity of these enzymes is extremely restricted because in addition to the homologous DNA only poly(dG·dC) serves efficiently as a template.

A more complex developmental pattern is exhibited by *B. subtilis* phage SP01. Three major temporal classes of phage transcripts can be distinguished (122), but in contrast to coliphage T7 the host core polymerase $\alpha_2\beta\beta'$ is conserved and required throughout phage development. Early phage RNA is transcribed almost immediately after infection. In vitro early RNA transcription, like that of phage T7, is directed by the host σ subunit. Next phage "middle" genes are transcribed at 4–5 min into the lytic cycle. This transcription depends on the product of SP01 gene 28 (123) which codes for a phage protein of 26,000 daltons (124). This polypeptide binds to core RNA polymerase and apparently directs the specific transcription of middle RNA in the absence of host σ factor (125, 126). Another host protein, δ (21,000 daltons), improves the accuracy of middle gene transcription by phage-modified polymerase (26).

Although gene 28 product binds to *B. subtilis* core polymerase and RNA polymerase containing this protein asymmetrically transcribes middle genes in vitro, proof that the gene 28 product itself directly influences the transcriptional specificity of RNA polymerase is still lacking. Indeed, a new protein species, apparently distinct from the product of gene 28, that promotes asymmetric transcription of phage middle genes in vitro has been identified (127).

Activation of SP01 late genes occurs between 8 and 14 min into the lytic cycle and is controlled in vivo by the products of genes 33 and 34 (123). These genes code for polypeptides of 13,500 and 24,000 daltons (128), respectively. Both of these proteins bind to core RNA polymerase and together with host δ protein act synergistically to direct the synthesis of late RNA by core RNA polymerase (129).

The pattern of transcriptional control during the development of T4 (see Figure 3), a coliphage of similar complexity to phage SP01, differs in several respects from that of the *B. subtilis* phage. At least four classes of transcripts, termed immediate early, delayed early, quasi-late, and late RNA appear sequentially during T4 development (130). Of these classes of RNA, only immediate early RNA is synthesized when cells are infected in the presence of chloramphenicol, suggesting that phage-specific protein synthesis is required for synthesis of most, if not all, the other classes of T4 RNA (131). Delayed early RNA synthesis begins at 2 min after infection at 30°C and is not inhibited by the addition of rifampicin to a sensitive cell immediately after infection. Thus,

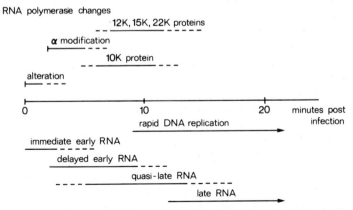

Figure 3. Patterns of RNA synthesis and changes in polymerase structure during the development of phage T4.

further initiation is not required for the appearance of delayed early RNA sequences. However, both immediate early and delayed early RNA sequences are transcribed from the same DNA strand and are interspersed and contiguous on the DNA chromosome (132). Consequently, it has been suggested that delayed early transcription could result from the inhibition of ρ-mediated termination which in vitro restricts transcription by host holoenzyme to immediate early transcription (133, 134). Such inhibition would allow polymerases which had initiated at immediate early promoters to read through into delayed early regions and would result in delayed early RNA sequences being found predominantly in long RNA molecules. This last prediction has been experimentally verified (132). By contrast, quasi-late RNA synthesis requires further initiation of RNA chains (135) and depends on the product of the *mot* gene (136). This gene may be identical with the *sip* (137) and *far* (138) loci identified by other criteria. Late RNA synthesis depends on the products of T4 genes 33, 45, and 55 (139). In addition, the level of late RNA synthesis is substantially enhanced by T4 DNA replication, although some late RNA synthesis does occur in the absence of detectable DNA replication (140). A further change in the pattern of transcription during T4 infection is the shut-off of host RNA synthesis during the first 5 min of the lytic cycle (141, 142). This process also requires the function of a T4 early gene product.

This changing pattern of gene expression is paralleled by structural changes in the host RNA polymerase because immediately after phage absorption the polymerase is "altered" (143). This step is independent of protein synthesis and involves the addition of adenosine diphosphoribose residues from NAD (144) to a small fraction (145) of the α subunits of the enzyme. In addition, a proportion of the σ and β or β' subunits may contain similar covalent additions. The polypeptide responsible for this ADP-ribosylation is injected with the phage DNA, but the functional significance (145), if any, of these changes in RNA polymerase remains obscure.

At about 4–6 min after infection, the subunit undergoes a similar stoichiometric ADP-ribosylation termed "modification" (146, 147). This change is blocked by treatment of infected cells with chloramphenicol. Phage which lack the modification function have no detectable lesions in the developmental program under normal laboratory conditions (145). However, reconstitution experiments in vitro have suggested that RNA polymerase containing modified α subunits is defective in the transcription of *lac* mRNA (148). A possible function for modification is thus the turn-off of host mRNA synthesis.

From 5 min after infection, small T4-coded polypeptides of molecular weights of 22,000, 15,000, and 12,000 become associated with core RNA polymerase (24, 149, 150). The 22K and 12K polypeptides have been identified on the products of genes 55 (151) and 33 (152), respectively, whereas the 15K polypeptide is missing from polymerase isolated from cells infected with a mutant T4 phage, *alc⁻* or *unf⁻*, which fails both to unfold the host genome normally and shut off host rRNA transcription (153, 154). This suggests that the 15K polypeptide may be involved in the shut-off of host transcrirtion, a finding which is supported by the in vitro observations that core polymerase from T4-infected cells supplemented with host σ factor has a lower affinity for *E. coli* rRNA promoters than *E. coli* holoenzyme (84). Furthermore, the specificity of the enzyme from infected cells is unaltered by ppGpp (84). By elimination, the 15K polypeptide is thought to be responsible for this change.

The identification of the genes coding for the 22K and 12K polypeptide suggests that these proteins are required for late RNA transcription. In addition, a role for at least two other proteins in late RNA transcription has been suggested. The T4 gene 45 product, a protein of 27,000 daltons, binds to a T4 core polymerase-agarose column but not to a similar column prepared with *E. coli* core polymerase (24). In addition, *alc*+ strains of T4 fail to synthesize late RNA under conditions in which hydroxymethylcytosine in the phage DNA is replaced by cytosine (155). This suggests that the 15K polypeptide normally functions in the synthesis of late RNA.

One other phage-coded polypeptide of 10,000 daltons which is genetically undefined binds to (149) and apparently alters the function of the host σ subunit so that the affinity of the holoenzyme for T4 promoters is reduced (156, 157). A possibly related observation is the identification of a transcription factor activity in crude extracts of T4-infected cells which directs *E.coli* or T4 core polymerase to synthesize RNA with the properties of quasi-late RNA (158). The factor could function by binding either to RNA polymerase (158) or to DNA (159).

Another mechanism of control of the sequences transcribed by *E. coli* holoenzyme is apparent during development of phage λ. Positive control of early gene expression in this phage requires the function of the λ N protein. This regulator permits the transcription of λ early genes, yet its sites of action are clearly distal to and distant from the early λ promoters. There is, however, a striking correlation between these sites and transcriptional terminators requiring ρ function. From this evidence, Roberts (37) proposed that protein N antagonized ρ function and thus allowed these termination signals to be overridden with a consequent extension of RNA chains initiated at early promoters. A block in λ development resulting from a lack of N function can be overcome by mutations in RNA polymerase (160, 161), suggesting that this mode of control might directly involve the transcribing enzyme. Indeed, recent evidence suggesting that the N protein is coprecipitated with RNA polymerase by antiserum to the enzyme (162) supports the concept that the termination specificity of RNA polymerase can be controlled in addition to the initiation specificity.

PROMOTER STRUCTURE AND RECOGNITION

What are the structural features within the promoter region which constitute the recognition parameters for RNA polymerase? Recognition of promoters by *E. coli* polymerase holoenzyme must satisfy at least two functional criteria. First, the σ polypeptide is required for the recognition of all *E. coli* promoters so far tested and also of phage promoters active in the absence of polymerase modification, suggesting that all σ-dependent promoters contain a common recognition element. This element could be a unique sequence or a family of closely related sequences. Second, the ability of *E. coli* holoenzyme to discriminate between different types of promoter (2) suggests that there must be a second recognition element which varies between promoters.

In functional and molecular terms, a promoter is a region along a DNA template at which RNA polymerase first recognizes a sequence and then forms a tight binary complex with the DNA. Within this latter complex, a stretch of DNA of 40–45 base pairs astride the mRNA start point is protected from DNase digestion (162–166). In several promoters, this domain contains a sequence in the coding strand closely related to -T-A-T-R-A-T-G, preceding the transcription start point by 6 or 7 residues (167, 168). This observation led to the suggestion that this heptanucleotide sequence is important for the formation of a stable RNA polymerase preinitiation complex with the DNA (167, 168). The biological importance of this sequence is demonstrated by the finding that the UV5 promoter mutation in the *lac* promoter region changes the heptanucleotide sequence of the wild type promoter so that it exactly matches the proposed common sequence and concomitantly abolishes the requirement for the auxiliary *crp* protein for efficient initiation (169). Nevertheless, a comparison of all known promoter sequences, including the recently determined *gal* (170), φX A, φX B, and φX D (171) promoters, reveals that at no position in this sequence is a residue wholly conserved. This variability may be compared with the recognition of mRNA translation initiation sites by the complementary interaction of a 16 S rRNA sequence and an mRNA sequence. Different initiation sites exhibit different extents of complementarity to 16 S rRNA (172, 173). The extent of complementarity is inversely correlated with the dependence of initiation on the presence of protein initiation factors (174), those binding sites which form the weakest mRNA-rRNA interaction having the greatest dependence on factors. By analogy, a requirement for transcription initiation factors may follow deviation from the most favorable heptanucleotide sequence in the conserved "box."

Although the conserved box is undoubtedly a crucial component of promoter activity, several arguments suggest that the sequence information in the polymerase-protected fragment is insufficient for overall promoter function. First, RNA polymerase does not form a stable complex with promoter fragments containing 20–25 residues on either side of the transcription start point (167, 168,

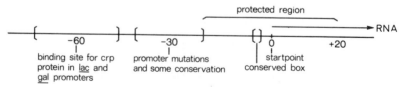

Figure 4. Promoter structure in prokaryotes.

175, 176). Second, a strong promoter mutation, *sex* (177), and two *lac* promoter mutations (178, 179) fall outside the protected domain in a region about 35 base pairs to the left of the corresponding mRNA start point. Sequence conservation is also apparent in the area. Thus, the pentanucleotide -T-G-T-T-G is present in two λ promoters and an SV40 "promoter" (180). Furthermore, the sequence -A-C-A-C-T-T-T is present in *gal* (170), *lac* (179), and tRNA$_{III}^{Tyr}$ (181) promoters.

A further feature of promoter sequences is a region of 2-fold rotational symmetry centered close to the transcription start point. This is particularly prominent in *gal* (170) and tRNA$_{III}^{Tyr}$ (181) promoters. It remains uncertain whether these "palindromic" sequences play any functional role in DNA transcription.

To what extent can the observed conservation of promoter structure be correlated with the requirements for promoter function? The most effective competitor template in template competition experiments is poly(d(A-T)). This copolymer competes effectively with both rRNA and phage mRNA promoters. Furthermore, this is the sole template on which core polymerase has been observed to form strongly temperature-dependent preinitiation complexes (53). These observations suggest first that a common recognition region for promoters is A-T rich and that this region is covered by core polymerase. The conserved heptanucleotide box is clearly a candidate for this function.

Clues concerning the region involved in discrimination may be gleaned from the differing degrees of inhibition of transcription of synthetic templates by ppGpp. Thus, transcription of poly(d(I-C)) is inhibited, whereas that of poly(d(A-T)) is not (100). One possibility then is the discrimination between G·C base pairs for the former and A·T base pairs for the latter. The fact that substitution of bromodeoxyuridine for thymine in T4 promoters alters the response of T4 transcription to ppGpp (79) and hence presumably the discriminatory recognition parameters of the promoters emphasizes that the distinction between A·T and G·C recognition must be positive.

The conclusion that discrimination by RNA polymerase involves the distinction between G·C and A·T base pairs is reinforced by the properties of the T4 *alc* mutants (157). The *alc*+ T4 strains are unable to synthesize late T4 RNA on T4 DNA containing cytosine in place of hydroxymethylcytosine (153). Mutation to *alc*− restores the capacity to use cytosine-containing DNA but in addition results in the failure to shut off host rRNA synthesis (ref. 155 and A. A. Travers, unpublished observation). Because the structure of hydroxymethylcytosine mimics thymine to the extent that both contain a group in the major groove of the DNA duplex this suggests that the putative product of the *alc* gene prevents recognition of cytosine in DNA. The obvious corollary is that recognition of G·C base pairs is required for rRNA synthesis.

An important consequence of promoter discrimination by multiple forms of RNA polymerase is the possibility that the relative affinity of different forms of the enzyme for a promoter can vary. At one extreme, a promoter may bind tightly only one polymerase isomer, whereas another promoter may interact equally well with two or more forms of the enzyme. In this way, a virtually continuous spectrum of recognition characteristics can be generated so that the promoter for any particular transcription unit is adapted to the control of polymerase specificity in a manner appropriate to the requirement of the cell. Such a model predicts that the responses of the synthesis of particular RNA species to ppGpp will not fall into a number of distinct classes but rather that the synthesis of each RNA species will have a characteristic individual response.

The role of the σ polypeptide is unclear. Although this protein does not bind to DNA-cellulose columns, it is retained by a heparin-agarose column (182). Perhaps the polypeptide directly recognizes a promoter sequence proximal to the region at which the core enzyme binds. Correspondingly phage-specific polypeptides which replace the σ factor might bind to different sequences and consequently alter the initiation specificity of the enzyme (1). The role of an auxiliary factor such as *crp* protein in promoter recognition is also obscure. Sequence determination and analysis of *crp* binding mutations suggest that the *crp* binding is centered at about 60 residues proximal to the transcription start point (170, 179). Dickson et al. (179) have proposed that the factor may facilitate initiation by acting as a DNA melting protein, but there is no direct experimental evidence to support this view. Another possibility is that *crp* protein interacts directly with polymerase holoenzyme and thereby increases the affinity of the enzyme for *crp*-dependent promoters (50, 51). Such an interaction could be with either the core or σ components of the polymerase.

In contrast to the intricacies of promoter recognition by *E. coli* polymerase, promoter recognition by T7 RNA polymerase seems simple. For efficient promoter utilization, the enzyme requires GC-rich sequences. Substitution of either G or A base analogues which no longer contain a group which projects into the minor groove of the DNA double helix completely blocks promoter utilization by the enzyme (183). Conversely, the presence or absence of a 5-methyl residue on the pyrimidine base which is exposed in the major groove of the duplex does not appear to alter promoter utilization significantly. These experiments show, therefore, that promoter utilization can be abolished without altering the base-pairing specificities within the promoter region. These findings suggest that the initial recognition of a promoter by the T7 polymerase may depend on the exposure of groups in the minor groove of the DNA complex. In contrast, the *E. coli* polymerase has a requirement for groups in the major groove as indicated by the change in promoter recognition consequent upon bromouracil substitution (79).

EUKARYOTIC TRANSCRIPTION

To what extent are the principles of control of transcriptional specificity in prokaryotes applicable to eukaryotes? The transcriptional machinery of eukaryotes differs in at least one major respect from that of prokaryotes. In place of a single eukaryotic polymerase, the nucleus of higher eukaryotes contains three structurally distinct enzyme types. One, polymerase I (or A), is located in the nucleolus and synthesizes rRNA. A second, polymerase II (or B), is located in the nucleoplasm where it synthesizes HnRNA, whereas the third, polymerase III (or C), synthesizes 5 S and tRNA. These enzymes are all heteromultimers whose subunit structures, although more complex, bear strong resemblances to those of the bacterial polymerases (184). In addition to the transcription systems within the nucleus, both mitochondria and chloroplasts contain DNA and RNA polymerases. The enzyme from the former is a single polypeptide chain of $\sim$60,000 daltons (90, 185, 186), whereas that from chloroplasts (187, 188) has a very close structural resemblance to the normal bacterial holoenzyme.

Eukaryotic Polymerases

The multiple forms of RNA polymerase in the eukaryotic nucleus are classified on the basis of their chromatographic and catalytic properties and their sensitivity to the toxin α-amanitin isolated from the fungus *Amanita phalloides*. Polymerase II is inhibited by low concentration (≤ 2 μg/ml), whereas polymerase III is inhibited only at higher concentrations ($\sim$200 μg/ml). By contrast polymerase I from higher eukaryotes is virtually insensitive to the toxin (184). Yeast polymerase I does, however, exhibit a sensitivity comparable to that of polymerase III from higher eukaryotes (189).

Within each type of nuclear polymerase, further heterogeneities have been observed. Multiple species of polymerase I have been described in rat liver (190, 191), calf thymus, *Xenopus laevis* (192), and yeast (189). In rat liver and calf thymus, these forms, termed "AI" and "AII," differ structurally. AI contains six types of subunits with molecular weights of 197,000, 126,000, 44,000, 25,000, and 16,500, present in the ratio 1:1:1:2:2, respectively (193), not including subunit SA3. AII contains an additional polypeptide of ~60,000 molecular weight. These forms of the enzyme can be related to the in vivo pools of "free" and "engaged" or actively transcribing enzyme (194). The latter consists entirely of form AII, whereas form AI is found only in the free fraction. A similar heterogeneity is observed in yeast (189). In this case, form AI (or A*) contains 10 types of polypeptide with molecular weights of 190,000, 135,000, 44,000, 40,000, 29,000, 24,000, 20,000, 16,000, 14,000, and 10,000, whereas form AII contains two additional apparent subunits of 48,000 and 37,000 daltons. The removal of these two subunits is associated with an increase in sensitivity to α-amanitin and with a decrease in the RNase H activity of the preparation (195).

The functional difference, if any, between the multiple forms of polymerase A remains obscure. Polymerase AI from *Xenopus* possesses transcriptional selectively, transcribing selectively the ribosomal RNA coding sequences in partially purified *Xenopus* ribosomal RNA (196). Nevertheless, for efficient transcription of yeast rDNA sequences an enzyme which has not been purified to the same degree is necessary (197).

Heterogeneity is also apparent in preparations of polymerase II (198). Three such forms, II_0, II_A, and II_B, have been isolated from a mouse plasmocytoma

(199). Each of these forms contains subunits with reported molecular weights of 140,000, 41,000, 30,000, 25,000, 22,000, 20,000, and 16,000. The molar ratios of all the subunits are unity except for the 20K polypeptide which has a molar ratio of 2. Each enzyme form is distinguished by its largest subunit, which has a molecular weight of 240,000, 205,000, and 170,000 in forms II_O, II_A, and II_B, respectively. In soybean, form IIA is the dominant form in ungerminated embryos comprising >95% of the class II enzyme (200). This form of the enzyme is free. However, as germination progresses increasing amounts of polymerase II can be recovered in nuclear preparations as a chromatin-bound enzyme. This enzyme exists exclusively as form II_B. One possibility is that in soybean form II_A enzyme may represent a storage or precursor form of polymerase II which becomes active only after its conversion via proteolytic cleavage to form II_B. Such proteolysis has been demonstrated in yeast (201) and in *Drosophila* (202), but in these cases the biological relevance of the phenomenon is far from clear.

In vitro polymerase II exhibits a strong preference for denatured DNA templates, but in highly purified transcription systems little or no transcriptional selectivity has so far been observed. Therefore, the molecular mechanism of initiation by this enzyme in vivo is unclear.

Polymerase III preparations, like those of polymerases I and II, exhibit structural heterogeneity (203). Two forms of the enzyme can be isolated from a mouse plasmocytoma line. Form III_A contains subunits of molecular weights of 155,000, 138,000, 89,000, 70,000, 53,000, 49,000, 41,000, 32,000, 29,000, and 19,000. The subunits of form III_B are identical with those of form III_A except for the replacement of the 32K polypeptide by a subunit of 33,000 daltons. Molar ratios of these subunits are close to unity for all except the 19K polypeptide which is present in molar excess.

Highly purified polymerase III preparations show high transcriptional selectivity in vitro. The enzyme preferentially transcribes low molecular weight RNA species from adenovirus DNA (204), an activity which is also apparent in the nuclei of adenovirus-infected cells. However, the requirements for the selective synthesis of 5 S RNA sequences by polymerase III appear to be more stringent (205). Both chromatin-associated components and the enzyme are necessary and sufficient for accuracy. Thus, the 5 S genes in *Xenopus* oocyte chromatin are transcribed by exogenous enzyme in a highly selective (3,000-fold above random) and predominantly asymmetric fashion. By contrast, no such selective transcription was observed with polymerases I or II or *E. coli* RNA polymerase. Similarly, all tested polymerases transcribed deproteinized 5 S DNA with only a slight degree of selectivity and both the sense and antisense strands of the gene were transcribed.

Although the structures of the three types of RNA polymerase seem to be complex, a few polypeptides seem, on the basis of molecular size, to be common to different enzyme classes (206). In the mouse plasmocytoma, these include a 52,000-dalton polypeptide common to classes I and III, a 41,000-dalton polypeptide common to classes II and III, and 29,000- and 19,000-dalton polypeptides common to classes I, II, and III. Immunological studies also suggest some common antigenic determinants, at least in the class I and II enzymes of calf thymus (207, 208) and yeast (209). In yeast this observation is supported by the presence of three polypeptides of apparently identical size in polymerases I and II (210).

Control of Eukaryotic Transcription

The molecular mechanisms that control transcription in eukaryotes are, to a large extent, poorly characterized. Studies with bacterial RNA polymerase probes have indicated that chromatin structural modifications are involved in regulating the transcription of the globin (211, 212), ovalbumin (3), and histone genes (213). However, the nature and mechanism of action of the putative regulatory proteins is not clear, nor is it clear whether chromatin structural modifications are the sole means by which these or other genes are regulated. Nevertheless, if the prokaryotic and eukaryotic polymerases are indeed homologues, it would seem unlikely that the property of specificity modulation acquired by the prokaryotic polymerase would be entirely lost during the emergence of the eukaryotic line. The separation of eukaryotic polymerases into three distinct types, each with apparently different functions, effectively releases the eukaryote from the prokaryotic restriction of balancing the use of a single enzyme. However, despite arguments of this type, evidence for structural changes in the eukaryotic polymerases which correlate with changes in transcription patterns remains meager. In one case changes in   hromatographic and catalytic properties of fibroblast polymerase I are associated with a change from a resting to an actively growing state (214). The phosphorylation of polymerase I in yeast may or may not be an example of a similar phenomenon (215). Thus, in eukaryotes the biological importance of the control of polymerase activity and specificity remains an open question.

REFERENCES

1. Losick, R., and Pero, J. (1976). RNA Polymerase, p. 227. Cold Spring Harbor Laboratory, New York.
2. Travers, A. (1976). Nature (Lond.) 263:641.
3. Tsai, M., Towle, H., Harris, S., and O'Malley, B. (1976). J. Biol. Chem. 251:1960.
4. Gall, J. G. (1968). Proc. Natl. Acad. Sci. USA 60:553.
5. Brown, D. D., and Dawid, I. B. (1968). Science 160:272.
6. Evans, D., and Birnstiel, M. L. (1968). Biochim. Biophys. Acta 166:274.
7. Burgess, R. R., Travers, A. A., Dunn, J. J., and Bautz, E. K. F. (1969). Nature (Lond.) 221:43.
8. Scrutton, M. C., Wu, C. W., and Goldthwait, D. A. (1971). Proc. Natl. Acad. Sci. USA 68:2497.
9. Burgess, R. R. (1969). J. Biol. Chem. 244:6168.
10. Scaife, J. (1976). RNA Polymerase, p. 207. Cold Spring Harbor Laboratory, New York.

262   Travers

11. Nomura, M., and Jaskunas, S. R. (1976). Alfred Benzon Symposium IX, Control of Ribosome Synthesis, p. 191. Munksgaard, Copenhagen.
12. Lindahl, L., Yamamoto, M., Nomura, M., Kirschbaum, J. B., Allet, B., and Rochaix, J. D. (1977). J. Mol. Biol. 109:23.
13. Austin, S. J., Tittawella, I. P. B., Hayward, R. S., and Scaife, J. G. (1971). Nature (New Biol.) 232:133.
14. Panny, S. R., Heil, A., Mazus, B., Palm, P., Zillig, W., Mindlin, S. Z., Ilyina, T. S., and Khesin, R. B. (1974). FEBS Lett. 48:241.
15. Fukuda, R., Iwakura, Y., and Ishihama, A. (1974). J. Mol. Biol. 83:353.
16. Nakamura, Y., Osawa, T., and Yura, T. (1977). Proc. Natl. Acad. Sci. USA. 74:1831.
17. Reeh, S., Pedersen, S., and Friesen, J. D. (1976). Mol. Gen. Genet. 149:279.
18. Losick, R., Sonenshein, A. L., Shorenstein, R. G., and Hussey, C. (1970). Cold Spring Harbor Symp. Quant. Biol. 35:443.
19. Stetter, K. O., and Zillig, W. (1974). Eur. J. Biochem. 48:527.
20. Nusslein, C., and Heyden, B. (1972). Biochem. Biophys. Res. Commun. 47:282.
21. Louis, B. G., and Fitt, P. S. (1971). FEBS Lett. 14:143.
22. Maitra, U. (1971). Biochem. Biophys. Res. Commun. 43:443.
23. Chamberlin, M., McGrath, J., and Waskell, L. (1970). Nature (Lond.) 228:227.
24. Ratner, D. (1974). J. Mol. Biol. 88:373.
25. Friesen, J. D., Parker, J., Watson, R. J., Bendiak, D., Reeh, S. V., Pedersen, S., and Fiil, N. P. (1976). Mol. Gen. Genet. 148:93.
26. Pero, J., Nelson, J., and Fox, T. D. (1975). Proc. Natl. Acad. Sci. USA 72:1589.
27. Pongs, O., and Travers, A. Manuscript in preparation.
28. Pongs, O., and Ulbrich, N. (1976). Proc. Natl. Acad. Sci. USA 73:3064.
29. Davison, J., Brookman, J., Pilarski, L. M., and Echols, H. (1970). Cold Spring Harbor Symp. Quant. Biol. 35:95.
30. Ramakrishnan, A., and Echols, H. (1974). J. Mol. Biol. 78:675.
31. Crimaldi, A., and Ihler, G. (1976). Eur. J. Biochem. 71:201.
32. Cukier-Khan, R., Jacquet, M., and Gros, F. (1972). Proc. Natl. Acad. Sci. USA 69:3643.
33. Wu, A. M., Ghosh, S., Echols, H., and Spiegelman, W. G. (1972). J. Mol. Biol. 67:407.
34. Ghosh, S., and Echols, H. (1972). Proc. Natl. Acad. Sci. USA 69:3660.
35. Crepin, M., Cukier-Khan, R., and Gros, F. (1975). Proc. Natl. Acad. Sci. USA 72:333.
36. Travers, A., and Cukier-Khan, R. (1974). FEBS Lett. 43:86.
37. Roberts, J. W. (1969). Nature (Lond.) 224:1168.
38. Schafer, R., and Zillig, W. (1973). Eur. J. Biochem. 32:201.
39. Richardson, J. P., Grimley, C., and Lowery, C. (1975). Proc. Natl. Acad. Sci. USA 72:1725.
40. Ratner, D. (1976). Nature (Lond.) 259:151.
41. Das, A., Court, D., and Adhya, S. (1976). Proc. Natl. Acad. Sci. USA 73:1959.
42. Morse, D. E., and Primakoff, P. (1970). Nature (Lond.) 226:28.
43. de Crombrugghe, B., Adhya, S., Gottesman, M., and Pastan, I. (1973). Nature (New Biol.) 241:260.
44. Lowery-Goldhammer, C., and Richardson, J. P. (1974). Proc. Natl. Acad. Sci. USA 71:2003.
45. Das, A., Court, D., and Adhya, S. (1976). Genetics 83:s19.
46. Gilbert, W., Maxam, A., and Mirzabekov, A. (1976). Alfred Benzon Symposium IX, Control of Ribosome Synthesis, p. 139. Munksgaard, Copenhagen.

47. Ptashne, M., Backman, K., Humayun, M. Z., Jeffrey, A., Maurer, R., Meyer, B., and Sauer, R. T. (1976). Science 194:156.
48. Lee, N., Wilcox, G., Gielow, W., Arnold, J., Cleary, P., and Englesberg, E. (1974). Proc. Natl. Acad. Sci. USA 71:634.
49. Dickson, R. C., Abelson, J., Johnson, P., Reznikoff, W. S., and Barnes, W. M. (1977). J. Mol. Biol. 111:65.
50. Silverstone, A. E., Magasanik, B., Reznikoff, W. S., Miller, J. H., and Beckwith, J. (1969). Nature (Lond.) 221:1012.
51. Gilbert, W. (1976). RNA Polymerase, p. 193. Cold Spring Harbor Laboratory, New York.
52. Fuchs, E., Millette, R. L., Zillig, W., and Walter, G. (1967). Eur. J. Biochem. 3:183.
53. Bautz, E. K. F., Bautz, F. A., and Beck, E. (1972). Mol. Gen. Genet. 118:199.
54. Chamberlin, M. (1974). Annu. Rev. Biochem. 43:721.
55. Mangel, W. F., and Chamberlin, M. J. (1974). J. Biol. Chem. 249:3002.
56. Travers, A. (1973). Nature (Lond.) 244:15.
57. Saucier, J. M., and Wang, J. C. (1972). Nature (New Biol.) 239:167.
58. Hayashi, M. (1965). Proc. Natl. Acad. Sci. USA 54:1736.
59. Mangel, W. F., and Chamberlin, M. J. (1974). J. Biol. Chem. 249:3007.
60. Hayashi, Y., and Hayashi, M. (1971). Biochemistry 10:4212.
61. Richardson, J. P. (1975). J. Mol. Biol. 91:477.
62. Nakanishi, S., Adhya, S., Gottesman, M., and Pastan, I. (1974). J. Biol. Chem. 249:4050.
63. Travers, A. (1974). Eur. J. Biochem. 47:435.
64. Levine, L., Gordon, J. A., and Jencks, W. P. (1963). Biochemistry 3:168.
65. Hinkle, D. C., and Chamberlin, M. J. (1972). J. Mol. Biol. 70:157.
66. Remold O'Donnell, E., and Zillig, W. (1969). Eur. J. Biochem. 16:152.
67. Khesin, R. B., Nikiforov, V. G., Zograff, Y. N., Danilerskaya, O. N., Kalyaeva, E. S., Lipkin, V. M., Modyanov, N. N., Dmitrier, A. D., Velkov, V. V., and Gintsburg, A. L. (1976). RNA Polymerase, p. 629. Cold Spring Harbor Laboratory, New York.
68. Travers, A. (1976). Mol. Gen. Genet. 147:225.
69. Giacomoni, P. U. (1977). FEBS Lett. 72:83.
70. Rabussay, D., and Zillig, W. (1969). FEBS Lett. 5:104.
71. Sippel, A., and Hartmann, G. (1968). Biochim. Biophys. Acta 157:218.
72. Sippel, A. E., and Hartmann, G. R. (1970). Eur. J. Biochem. 16:152.
73. Johnston, D. E., and McClure, W. R. (1976). RNA Polymerase, p. 413. Cold Spring Harbor Laboratory, New York.
74. Kessler, C., and Hartmann, G. R. (1977). Biochem. Biophys. Res. Commun. 74:50.
75. Bahr, W., Standen, W., Scheit, K. H., and Jovin, T. M. (1976). RNA Polymerase, p. 329. Cold Spring Harbor Laboratory, New York.
76. Wehrli, W., Handschin, J., and Wunderli, W. (1976). RNA Polymerase, p. 397. Cold Spring Harbor Laboratory, New York.
77. Nakamura, Y., and Yura, T. (1976). Mol. Gen. Genet. 145:227.
78. Rhodes, G., and Chamberlin, M. J. (1975). J. Biol. Chem. 250:9112.
79. Friedman, E. Y., and Travers, A. A. Manuscript in preparation.
80. Beckmann, J. S., and Daniel, V. (1974). Biochemistry 13:4058.
81. Debenham, P., and Travers, A. (1977). Eur. J. Biochem. 72:515.
82. Matsukage, A. (1972). Mol. Gen. Genet. 118:11.
83. Debenham, P. (1977). Ph.D. thesis. University of Cambridge.

84. Baralle, F. E., and Travers, A. (1976). Mol. Gen. Genet. 147:291.
85. Thomas, R. (1954). Biochim. Biophys. Acta 14:231.
86. Richardson, J. P. (1966). Proc. Natl. Acad. Sci. USA 55:1616.
87. Wu, F. Y.-H., Yarbrough, L. R., and Wu, C. W. (1976). Biochemistry 15:3254.
88. Travers, A. (1976). FEBS Lett. 69:195.
89. Stevens, A., and Rhoton, J. (1975). Biochemistry 14:5074.
90. Kuntzel, H., and Schafer, K. P. (1971). Nature (New Biol.) 231:265.
91. Primakoff, P., and Berg, P. (1970). Cold Spring Harbor Symp. Quant. Biol. 35:391.
92. Cashel, M. (1969). J. Biol. Chem. 244:3133.
93. Cashel, M., and Gallant, J. (1969). Nature (Lond.) 221:838.
94. Fiil, N. P., Willumsen, B. M., Friesen, J. D., and von Meyenburg, K. (1977). Mol. Gen. Genet. 150:87.
95. Reiness, G., Yang, H. L., Zubay, G., and Cashel, M. (1975). Proc. Natl. Sci. USA 72:2881.
96. Block, R. (1976). Alfred Benzon Symposium IX, Control of Ribosome Synthesis, p. 226. Munksgaard, Copenhagen.
97. Lindahl, L., Post, L., and Nomura, M. (1976). Cell 9:439.
98. van Ooyen, A. J. J., Gruber, M., and Jorgensen, P. (1976). Cell 8:123.
99. Fiil, N. P., von Meyenburg, K., and Friesen, J. D. (1972). J. Mol. Biol. 71:769.
100. Cashel, M., Hamel, E., Shapshak, P., and Boquet, M. (1976). Alfred Benzon Symposium IX, Control of Ribosome Synthesis, p. 279. Munksgaard, Copenhagen.
101. Goldthwait, D. A., Anthony, D. D., and Wu, C. W. (1970). RNA Polymerase and Transcription, p. 10. North Holland, Amsterdam.
102. Pedersen, F. S., Lund, E., and Kjeldgaard, N. O. (1973). Nature (New Biol.) 243:131.
103. Haseltine, W. A., and Block, R. (1973). Proc. Natl. Acad. Sci. USA 70:1564.
104. Friesen, J. D., Fiil, N. P., and von Meyenburg, K. (1975). J. Biol. Chem. 250:304.
105. Hansen, M. T., Pato, M. L., Molin, S., Fiil, N. P., and von Meyenburg, K. (1975). J. Bacteriol. 122:585.
106. Chakrabarti, S. L., and Gorini, L. (1975). Proc. Natl. Acad. Sci. USA 72:2084.
107. Muto, A., Kimura, A., and Osawa, S. (1975). Mol. Gen. Genet. 139:321.
108. Debenham, P., and Travers, A. Manuscript in preparation.
109. Heimark, R. L., Hershey, J. W. B., and Traut, R. R. (1976). J. Biol. Chem. 251:7779.
110. Kimura, A., Muto, A., and Osawa, S. (1974). Mol. Gen. Genet. 130:203.
111. Kari, C. Manuscript in preparation.
112. Maaloe, O. (1969). Dev. Biol. Suppl. 3:303.
113. Rose, J. K., and Yanofsky, C. (1972). J. Mol. Biol. 69:103.
114. Willumsen, B. Cited in K. von Meyenburg, Alfred Benzon Symposium IX, Control of Ribosome Synthesis, Munksgaard, Copenhagen, 1976, p. 448.
115. Furano, A. V. (1975). Proc. Natl. Acad. Sci. USA 72:4780.
116. Krauss, S. W., and Leder, P. (1975). J. Biol. Chem. 250:3752.
117. Brunovskis, I., and Summers, W. C. (1972). Virology 50:322.
118. Simon, M. N., and Studier, F. W. (1973). J. Mol. Biol. 79:249.
119. Ponta, H., Altendorf, K. H., Schweiger, M., Hirsch-Kaufmann, M., Pfennig-Yeh, M. L., and Herrlich, P. (1976). Mol. Gen. Genet. 149:145.
120. Studier, F. W., and Maizel, J. V. (1969). Virology 39:575.
121. Summers, W. C., and Siegel, R. B. (1970). Nature (Lond.) 228:1160.
122. Gage, L. P., and Geiduschek, E. P. (1971). J. Mol. Biol. 57:279.

123. Fujita, D. J., Ohlsson-Wilhelm, B. M., and Geiduschek, E. P. (1971). J. Mol. Biol. 57:301.
124. Fox, T. D., Losick, R., and Pero, J. (1976). J. Mol. Biol. 101:427.
125. Pero, J., Tjian, R., Nelson, J., and Losick, R. (1975). Nature (Lond.) 257:4530.
126. Duffy, J. J., and Geiduschek, E. P. (1975). J. Biol. Chem. 250:4530.
127. Tjian, R., Pero, J., Losick, R., and Fox, T. D. (1976). Molecular Mechanisms in the Control of Gene Expression. Academic Press, New York.
128. Fox, T. D. (1976). Nature (Lond.) 262:748.
129. Tjian, R., and Pero, J. (1976). Nature (Lond.) 262:753.
130. Guha, A., Szybalski, W., Salser, W., Bolle, A., Epstein, R. H., Geiduschek, E. P., and Pulitzer, J. F. (1971). J. Mol. Biol. 59:329.
131. Grasso, R. J., and Buchanan, J. M. (1969). Nature (Lond.) 224:882.
132. Brody, E., Sederoff, R., Bolle, A., and Epstein, R. H. (1970). Cold Spring Harbor Symp. Quant. Biol. 35:203.
133. Richardson, J. P. (1970). Cold Spring Harbor Symp. Quant. Biol. 35:127.
135. Travers, A. (1970). Cold Spring Harbor Symp. Quant. Biol. 35:225.
135. O'Farrell, P. Z., and Gold, L. M. (1973). J. Biol. Chem. 248:5512.
136. Mattson, T., Richardson, J., and Goodin, D. (1974). Nature (Lond.) 250:1014.
137. Freedman, R., and Brenner, S. (1972). Genet. Res. Camb. 19:165.
138. Homyk, T., Rodriguez, A., and Weil, J. (1976). Genetics 83:477.
139. Wu, R., Geiduschek, E. P., Rabussay, D., and Cascino, A. (1973). Virus Research, p. 181. Academic Press, New York.
140. Riva, S., Cascino, A., and Geiduschek, E. P. (1970). J. Mol. Biol. 54:83.
141. Adesnik, M., and Levinthal, C. (1970). J. Mol. Biol. 48:187.
142. Nomura, M., Okamoto, K., and Asano, K. (1962). J. Mol. Biol. 4:376.
143. Seifert, W., Qasba, P., Walter, G., Palm, P., Schachner, M., and Zillig, W. (1969). Eur. J. Biochem. 9:319.
144. Zillig, W., Mailhammer, R., and Rohrer, H. (1976). Alfred Benzon Symposium IX, Control of Ribosome Synthesis, p. 43. Munksgaard, Copenhagen.
145. Horvitz, H. R. (1974). J. Mol. Biol. 90:739.
146. Walter, G., Seifert, W., and Zillig, W. (1968). Biochem. Biophys. Res. Commun. 30:240.
147. Goff, C. G. (1974). J. Biol. Chem. 249:6181.
148. Mailhammer, R., Yang, H. L., Reiness, G., and Zubay, G. (1975). Proc. Natl. Acad. Sci. USA 72:4928.
149. Stevens, A. (1972). Proc. Natl. Acad. Sci. USA 69:603.
150. Stevens, A. (1970). Biochem. Biophys. Res. Commun. 41:367.
151. Ratner, D. (1974). J. Mol. Biol. 89:803.
152. Horvitz, H. R. (1973). Nature (New Biol.) 244:35.
153. Sirotkin, K., Wei, J., and Snyder, L. (1977). Nature (Lond.) 265:28.
154. Snustad, D. P., Tigges, M. A., Parson, K. A., Bursch, C. J. H., Caron, F. M., Koerner, J. F., and Tutas, D. J. (1976). J. Virol. 17:622.
155. Snyder, L., Gold, L., and Kutter, E. (1976). Proc. Natl. Acad. Sci. USA 73:3098.
156. Stevens, A. (1974). Biochemistry 13:493.
157. Khesin, R. B., Bogdanova, E. S., Goldfarb, A. D., and Zograff, Y. N. (1972). Mol. Gen. Genet. 119:299.
158. Travers, A. (1970). Nature (Lond.) 225:1009.
159. Thermes, C., Daegelen, P., de Franciscis, V., and Brody, E. (1976). Proc. Natl. Acad. Sci. USA 73:2569.
160. Georgopoulos, C. P. (1971). Proc. Natl. Acad. Sci. USA 68:2977.
161. Ghysen, A., and Pironio, M. (1972). J. Mol. Biol. 65:259.

162. Epp, C., and Pearson, M. L. (1976). RNA Polymerase, p. 667. Cold Spring Harbor Laboratory, New York.
163. Le Talaer, J., and Jeanteur, P. (1971). Proc. Natl. Acad. Sci. USA 68:3211.
164. Okamoto, T., Sugiura, M., and Takanami, M. (1972). Nature (New Biol.) 237:108.
165. Ruger, W. (1971). Biochim. Biophys. Acta 238:202.
166. Heyden, B., Nusslein, C., and Schaller, H. (1972). Nature (New Biol.) 240:9.
167. Schaller, H., Gray, C., and Herrmann, K. (1975). Proc. Natl. Acad. Sci. USA 72:737.
168. Pribnow, D. (1975). Proc. Natl. Acad. Sci. USA 72:784.
169. Gilbert, W., Gralla, J., Majors, J., and Maxam, A. (1975). Protein-Ligand Interactions, p. 193. Walter de Gruyter and Co., Berlin.
170. Musso, R., di Lauro, R., Rosenberg, M., and de Crombrugghe, B. (1977). Proc. Natl. Acad. Sci. USA 74:106.
171. Barrell, B. Chapter 5, this volume.
172. Shine, J., and Dalgarno, L. (1974). Proc. Natl. Acad. Sci. USA 71:1342.
173. Steitz, J. A., and Jakes, K. (1975). Proc. Natl. Acad. Sci. USA 72:4734.
174. Steitz, J. A., Wahba, A. J., Laughren, M., and Moore, P. B. (1977). Nucleic Acids Res. 4:1.
175. Walz, A., and Pirotta, V. (1975). Nature (Lond.) 254:118.
176. Maurer, R., Maniatis, T., and Ptashne, M. (1974). Nature (Lond.) 249:221.
177. Gottesman, M. E., and Weisberg, R. A. (1971). The Bacteriophage Lambda, p. 113. Cold Spring Harbor Laboratory, New York.
178. Maniatis, T., Ptashne, M., Backman, K., Kleid, D., Flashman, S., Jeffrey, A., and Maurer, R. (1975). Cell 5:109.
179. Dickson, R. C., Abelson, J., Barnes, W. M., and Reznikoff, W. S. (1975). Science 187:27.
180. Dhar, R., Weissman, S. N., Zain, B. S., and Pon, J. (1974). Nucleic Acids Res. 1:595.
181. Sekiya, T., Grit, M. J., Noris, K., Ramamoorthy, B., and Khorana, H. G. (1976). J. Biol. Chem. 251:4481.
182. Sternbach, H., Engelhardt, R., and Lezius, A. (1975). Eur. J. Biochem. 60:51.
183. Stahl, S., and Chamberlin, M. J. (1976). RNA Polymerase, p. 429. Cold Spring Harbor Laboratory, New York.
184. Roeder, R. G. (1976). RNA Polymerase, p. 285. Cold Spring Harbor Laboratory, New York.
185. Wu, G. J., and Dawid, I. B. (1972). Biochemistry 11:3589.
186. Reid, B. D., and Parsons, P. (1971). Proc. Natl. Acad. Sci. USA 68:2830.
187. Smith, H. J., and Bogorod, L. (1974). Proc. Natl. Acad. Sci. USA 71:4839.
188. Surzycki, S. J., and Shellenbarger. (1976). Proc. Natl. Acad. Sci. USA 73:3961.
189. Huet, J., Buhler, J. M., Sentenac, A., and Fromageot, P. (1975). Proc. Natl. Acad. Sci. USA 72:3034.
190. Chesterton, C. J., and Butterworth, P. H. W. (1971). Eur. J. Biochem. 19:232.
191. Matsui, T., Onishi, T., and Muramatsu, M. (1976). Eur. J. Biochem. 71:351.
192. Roeder, R. G. (1974). J. Biol. Chem. 249:241.
193. Gissinger, F., and Chambon, P. (1975). FEBS Lett. 58:53.
194. Kellas, B. L., Austoker, J. L., Beebee, T. J. C., and Butterworth, P. H. W. (1977). Eur. J. Biochem. 72:583.
195. Huet, J., Wyers, F., Buhler, J. M., Sentenac, A., and Fromageot, P. (1976). Nature (Lond.) 261:431.
196. Beebee, T. J. C., and Butterworth, P. H. W. (1974). Eur. J. Biochem. 44:115.
197. Hager, G., Holland, M., Valenzuela, P., Weinberg, F., and Rutter, W. J. (1976). RNA Polymerase, p. 745. Cold Spring Harbor Laboratory, New York.

198. Kedinger, C., and Chambon, P. (1972). Eur. J. Biochem. 28:283.
199. Schwartz, L. B., and Roeder, R. G. (1975). J. Biol. Chem. 250:3221.
200. Guilfoyle, T. J., and Key, J. L. (1977). Biochem. Biophys. Res. Commun. 74:308.
201. Dezelee, S., Wyers, F., Sentenac, A., and Fromageot, P. (1976). Eur. J. Biochem. 65:543.
202. Greenleaf, A. L., Haars, R., and Bautz, E. K. F. (1976). FEBS Lett. 71:205.
203. Sklar, V. E. F., and Roeder, R. G. (1976). J. Biol. Chem. 251:1064.
204. Jaehning, J. A., Weinmann, R., Brendler, T. G., Raskas, H. J., and Roeder, R. G. (1976). RNA Polymerase, p. 819. Cold Spring Harbor Laboratory, New York.
205. Parker, C. S., and Roeder, R. G. (1977). Proc. Natl. Acad. Sci. USA 74:44.
206. Sklar, V. E. F., Schwartz, L. B., and Roeder, R. G. (1975). Proc. Natl. Acad. Sci. USA 72:348.
207. Kedinger, C., Gossinger, F., and Chambon, P. (1974). Eur. J. Biochem. 44:421.
208. Ingles, C. J. (1973). Biochem. Biophys. Res. Commun. 55:364.
209. Sebastian, J., Hildebrandt, A., and Halvorson, H. D. (1973). Fed. Proc. 32:646.
210. Buhler, J. M., Iborra, F., Sentenac, A., and Fromageot, P. (1976). J. Biol. Chem. 251:1712.
211. Gilmour, R. S., and Paul, J. (1973). Proc. Natl. Acad. Sci. USA 70:3440.
212. Axel, R., Cedar, H., and Felsenfeld, G. (1973). Proc. Natl. Acad. Sci. USA 70:2029.
213. Stein, G., Park, W., Thrall, C., Mans, R., and Stein, J. (1975). Nature (Lond.) 257:764.
214. Mauck, J. C. (1977). Biochemistry 16:793.
215. Bell, G. I., Valenzuela, P., and Rutter, W. J. (1976). Nature (Lond.) 261:429.

Index